UNDERSTANDING RELATIVITY

· *Stanley Goldberg* ·

Birkhäuser

Boston · Basel · Stuttgart

UNDERSTANDING RELATIVITY

Origin and Impact of a Scientific Revolution

Library of Congress Cataloging in Publication Data

Goldberg, Stanley.
Understanding relativity.

Bibliography: p.
Includes index.
1. Special relativity (Physics) 2. Science — Methodology.
3. Science — — History.
I. Title.
QC173.65.G65 1984 530.1′1 83-22368
ISBN 0-8176-3150-X

CIP-Kurztitelaufnahme der Deutschen Bibliothek

Goldberg, Stanley:
Understanding relativity / by Stanley Goldberg.
- Basel ; Boston ; Stuttgart : Birkhäuser,
1984.
ISBN 3-7643-3150-X

Printed in the United States of America.

ISBN 0-8176-3150-X

To my mother, Sarah Belle Talisman Goldberg.

And to Frank Weissman, Ollie Loud and Trevor Coombe:
Though miles, even light years apart, they have kept their
eyes on the same star:
The Dignity of People.

Contents

Preface

The central subject matter of this book is Einstein's special theory of relativity. While it is a book that is written primarily for a lay audience this does not necessarily mean an audience not versed in the ways of doing science. Rather, this book is written for anyone wishing to consider the nature of the scientific enterprise: where ideas come from, how they become established and accepted, what the relationships are among theories, predictions, and measurements, or the relationship between ideas in a scientific theory and the values held to be important within the larger culture.

Some readers will find it strange that I raise any of these issues. It is a common view in our culture that the status of knowledge within science is totally different from the status of knowledge in other areas of human endeavor. The word "science" stems from the Latin word meaning "to know" and indeed, knowledge which scientists acquire in their work is commonly held to be certain, unyielding, and absolute. Consider how we use the adjective "scientific." There are investors and there are scientific investors. There are socialists and there are scientific socialists. There are exterminators and there are scientific exterminators. We all know how the modifier "scientific" intrudes in our daily life. It is the purpose of this book to challenge the belief that scientific knowledge is different from other kinds of knowledge.

Other readers are no doubt aware that scientific theories that at one time are generally accepted are later rejected as insufficient or inaccurate. This does not affect the commitment of many of these readers to the notion that scientific knowledge is special and more certain than other kinds of knowledge, for they believe that the reason for the replacement of one theory by another is the result of the improvement in measurement or the discovery of hitherto unknown phenomena. The bedrock on which theories rest is experiments and measurements. The results of the experiments and the measurements persist regardless of how theories change. Thus the succession of one theory by another is seen as part of an evolutionary process in which agreement between prediction and measurement become closer and closer and, in the process, an ever-growing number of phenomena are included within the widened perimeter of that which the new theory explains. These views will also be challenged in this book.

The premise underlying the views that are being challenged here is the belief that there is such a thing as "scientific thinking," which is essentially different from thinking in other spheres of human activity. Scientific thinking is different, according to this view, because of the use of something called "the scientific method." This scientific method is supposed to make knowledge in science logical and inevitable. In fact, in some circles science is considered a branch of logic. The premise that "scientific thinking" is different from other kinds of thinking by virtue of the application of the scientific method, will be challenged as a myth in this book. There is no magic formula such as the scientific method and science is not a branch of logic. In fact, when formalized, the relationship between measurements and the scientific theories which are supposed to explain those measurements contains a logical fallacy which Aristotle recognized as "affirming the consequent." Whereas it is true that in practicing science one should be logical, being logical is not the same as using formal logic. Regardless of which problem a person is working on, be it a problem in science, cooking, banking, driving an automobile or train, one should always be logical. To say that in doing science one should be logical is to say little.

Not only are the generally accepted views about the status of scientific knowledge challenged in this book, but also called into question are the commonly held notions about the niche occupied by science as a social institution within the matrix of institutions that make up a culture. For example, the glib claim that it is necessary to do science in order to fuel the fires of technological progress will be scrutinized and found wanting. In fact, it seems to be the case that more often than not, it is technological innovation that suggests the questions on which scientists should be working and which make it possible to ask questions that are unanswered. Even today, technological innovation is largely *sui generis,* requiring only prior technological innovation.

At this point it might well be asked what these points have to do with the special theory of relativity. The ideas and experiments associated with the special theory of relativity, their development, early reception, and assimilation, will be used as a case study to illustrate and support the general argument. In many ways, the development of the special theory of relativity is typical, it is claimed, of the development of any scientific theory. But there are features of the special theory of relativity and its development which make it particularly suited for our purposes. Although there has been an enormous mystique surrounding the theory and its creator, the special theory of relativity is an exceedingly simple theory. The core of the theory is understandable without great mathematical development. This understanding will allow an examination of the relationship between the ideas in the theory and the mathematical development which is normally associated with the theory and will help to give some insight into the role of a formal language such as mathematics in scientific theories.

The special theory of relativity is a modern theory. Very often, when the issues that are raised here have been discussed, the illustrative examples have arisen from earlier periods in the history of science: for example, the Copernican revolution of the sixteenth century, the Newtonian synthesis of the late seventeenth and early eighteenth century, or Dalton's atomic theory in the early nineteenth century. This has allowed some to argue that, whereas social factors might have been a consideration earlier, they are no longer and the nature of science has changed since the early modern period. That this is not the case will be one of the major focuses of this study.

Among the motivations for writing this book is the belief that these issues have a direct bearing on public policy in the sciences. If the public is to actively participate in policy decisions, there must be an understanding of the issues about which policy is being made. Regarding the sciences, this means understanding substantive issues in science. We are investigating the content and reaction to the special theory of relativity not only because it represents interesting and beautiful intellectual puzzles, but also because the contemplation of the issues can provide insight into current social questions about which science has a bearing. The first part of this book, therefore, is devoted to describing the intellectual content of the theory and making it understandable and accessible to anyone who can read and wishes to work at thinking about the problems that the theory addresses. The first part of this work is intended to *demystify* the substantive content of the special theory of relativity. To that end, the development of the ideas has been placed into an historical context. In order not to break the flow of ideas, technical asides, generally the formal developments that one might want to examine for completeness, have been placed in a set of appendices. They are not crucial to an understanding of the intellectual content of the special theory of relativity nor the contents of the book.

The second part of the book is devoted to an examination of how the special theory of relativity was received in the four cultures responsible for more than ninety-nine percent of the literature about it in the years following publication by Albert Einstein in September, 1905. We can identify within the matrix of the responses of the scientists in the four countries, something which we will call "national styles" about how the theory was understood. We can also learn how those national styles are comprehensible within traditional ways of practicing science in those cultures as well as being compatible with other facets of the social institutions of those cultures.

In the third part of this work, we raise the question of how the theory, which was initially found unacceptable in all four of the cultures, was assimilated not only within the scientific communities but within the general societies. Rather than examine the history of that assimilation in each of the four countries, the assimilation of relativity in the United States is used as an exemplar. The claim is not that the substantive understanding of the special theory of relativity acquired within American culture is typical but that the process of assimilation is representative. This study reveals that one of the purposes of scientific social institutions is to provide mechanisms for interpreting new scientific theories to insure the survival of traditional cultural views about the nature of science and its value to the society.

This raises the question of the precise nature of scientific revolutions. Do we toss out old theories that have been found wanting and, albeit with a significant struggle, embrace newer revolutionary views? On the other hand, if the role of our social institutions in science is to shield us from changing our minds about the nature of the universe, how are new ideas introduced?

Throughout, although the style suggests that all the answers are here, it should be emphasized that the intent is exploratory, tentative and heuristic. The issues are important for our culture and they have always been important. There is no illusion that this work represents the final word. On the contrary, the answers are not as important as the elaborative process that we undertake to address the questions that are being asked.

Acknowledgments

There are three mentors I wish to cite for shaping my outlook on these matters. I am sure we no longer agree on the details, yet that, after all, is part of the fun.

Vernon C. Cannon was my first teacher of physics. He made an incredibly difficult although beautiful subject acceptable and understandable. It was watching him build physical models out of the air between his hands that gave me the clue that physics modelled the universe of our experience. And it was by standing at the blackboard with him, sometimes for hours, trying to understand the most trivial formal statement about the motion of an hypothesized mass point, that made me realize the joy to be gained from considering abstractions.

Gerald Holton was the director of my doctoral work in the history of the early reception of Einstein's special theory of relativity. This subject is considered mostly in the second part of this book. That work was completed about fourteen years ago. It brings to mind something that Holton said to me as we began the project: "Whatever you do," he said, "there are two things to bear in mind. It should be something you love because you are going to be working on it for some time. And it should be considered a beginning and not an end." No teacher ever gave better, sounder, more sensible advice. And no teacher was ever more generous in introducing an important part of the garden of material that was to be part of the work. In addition, Holton's intellectual prowess has been a source of inspiration and guidance. Even when we disagreed, his critiques have been instrumental in tempering my outlook.

Leonard K. Nash taught me the importance of teaching honestly, without talking down, and without treating the student as if he or she were the village idiot. He is a master. Although I can never hope to be the kind of teacher Nash is, the model is always in front of me. It was Nash who introduced me to the concept of "case study" in a way that I had not considered earlier, and it was Nash who taught me to say *aloud,* "I don't know, but I know how to find out."

In addition, I am grateful to the following people who have been helpful with this project:

Herb Bernstein, I. B. Cohen, Peter Donahue, Paul Forman, Beth Ann Friedman, Kurt Gordon, John Heilbron, Evelyn Fox Keller, Patty Medor, Arthur Molella, Lewis Pyenson, Albert B. Stewart, Terry Roth Schaff, Monique Vincent all read parts of the manuscript and made valuable suggestions for improvements.

I am indebted to John Stachel, Editor of the Einstein Papers, for engaging with me over the last several years in an intense, lively, demanding correspondence on the epistemological implications of the structure of the special theory of relativity. While we do not agree, one of the outcomes has been a more profound understanding of the questions and issues to be addressed. For me the experience represents the epitome of what counts as educational. I hope that Stachel has also found the experience worthwhile.

For months before publication of this book, my once-a-week intellectual tussle with Michael Dennis was a wonderful stimulant. There is no way to convey just how important those discussions have been to me, nor is there any way to adequately thank him.

Fran Duda, Bobbie Rosenau, and Andrea Barrett typed drafts of some of the chapters.

I am indebted to my son David for his perceptive critique of part of the text and for being, along with his sisters Ruth and Eve, irreverently and lovingly supportive. My wife, Susan Galloway Goldberg, has been grand through it all.

I acknowledge the help and cooperation of Judith Goodstein, Archivist at The California Institute of Technology for permission to use materials from the Tom Lauritsen papers; Spencer Weart and Joan Warnow of The Center for the History of Physics at the American Institute of Physics for permission to quote from the E. H. Kennard papers; and Clark Elliott, Harvard University Archives for permission to quote from the Percy W. Bridgman papers.

My editor Angela von der Lippe has been a joy to work with. The copyediting by Pat Eden was an inspired work of art. Tad Gaither and Helane Manditch-Prottas oversaw production with calming competence. I am indebted to Mike Prendergast for transforming my chicken scratchings into recognizable drawings.

The final draft of this book was completed on "Deck D" (now "Deck A") of The Library of Congress. I am indebted to the Research Facilities Office of the Library, in particular to Suzanne Thorin and her staff. And I take this opportunity to acknowledge the wonderful help given to us all by the long-suffering, patient reference librarians at The Library of Congress: They are miracle workers.

The work in this book was partially supported by grant #8206019 from the National Science Foundation.

In citing the help of these individuals and institutions I am not suggesting that they will find this work satisfactory, or even acceptable. It is I who bear responsibility for what follows. May He or She who is responsible for these sorts of things have mercy when it comes time for The Great Accounting.

·I·

THE CREATION OF THE THEORY OF RELATIVITY

One of the purposes in writing this book is to use the theory of relativity to demonstrate that ideas in science are like any other abstractions which human beings create. Scientific knowledge is neither better nor more certain than other kinds of knowledge; it is different. But in order to talk *about* the theory of relativity it is necessary to know something about the content of the theory. The first part of the book is devoted to that subject.

While the ideas associated with the theory of relativity are developed here in a historical context, Part I of this book is not a history. It is a case study. Those familiar with ideas in physics will have no difficulty incorporating the story told here into what they already know. Those who are novices in physics or who have up to now found the subject too forbidding will not find the usual mathematical development. That is not to say that mathematical statements are not used. But they are not an integral part of the development. They serve only as a benchmark for historically important mathematical formulations of key physical ideas. For those who prefer a more formal analysis a series of appendices is provided with a mathematical treatment of key ideas. The appendices are not required to understand the intellectual issues.

The organization of Part I is very straightforward. Chapter 1 examines the question of how ideas in science compare to ideas in other disciplines and raises the issue of the social nature of scientific knowledge. Einstein's theory of relativity relates how the description of physical events by observers in different states of motion compare. The special theory of relativity deals only with observers moving at a uniform speed. In the general theory of relativity that restriction is relaxed. Chapter 2 explores the ideas in physics prior to the twentieth century which led to the development of the theory of relativity. In Chapter 3, the foundations of Einstein's special theory of relativity are developed in the context of the problems he and his contemporaries considered important in physics. Chapter 4 discusses some of the physical ramifications of those ideas and Chapter 5 presents a synopsis of the domain of the general theory of relativity.

As I stated in the Preface, the ideas we associate with the theory of relativity are exceedingly simple. That is not the same as saying it is an easy subject. There are no easy subjects. As with any other subject, simply reading about the theory will not bring an understanding of the theory's content. Understanding requires thought, argument, reformulation; in short, the internalization of the ideas being studied. A subject becomes easy when a person's intuitions about the subject become honed to the point that he or she no longer has to think about the structure of the argument. This is not the same as memorizing or learning by rote.

While the treatment in this book is not "baby science," and does not condescend to oversimplification, it is my hope that the first part of this book will make the special theory of relativity accessible to those who identify themselves as "science shy" or who profess to being so frightened by science, they flee from it in terror. I do not know if this book will help them. We understand very little about why a significant segment of the population is so frightened of science. For those people, science is not just hard, it is impenetrable. I am speaking here of people who simply tune out when presented with one equation in one unknown. Whatever leads to this state of mind, it happens at quite an early age. It happens more often to women than to men. And yet these people are not incapable of understanding the issues. When the same types of abstractions are translated into a familiar context, supermarket problems for example, these people have no difficulty confronting the puzzle and obtaining the answer. This suggests, though does not prove, that we are dealing here with a social phenomenon.

I urge those with this kind of block to bear in mind that the relationship of evidence to belief in science is no different than in any other area of human intellect. There is no special kind of thinking that is "scientific." There are no magic exercises that will allow a person to penetrate the fog. Some of the people I have worked with who have this kind of block about science have found it helpful to practice translating back and forth between ordinary language and formal arguments and to do so over and over again.

If it is going to be so much work, why bother? There are any number of aesthetic and practical rationales I could produce to answer this question. You know them already. Some of them will be found in the following chapters. But there is one other reason you might consider: I and many other people love science. We find it exciting, challenging, beautiful to the point of pain. You might want to hang around long enough to find out why.

·1·

Science, Logic, and Objectivity

WE EXAMINE SEVERAL GENERAL QUESTIONS about the nature of science and its place in the matrix of social institutions in a society in this chapter. Our discussion begins with the question of the role that mathematics plays in science. Because the issues are important for understanding developments in later chapters, and because there is much confusion in the literature about what mathematics contributes to science, the discussion is extended. In subsequent sections, the relationship of science to other abstract disciplines like sociology, history, or art is examined in the context of the canons of organization that communities have used in defining their role in our culture.

THE ROLE OF MATHEMATICS AND OTHER FORMAL SYSTEMS IN SCIENCE

Mathematics is a branch of formal logic. When a mathematician is doing mathematics, it hardly matters for the quality of the mathematics that he or she is doing, whether it relates to experience or a theoretical system that describes experience. Mathematical systems are self-contained and, to the mathematician, of interest for their own sakes.

Science is not like that. First, science is not a branch of formal logic. If it were, it would not be necessary to perform experiments. It wouldn't even be necessary to stick one's finger in the air to see which direction the wind was blowing because whether the wind blows or not has no bearing on the outcome of the structure of formal systems. There must be a relationship between science and experience. The nature of that relationship will be taken up in this book.

To say that mathematics is a branch of formal logic, is to say that it has a definite structure. That structure consists of a set of premises, a set of statements in which those premises are manipulated according to definite rules which lead to a conclusion. Different branches of mathematics have different rules but in all branches, since the rules are predetermined, the conclusion is actually a restatement, in a new form, of the premises. Mathematics, like all formal logic, is tautological. That is not to say that it is uninteresting or that it doesn't contain many surprises.

Here is a well worn and well understood example that you may have encountered in other guises:

(a) All people are mortal.
(b) Emma Goldman is a person.
(c) Therefore, Emma Goldman is mortal.

Statements (a) and (b) are called "premises" of the argument and (c) is the conclusion. There is no question that it is a correct argument according to the rules. In fact, the correctness of the argument can be better seen if we generalize the situation:

(a′) All *A* are *B*
(b′) *C* is an *A*
(c′) Therefore, *C* is a *B*

It is obvious that the conclusion (c′) follows from the premises since according to the premises anything that is an *A* is also a *B* and *C* happens to be an *A*.

What does it mean to say that the argument is a correct argument? It means that the conclusions follow from the premises, nothing more. Another question might be asked however. "Is the argument true?", that is, does it conform to experience? This is a different kind of question. The answer is not at all obvious. For example, consider the following case:

(a″) All people are mortal.
(b″) Jesus is a person.
(c″) Therefore, Jesus is mortal.

The argument is still correct. Is the argument true? If you are a Christian, the answer is "no."

"Although the argument is correct," you might say, "one of the premises of the argument is not acceptable to me. I agree that all men are mortal, but I do not agree that Jesus is a man."

And so, although the argument is correct, it is not true. It is possible to construct a correct argument that satisfies the rules of formal logic, and at the same time, draw conclusions that are not acceptable and do not represent the world of your experience.

There are certain features of formal logical systems which bear emphasis. No argument can be self-contained. That is, it is not possible *from within the system* to prove the postulates that are used to construct arguments. Thus, within the system we have been using it is not possible to prove either (a′) or (b′). One might want to construct a different argument in which (a′) or (b′) are the conclusion of an argument based on other premises. For example, one might want to construct an independent argument with the conclusion that "All men are mortal."

Not only are the premises of a logical system unprovable within the system, key terms in those premises are also undefined. Thus, within the system that we first gave, the terms "people," and "mortal," are undefined. They are sometimes referred to as the primitives in the system.

THE CASE OF EUCLIDEAN GEOMETRY

All of this applies, not only to the simple syllogistic-like systems we have been considering, but to any system of formal logic, including mathematical systems. Consider, for example, the system that you learned in secondary school, Euclidean plane geometry. Euclid did not invent geometry. His major contribution was to weld together a series of theorems by many individuals into a unified system which could be developed from five postulates. The postulates (themselves unproveable) are:

1. It is possible to draw a straight line from any point to any point.
2. It is possible to extend a finite straight line continuously.
3. It is possible to describe a circle with any center.
4. All right angles are equal to one another.
5. If a straight line falling on two straight lines makes the interior angles on the same side less than two right angles, the two straight lines, if extended indefinitely, meet on that side on which the angles are less than the two right angles.*

In addition to the five postulates, Euclid introduced a series of five *common notions:* postulates not specific to geometry but which are required for any mathematical system:

1. Things which are equal to the same thing are also equal to one another.
2. If equals be added to equals, the wholes are equal.

* Morris Kline, *Mathematical Thought from Ancient to Modern Times* (New York, 1972), p. 59.

3. If equals be subtracted from equals, the remainders are equal.
4. Things which coincide with one another are equal to each other.
5. The whole is greater than the part.

The system of Euclidean plane geometry contains a number of undefined terms; for example, "point," "line," "straight," "angle," etc. even though Euclid had provided an auxiliary set of definitions for those things. The definition of a point is "that which has no parts." A line is "a breadthless length." All told, there are twenty-three definitions.

From this starting point, all of the theorems of plane geometry are derivable.

As with the case of the syllogism, as long as no logical errors are made, the system is correct. The question of whether it is a true system or not is an entirely different matter. For at least two millenia, it was generally believed that the space in which we live is Euclidean. Since the theorems which are proven by using the postulates appeared to conform to our experience, this was generally interpreted to mean that the postulates must be true even though they are not testable.

For example, recall that one of the theorems of Euclidean geometry states that the sum of the angles in any triangle must be equal to two right angles, or 180 degrees. As near as any one has been able to measure, the sum of the angles in any actual plane triangle has never been anything else.

Over the centuries, however, there was some disquietude regarding the structure of Euclidean geometry. It was often argued that the fifth postulate, the so-called "parallel line postulate," seemed to be overly complicated. Many efforts were made to prove the fifth postulate on the basis of the other four postulates. Had such an effort been successful, the fifth postulate would have no longer been a *postulate,* but a *theorem.* The overall effect would have been to simplify the geometry in the sense that, rather than being based on five postulates, it would have been based on four. Such attempts were never successful. Finally, in the nineteenth century, several mathematicians derived an alternative method of dealing with the problem. Rather than trying to prove the fifth postulate of Euclid on the basis of the other four, they denied the fifth postulate. What would the effect on the resulting theorems be? There are two obvious ways to deny the fifth postulate. One way would be to postulate that no line can be drawn parallel to another. The second is to postulate that an infinite number of lines can be drawn parallel to a given line. In either case, it was discovered that one could develop a perfectly consistent geometry, different from Euclidean geometry, but totally consistent with itself. These non-Euclidean geometries also appear to have applications in the world of experience; therefore the question of whether or not space is Euclidean no longer results in the obvious answer that it was once thought to have.

There is an implicit relationship between our experiences about the space in which we live and these various geometries. If we make the claim, for example, that space is Euclidean, Euclidean geometry becomes a language in which one may discuss and describe the space and the kinds of qualities that space should have. One can also use that language to derive relationships that were hitherto unsuspected. Suppose, for example, you wish to determine the height of a very tall building, say, an Egyptian pyramid. It is possible to perform by a simple technique if one makes the assumption that space is Euclidean. You then exploit the theorem of Euclid which states that corresponding parts of similar triangles are in proportion to each other.*

In order to measure the vertical height of the pyramid you must put a straight stick vertically in the ground, as in Figure 2. Knowing the height of the stick, and measuring the length of the shadow of the stick, and of the pyramid, allows you to calculate the height of the pyramid by simple ratio and proportion.

Before proceeding, you should read the caption associated with Figure 2. If you have read the caption but skipped the two equations that describe the relationship between the height of the pyramid and the shadow of the pyramid cast by the sun, reread the caption, and also *read* the equations. I say this because far too often, people who are not familiar with, or who are frightened by, mathematics do not realize that when mathematics is applied to physical situations, the mathematical statements are a language in which each term represents a physical quality. As mentioned earlier, the difference between applications of mathematics and considerations of mathematical relationships for their own sakes is that there is no requirement that the terms of a mathematical system apply to the physical world. Whereas internally, the mathematics remains a language, it can be a language without physical referents. This is not true when a mathematical system is applied

* The interpretation of the technical term "similar" of Euclidean geometry into ordinary language is as follows: The Euclidean theorem proves that two triangles are similar if they have the same angles. That means, colloquially, that they have the same shape. Triangles I and II in the accompanying figure do have the same angles. To say that the corresponding parts are in proportion to each other is to say that the ratio of length AB to BC is the same as the ratio of ab to bc, etc.

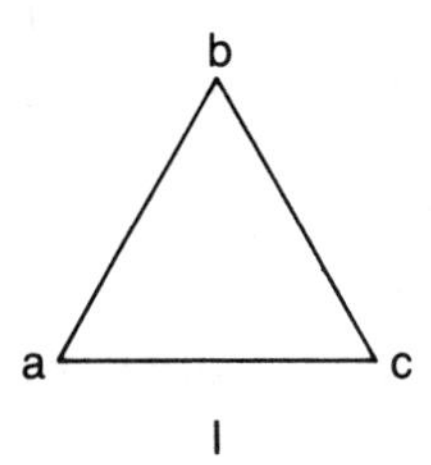

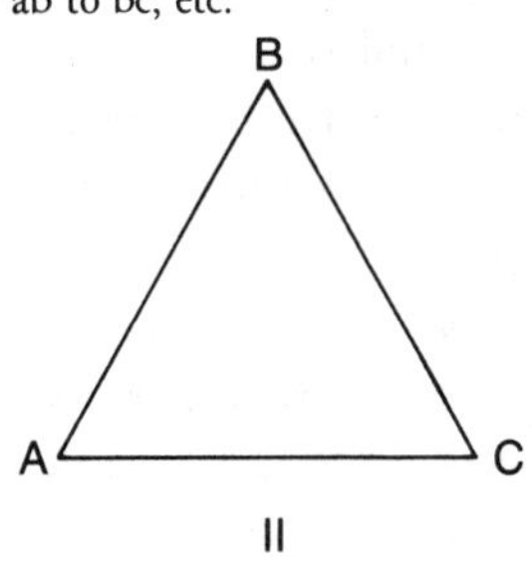

Figure 1.1

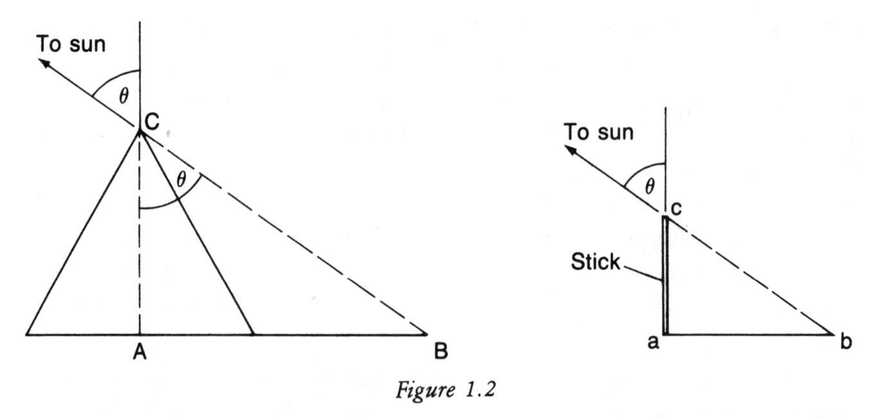

Figure 1.2

Length *ab* is the shadow of the stick. Length *AB* is the distance from the center of the pyramid to the end of the shadow. *ac* is the length of the stick. *AC* is the vertical height of the pyramid. Then it must be the case, (assuming that space is Euclidean) that triangle *abc* is similar to triangle *ABC*.

The two triangles, *ABC* and *abc*, are similar because they contain the same angles; the right angles (*CAB* and *cab*) at the base, the angles made by the sun at the top (*ACB* and *acb*) and finally, since the sum of the three angles must be the same, angle *ABC* must equal angle *abc*.

$$\frac{AC}{AB} = \frac{ac}{ab}$$

This may be rewritten as:

$$AC = \frac{ac}{ab} AB$$

which gives us the height of the pyramid.

to physical situations. The terms always mean something. The mathematics serves as a useful language for expressing fairly complex ideas. In each case the statements can be translated into ordinary language.

Similarly, problems that are encountered in ordinary language can often be translated into mathematical statements. In this case, a simple English rendition of the mathematical conclusion is: the height of the pyramid is ascertained by the product of the length of the shadow cast by the pyramid, and the ratio of the length of the shadow cast by the stick to the length of the stick. However, a less literal, more interpretive translation might be more informative. Since the sun is very far from the earth and light travels in straight lines, the angle between the vertical and the sun will be the same at the base of the pyramid and at the bottom of the stick. Therefore, as long as a line drawn from the peak of the pyramid to the ground and the stick are both at right angles to the ground, the ratio of the height of the pyramid to the length of the shadow cast by the pyramid must be the same as the ratio of the height of the stick to the shadow cast by the stick. As one can measure the height of the stick, and the length of the shadow cast by the pyramid, the only undetermined quantity in the system is the height of the pyramid. It can be calculated by simple ratio and proportion.

THE CASE OF NATURAL NUMBERS

As I indicated earlier, one of Euclid's major contributions was not the creation of plane geometry, but its organization into a unified system based on a small number of postulates. There are many other cases where the organization of the system into a formal one occurs long after the system itself has been established. It was not until the nineteenth century, for example, that the logical foundation was supplied for the natural numbers. "Natural numbers" are whole positive integers, 1, 2, 3, etc. Obviously, the fact that one and one are two, or that two hundred and thirty-four and twenty-six equals two hundred and sixty, or that twelve times six equals seventy-two were not revelations that required logical formulation. At some point in our lives many of us have realized that *this many* (say, one of anything) plus *this many* (say, one of anything) makes two. That appears to be a truth, not only independent of there being things, but it is also a truth that is independent of me, you or anyone.

The axiomatization of the natural numbers was accomplished by Giusepe Peano near the end of the nineteenth century when interest in the logical foundations of all mathematical systems was widespread. Peano began with the following undefined concepts: "successor" and "belong to." There follow five postulates for the natural numbers:

1. 1 is a natural number.
2. 1 is not the successor of any other natural number.
3. Each natural number a has a successor.
4. If the successors of a and b are equal then so are a and b.
5. If a set S of natural numbers contains 1, and if, when S contains any number, a, it also contains the successor of a, it also contains all the natural numbers.

Peano also employed a certain number of relational postulates:

$$a = a; \text{ If } a = b \text{ then } b = a; \text{ If } a = b \text{ and } b = c \text{ then } a = c$$

These are often referred to, respectively, as the reflexive, symmetrical and transitive axioms for equality. Peano then defined two operations: addition and multiplication. If we let the symbol a+ denote the next natural number after a, then addition is defined as:

$$a + 1 = a+$$
$$a + (b+) = (a + b)+$$

Multiplication is defined that to each pair of natural numbers a and b there is a unique product such that:

$$a \cdot 1 = a$$
$$a \cdot b+ = (a \cdot b) + a$$

From these axioms and definitions all of the familiar properties of natural numbers can be established. It is then only a short step to create, axiomatically, the properties of negative whole numbers, and, it follows, of all the rational numbers.

Peano did not invent the natural numbers nor their properties. He had shown the kind of logical foundation that is required to establish them. His set of postulates are not *the* postulates required. There are probably an infinite number of ways that one could establish the natural numbers, for example, by taking different undefined concepts and different postulates from those used by Peano. One might not be as efficient as Peano was in the sense of the number of concepts and postulates required, but efficiency is an arbitrary, aesthetic concept. In any case, all of these mathematical systems have a direct relationship to the syllogistic-like system we referred to above. The relationship between the undefined concepts and the postulates, on the one hand, and the theorems that are proved, on the other hand, may be put in the form:

If all *A* are *B*,
and if *C* is an *A*,
then *C* is a *B*

Science is never like that.

THE CASE OF FREE FALL

Whereas logic is employed in the development of theories in natural science, there is a further requirement in science that is never made within the confines of pure mathematics. It is always required that a connection be made between the results obtained by theoretical manipulation of formal logical systems and phenomena taking place in the world of experience. In other words, if natural science is concerned with anything, it must be concerned with observation and experiment. The question is, "How can I explain this or that?" or "How can I predict this or that?" or "Since my explanatory system predicts this or that, how can I arrange an experiment to show that it, in fact, occurs?"

For now, it is enough to say that, when one works in science, one is concerned with explaining the world of experience. It might be the case that the experiments and observations—the phenomena—have been recognized for some time; but it might also be the case that in the process of creating an explanatory pattern, hitherto unsuspected phenomena are suggested. Very often, in the process of creating certain kinds of technological processes, phenomena are uncovered that defy current theory and call for new explanatory systems. Whatever the order of events, the relationship between explanation or theory on the one hand, and evidence, that is, phenomena and experiments on the other hand, may be cast into the following form:

(d) If *A* then *B*,
(e) I observe *B*.
(f) Therefore *A*.

Many examples of this application in science will be provided in this book. But, before giving the first one, the law of free fall, let us be clear that this form is not logically correct. In fact it contains a logical fallacy known as "affirmation of the consequent." In statement (d) a logical connection is made between *A* and *B*. It cannot be shown, however, that it is the only set of statements, *A*, that yield the result, *B*. In other words, statement (d) represents a theoretical explanation of the type we have already encountered. It might be a statement in ordinary language or mathematical language, but the set of statements conforms to the rules of formal logic. It is impossible to show that the explanatory pattern represented by *A* is unique and the only one to yield the set of results labeled as B. There may be another system of logic, *C* or *D*, or *E*—indeed, there may be an infinite number of such systems—that would yield *B*. In order for the form presented in statements (d), (e), and (f) to be absolute, unique, and logical, statement (d) would have to be modified as follows:

(d′) If and only if *A*, then *B*

That can never be the case in natural science. There is an impenetrable barrier between the world of experience and the postulates chosen to make the theories to explain those experiences. It is always possible for an alternative accounting of the facts [the "*B*" in statement (e)]. We shall see later that it is this aspect of natural science that is the key to understanding why we change our views about the way the universe *actually* works. It is not possible to believe that knowledge in natural science is certain in a way that is untrue in other fields.

We also observe that, at any given time, there are entire classes of theory about which scientists are confident and sure. They would stake their lives on the claim that the universe is really *like that*. This happens in every generation of scientists.

And yet no theory remains immutable. Until now, many have given way to contrary evidence, to more comprehensive, theoretical structures, or because postulates that at one time were acceptable are no longer valid. In our investigation of the theory of relativity we will find examples of these circumstances.

For the moment I wish to illustrate this discussion with a familiar case: the transition from Aristotelian concepts of motion to Galileo's law of free fall. At the time that Galileo lived and worked, between the end of the sixteenth century and the middle of the seventeenth century, the predominant theory accounting for the motion of objects near the surface of the earth (including free fall) had its origins within Aristotelian physics. The basis of the patterns of explanation for motion was empirical, and directly tied to experience.

This is probably contrary to what you believe about Aristotelian concepts of motion and physical science. First, whereas we no longer see the world as the Greeks saw it and we no longer take Aristotelian physics seriously, there is no question that during his era it was respectable, and it made sense out of the world of experience for his contemporaries. And when Aristotelian concepts were modified and incorporated within medieval Catholic dogma, the relationship between the concepts of motion and the world of experience remained culturally acceptable.

While comparing Aristotelian and Galilean analyses of the motion of falling bodies, we will use, directly, more mathematics than in other chapters of this book. We do this to illustrate the relationship between mathematics and physical science.

THE ARISTOTELIAN ANALYSIS OF MOTION

When we say that Aristotelian concepts of motion were connected to experience we mean that his explanations and theories had very few abstractive qualities. Aristotle's system explained the observations simply by positing that the reasons events happen the way they do is that they actually happened that way. On the other hand, Galileo's explanation of motion in a sense explains *away* that which we observe.

The medieval Aristotelian system is based on the importance of position — the idea of natural place. Underlying the entire system is a set of general postulates that include:

1. The world is composed of four elements: earth, air, fire and water. The four elements represent combinations of four basic qualities: hot, cold, wet and dry.
2. Each element has its natural place.
3. Spontaneous change results from lack of perfection.

4. The most perfect of shapes is the sphere since it has an infinite number of planes of symmetry and is therefore indifferent with regard to motion.
5. The heavens contain the following objects: the moon, the sun, the five planets, and the stars. Since the heavens are perfect, change (for example, motion) is impossible. Since the sphere is indifferent about motion, the trajectories of all heavenly bodies must be circular, along the surface of a sphere.
6. Below the dividing line between the heavenly and non-heavenly parts of the universe, the natural order of the elements is the sphere of the earth (at the center of the universe) surrounded by the sphere of water, the sphere of air and the sphere of fire. Then comes the first heavenly sphere, the sphere of the moon.
7. Matter is continuous. There are no empty spaces. Beyond the sphere of the moon, the heavens are filled with a perfect fluid, the ether, having no weight, offering no resistance to either matter or light.

This universe is depicted in Figure 3 (p. 18).

The first conclusion that one may draw from these postulates is that below the sphere of the moon, things are imperfect. The evidence is that spontaneous change occurs. Such change is a kind of motion, be it a change in place, temperature, mood, etc. The hypothesized reason for the change is that the elements have been mixed. When spontaneous change occurs, it represents the return of the elements to their natural places. For example, lightning strikes. The explanation is that fire and air have been liberated from each other and are spontaneously returning to their natural places. Suppose the lightening strikes a tree that bursts into flame. The tree is composed of the four elements because now, as the tree burns, the elements can almost visibly be seen to go back to their natural places: the fire leaps up toward its sphere, the smoke (air) rises to its sphere. As the log burns, the sap (water) runs out and returns to its sphere and, when the burning process is complete, we are left with ash (earth), returned to its sphere.

We should caution that some of the words being used here—for example, "element"—did not have the same meaning to the Greeks and medieval Catholic intellectuals that they now have. First, the elements were mutable one into the other. They represented abstract principles as often as they described actual materials.

Earlier I pointed out that there is always an alternative accounting for the experiments and observations. Our observation of the burning tree would not be different from the observations made by the Greeks. Our explanation of an identical set of events would be entirely different. We would employ concepts like potential energy, a different concept of element, oxidation, reduction, and so on. Note that the phenomena are the same: The tree is burning.

The Aristotelian analysis of the motion of objects from place to place, "local motion'' (since contracted to *locomotion*) may conveniently be divided into two categories: forced and natural. Natural motion occurs spontaneously. For example, a seed falls from a tree and this represents the spontaneous return to a natural place. Earthy things fall to the earth. But, even if the earth were somehow moved out of the way, earthy things would continue to fall to the center of the universe and not to where the earth had been moved. Forced motion is motion that is imposed by a mover. The object can then be moved in a manner which is contrary to natural motion. Thus, by flinging it with the arm, a stone may be thrown up away from the earth, its natural place. The taut string of a bow can force an arrow on a trajectory contrary to its natural inclinations. The greater the force exerted on an object, the faster it moves. If no force is exerted it would not move. These statements conform precisely to observation.

When a force is exerted upon an object, there is a counterforce which tends to prevent the motion of the object. This counterforce is called *resistance*. The higher the resistance, the slower the object moves from the action of a given force. Should the force of the mover no longer be applied, the object would quickly come to rest not because of the resistance but because, without the force, the object would return to its natural state of rest. (Note that this describes our experience, and also note that currently we explain away the observation that things come naturally to rest. Our premises include the concept of inertia—that whatever the state of motion of an object, that state of motion persists in the absence of outside influences.) The role that the resistance of the medium played was to specify the minimum force necessary to get the object moving and to determine the speed of the object under the influence of a given force. Thus, if a person pushed a cart on solid ground, it would move faster than were it pushed with the same force through mud. If the person could not push hard enough, the resistance would not be overcome and the cart would not move.

The mathematical language that the Greek and early western cultures had available for the analysis of these kinds of problems was geometry. Algebra was not available. This meant that certain concepts, which we can easily express, could not then be so easily expressed. For example, one could not speak of the speed of an object in the same way that we do. Today we define the speed of an object as "the time rate of change of the position." Algebraically that is translated into $v = s/t$, where v is the speed, s is the distance traversed, and t is the time elapsed in traversing that distance. Given that the Greeks were limited to geometry, it is understandable that they could only compare length to length, time to time, force to force, etc. Three variables were at play in the Aristotelian analysis of motion: the speed of an object, the force exerted on the object and the resistance of the object. Hence, we might write an algebraic equation relating those variables:

$$\left(v = k\frac{F}{R}\right)$$

where v is the speed, F is the force, R is the resistance and k is a proportionality constant that allows us to scale the equality and make certain that it comes out with the right units (distance over time). But the Greeks would have been forced to consider no fewer than three, and probably six relationships (All the possibilities of three things, taken two at a time). For example:

The *quicker* would be defined as follows:

For a medium of a given resistance, the quicker of two objects is the one on which a larger force is exerted.

For two identical objects on which the same force is applied, the quicker is the object moving through the medium offering less resistance.

For two objects of equal quickness, the force must be greater on the one in the medium of higher resistance.

For a given force, the resistance must be less on one of two objects which is the quicker, and so on.

In natural science the role of mathematics is that of a language, and a given language favors the emergence of certain concepts over others.

Obviously there is much that the Aristotelian concept of motion could not explain. For centuries scholars and natural philosophers struggled with some of these problems. How do we explain that an arrow continues to move once it is no longer being forced by the string? Or what causes the motion of a heavy potter's wheel turning in place? And consider the fall of an object from a height. Although that fall is natural, how do we explain that, as it descends, it appears to move faster and faster?

In the case of the arrow, Aristotle originally proposed a kind of communion between the moving arrow and the surrounding air in which the quality of motion went back and forth between the air and the object. The quality was gradually attenuated until, finally, the arrow fell. This concept was interpreted in many different ways by those who came after Aristotle and is likely to have been the precursor of early notions of impetus, and finally the modern notions of inertia and momentum.

The case of free fall was more difficult. The role of force in *forced motion* had to be replaced by the *seeking of natural place*. Obviously as more of the element seeking its natural place was available, it moved more quickly. Therefore, it should follow that for earthy things, the heavier they are the quicker they should fall. As to why objects speed up as they fall, this seems to have been explained by a kind of anthropomorphization: the closer they get to their place, the more overjoyed they are at the prospect and the quicker they move. Hence, one should expect that heavier objects fall faster and that the speed of an object increases in proportion to the distance it falls.

Whereas there were obvious difficulties, this system of analysis was explanatory. The motion of objects in free fall is very fast and it was sufficient that objects increased their speed as they fell. Furthermore, there are many examples of objects that do not fall at the same rate. In nature, feathers and leaves fall much more slowly than rocks and bricks.

This is not to say that there were not criticisms of the Aristotelian system. From the time he advocated the system it was subject to severe criticism. During the Middle Ages there were several attempts to modify the theory, as well as reports of phenomena not in line with the predictions. For example, it was noted quite early that two objects of the same shape but of different weights fell in more or less the same amount of time.

It is not strange, nor is it surprising, that such observations did not lead to the immediate rejection of the Aristotelian theory of free fall. It does not mean that Aristotle, his disciples, nor those who came after him were ignorant. Different kinds of things were important to those early scientists, just as different kinds of things that are important to us may not be important to our successors a thousand years from now.

Contrary instances between prediction and observation, between theoretical outcomes and experimental measurements, do not *ipso facto* result in the rejection of a theory, especially when that theory is centrally imbedded in a large metaphysical scheme as was the Aristotelian concept of motion. The Aristotelian concept of motion stemmed from a beautifully elaborated system of postulates in medieval culture that were part of almost all beliefs about the meaning of life, the place of man in the universe, the relationship between man and god, and so forth. That the Aristotelian theory of motion did not predict certain aspects of the behavior of moving objects would be overlooked as something to be worked on later, a trivial discrepancy that would ultimately be settled satisfactorily.

The issue is probably even clearer when one considers predictive astronomy during the same period. According to the cosmology of Aristotle (see the postulates we listed earlier) and medieval culture, the planets, sun, and moon moved in perfect circles around a center occupied by the stationary earth. This theory defies observation, and there are no ways to reconcile the discrepancies. The system of predictive astronomy that was developed (synthesized by Ptolemy and modified by later commentors) worked rather well, and made use of circles whose centers moved in circular paths. Strictly speaking the earth was not at the center of the system and no other objects in the heavens moved in circular paths around the earth. (Compare Figure 4 to Figure 3). When scholars were studying predictive

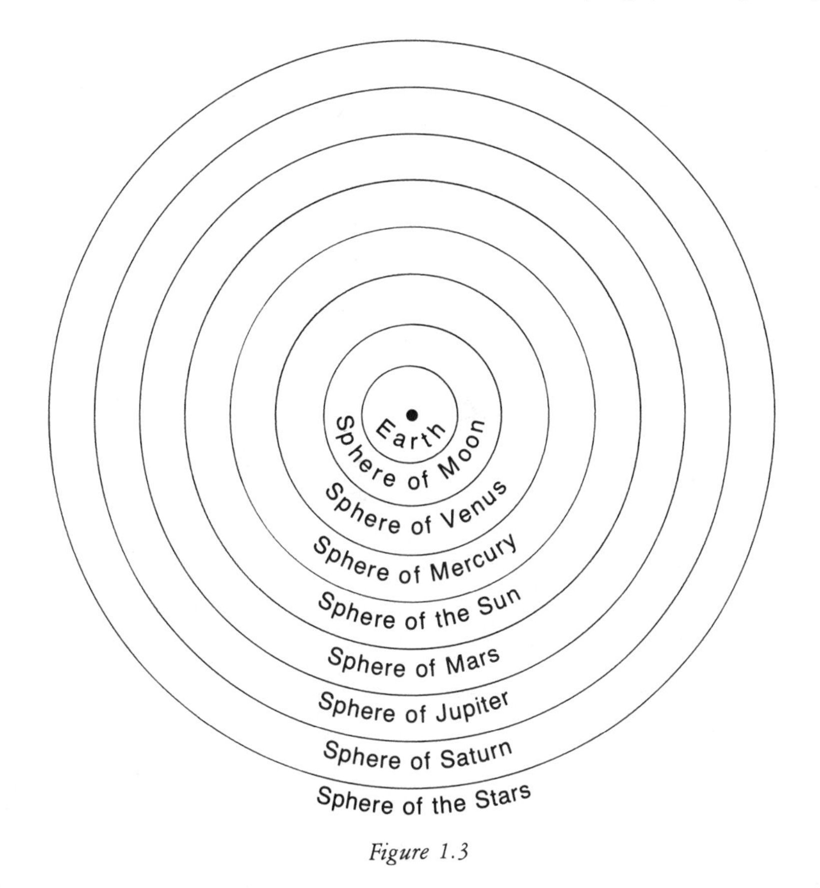

Figure 1.3

astronomy, they made use of Ptolemy. When they were studying cosmology they made use of Aristotle, and the two were never confused. Similarly, the ideal that explained the motion of objects near the earth was the Aristotelian system. Discrepancies between observations and the theoretical system in this area could not shake the cultural contentment with the system as a whole.

By the end of the sixteenth century all this had changed. In the atmosphere of the Renaissance, basic beliefs about the nature of the universe that had been static within western culture for ten centuries were being questioned because of many cultural and technological factors: the introduction of printing, the reemergence, in quantity, of the intellectual remnants of Greek and Roman culture, the progress in travel technology (navigation, sailing ships) and the technology of war. But in truth, the degree to which western culture was willing to entertain hitherto undreamed of views reflects how catastrophic and unsettling the era must have been. The phenomenon of upheaval in a culture, what causes it, why certain cultural institutions persist and others are tossed aside, is not completely under-

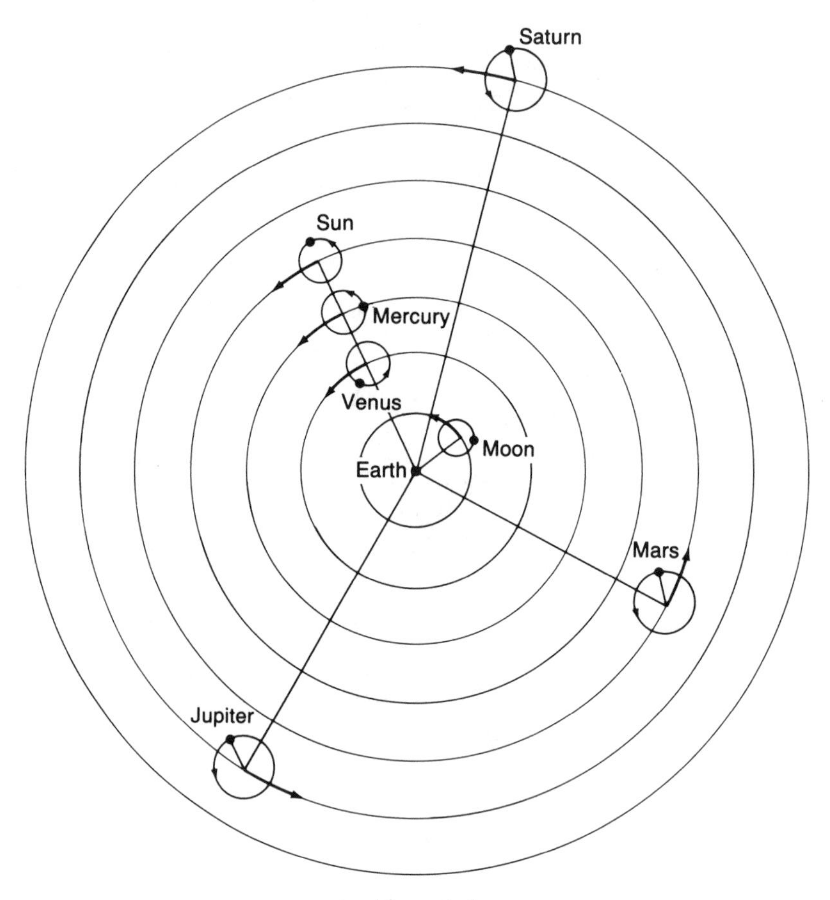

Figure 1.4

stood and may not be amenable to rational analysis. Even when dealing with rational subjects like science, people, even those acting in concert, may act irrationally. Furthermore, emerging questions and institutions are part of the catastrophic change so that *cause* and *effect* become difficult to sort out.

This seems to have been the case during the Renaissance. It was during this period that Protestantism was born. At the end of the transition between feudal and capitalistic economy Francis Bacon formalized, for the first time, a system in which science, technology and progress were linked by the concept of "useful knowledge." It was in this setting that Copernicus proposed his heliocentric system, which defied common sense and the hitherto entrenched Aristotelian concepts of the structure of the universe and of motion.

And yet a system centered on the Sun was eventually adopted. To make the heliocentric views acceptable did not mean exchanging the positions of the earth

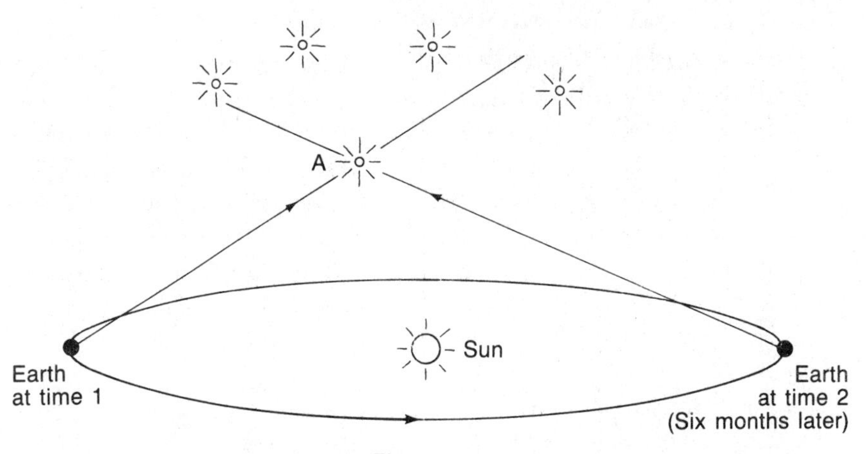

Figure 1.5

and the sun. Rather, a radical reformulation of cosmology, physics, religion, of the commonly held notions about the relationship of one class of men to another took place. All of this changed, and the heliocentric system was accepted, although at the beginning it was not a better predictive instrument and the kind of evidence that was necessary to support the system was wanting.*

THE GALILEAN ANALYSIS OF FREE FALL

It was in this period that the modern analysis of the motion of falling objects was born. The individual who is honored with the discovery of the analysis is Galileo. His reputation is well deserved but, in truth, at the time that Galileo worked, several other scientists were undertaking the same analysis in almost precisely the same terms. The idea was in the air. Galileo's unique contribution was his ability to proselytize and make the new point of view appear to be reasonable, appealing, and popular. It is significant that the mathematical formulation that Galileo used in analyzing the motion of objects had been known for almost two centuries. And, while it had been applied to the problem of motion in the fourteenth century, no fourteenth century scholar who developed the analysis had asked the question, "Does anything in the world actually move in this way?" This fact points up the importance of the cultural context in which such an analysis takes place.

* An example of the kind of evidence that was to be expected was the observation of parallax between the outer planets and the stars or between individual stars. Parallax is a phenomenon in which the apparent position of objects at different distances from the observer change, relative to each other. Thus, as the earth moves around the sun, the apparent position of the star A in Figure 5 will appear to shift, relative to the other stars.

Galileo was trained in the medieval Aristotelian tradition. His break with that tradition occurred about 1600. He began to defend the Copernican heliocentric system in the last decade of the sixteenth century, and his analysis of the motion of falling bodies was developed during the first decade of the seventeenth century. As we observed, this was a time of great social and intellectual ferment. During this period Galileo held professorships, first at Pisa, and later at Padua. He was surrounded by interesting, exciting, and stimulating colleagues and the atmosphere was intellectually open and challenging. It was under those circumstances, with others raising several of the same questions, that Galileo was able to make an analysis of the motion of falling objects and projectiles that has been incorporated in all subsequent theories of motion.

The mathematical language used by Galileo was geometry. Algebra was not to become the popular language of discourse in natural science until the eighteenth century. As a result, Galileo could not talk about mixed quantities, for example, the time-rate of change of position, or the time-rate of change of speed. He compared speed to speed, position to position, time to time, etc. We will not follow him faithfully in this, since we now find it easier to reason about these matters by using the language of algebra.

Galileo's first attempts to understand the motion of falling objects were not successful. There is a record of a letter he wrote in 1604 to his friend and colleague Sarpi that, if Sarpi were to allow him the assumption that as an object fell, its speed increased in proportion to the distance it had fallen, Galileo could prove that the distances traversed are proportional to the square of the elapsed time. Apparently, Galileo had earlier made a series of experiments to convince himself that indeed, when objects fall, the distance the object falls increases in proportion to the square of the time elapsed. He thought that this was true regardless of the nature, weight, shape, size, color, or texture of the object as long as one did not speculate about the resistive effects of the medium through which the object fell. Galileo's argument was defective but, before proceeding, let us translate his argument into mathematical form and compare the structure of the argument to other arguments we have already examined.

Galileo postulated that when an object falls, the speed of the object increases in proportion to the distance traversed. Since we will want to distinguish between speed, and change in speed, between time and change in time, between position and change in position (that is, distance traversed), we introduce a special mathematical symbol, Δ, to represent the notion of "change in" or "interval of." If we let t stand for time, Δt does *not* mean multiply Δ and t together. The symbol Δ (the Greek letter delta) is being used as an operator in the same way that $+$ means "add," $\times$ means "multiply," $-$ means "subtract" and $\div$ means "divide."

IL DIVINO GALILEO DI VINCENZIO GALILEI
PATRIZIO FIOR. FILOS. E MATEM.
DI FERDINANDO II. G. D. DI TOSCANA.

Galileo Galilei. Courtesy of the American Institute of Physics Niels Bohr Library.

Thus the symbol Δt is to be translated as "the change in the variable t." Since we are making t represent the time, the notation Δt means "the change in the time." Mathematically, this can be refined even further. If we specify two times, say a beginning time, t_1, and an ending time, t_2, where the subscripts denote given times, the interval of time may be specified as Δt. That is,

$$\Delta t = t_2 - t_1$$

Let us now carefully examine the structure of Galileo's argument, both in ordinary language and by using mathematical symbols, and compare its structure to the structure of earlier arguments. The symbol x represents the position of an object, the symbol t represents the time, and v represents the speed of an object.

> Postulate: If the speed of an object in freefall increases in proportion to the distance traversed [$\Delta v = b(\Delta x)$]

then

> Conclusion: the distance traversed should be proportional to the square of the time elapsed. [$\Delta x = c(\Delta t)^2$]
>
> Observation: the distance traversed when an object falls is proportional to the square of the time elapsed.

First, let us look at the structure of the argument and then we will examine the relationship between the English and mathematical statements (We again emphasize that this argument proved to be incorrect):

Galileo postulated that when something falls the speed increases in proportion to the distance. Call that assumption "*A*".

Galileo then utilized the logic of geometry (we utilize the equivalent logic of algebra) to reach the conclusion from the premise that the distance traversed should be proportional to the square of the time elapsed. Call that conclusion "*B*". It is *B* that Galileo observed. The argument may be summarized as:

If *A* then *B*
I observe *B*
therefore *A*

conforming to the earlier claim about the nature of argumentation in science when one attempts to connect theory with observation. Galileo could not prove that the speed of an object increases in proportion to the distance traversed. There may be any number of other sets of postulates that one might contrive to arrive at the conclusions that the distance traversed in free fall should be proportional to the square of the time elapsed.

At this point the reader and I might have challenged Galileo as follows: "Why didn't you measure the speed of the object at different points in its trajectory rather than measuring distances and times? Then you would know precisely what the relationship was between the speed of the object and the distance traversed."

The answer to that question is simple and crucial. We *never* measure speeds or changes in speed. Speed is an *abstract,* defined quantity. It is *defined* as the time rate of change of position.* We could define it many ways, but other definitions would not satisfy our intuitions about what we *want* "speed" to mean. We can only measure distances and times. Even speedometers in cars and other measuring devices which give us a "reading" of speed do not measure "speed." Rather, they measure the rate at which certain distances are covered, sometimes in *very* complicated ways. They measure distances and times and an engineer cleverly arranges it so that the pointer tells us directly, how those measurements translate into "time rate of change of position." So Galileo's argument is not limited by the technology of the time. No matter how complex we become and no matter that we now have computers and accurate high speed electronic clocks, high speed motion picture cameras, and highly accurate techniques for measuring distances, and lasers, and so forth, we can only measure distances and times. We never measure speeds. We make some assumption (postulate) about the relationship between speed and distance, or speed and time.

Earlier I said that this argument by Galileo was false. It was false in the logic that Galileo used in connecting his postulate A to his conclusion B. Without going into the details,† we can say that Galileo made two mistakes in his mathematical (geometric) argument and came to the false conclusion that his Aristotelian assumption that the speed of an object increased in direct proportion to the distance traversed, implied that the distance traversed is proportional to the square of the time elapsed. This result was obviously important to Galileo, because he was *convinced* that experiment told him that it was accurate. But he apparently was not yet prepared to break with the tradition of two millenia of collected wisdom. So intent was he on reaching the conclusion that the distance traversed is proportional to the square of the time elapsed that he made several mistakes in order to twist the premise of the logical argument to the conclusion he already knew must be the case.

Before examining the relationship between the mathematical statements in the argument and their ordinary language counterparts, let us see how Galileo eventually altered his argument. By 1609 Galileo had realized that something was wrong with his theory, for in that year he revised his argument in a way that has

* This is because distance and time are primitives in our system. We could build a system in which speed is a primitive. But given the kind of perceptual machinery with which we are equipped, we don't.

† This story is told forcibly and clearly in N. R. Hanson, *Patterns of Discovery* (Cambridge, 1958) chapter 2. It is significant, as Hanson points out, that the kind of mistake made by Galileo was made by others at about the same time and the steps that led Galileo to the proper formulation were also taken by others.

since remained unchallenged. Just what led him to see his error is not clear. In any event, the new argument (which was not made in full publicly until the year 1638) began with a different premise:

> Postulate: If the speed of an object in free fall increases in proportion to the time elapsed [$\Delta v = a(\Delta t)$]

then,

> Conclusion: the distance traversed by the object should be proportional to the square of the time elapsed [$\Delta x = \frac{1}{2}a\,(\Delta t)^2$]

(Note that the conclusion makes use of the same proportionality constant as the premise, but divided in half.)

Everything else follows as before. Galileo obviously thought that this conclusion was supported by experimental evidence. It is important to emphasize that the relationship between the postulate and the conclusion is mathematical. It is a matter of logic. One can, in fact, in an error free way, show that using the premise that the speed increases in proportion to the elapsed time correctly leads to the conclusion that the distance will increase in proportion to the square of the time. In other words, if, at the end of one second an object has fallen one unit of distance, at the end of two seconds it will have fallen four units of distance, at the end of three seconds it will have fallen nine units of distance, etc. Similarly, if, at the end of the first second the object, starting from rest had acquired one unit of speed, the postulate states that at the end of the second second, it would have acquired two units of speed, at the end of the third second, it would have acquired three units of speed, etc. The mathematical details of the argument are examined in Appendix 2. For now, let us make sure that the relationship between the mathematical statements and their ordinary language counterparts is clear.

Consider the postulate: To say that the speed increases in proportion to the time elapsed means simply that if the time elapsed is doubled, the speed is doubled. If the time elapsed is trebled, the speed is trebled; if the time elapsed is increased by a factor of 1.3, the speed is increased by a factor of 1.3.

The mathematical statement reflects this. The symbol "a" serves two functions: it is a proportionality constant that *scales* the relationship between changes in speed and changes in distance, and it provides a means for converting the units or dimensions of time into the units or dimensions of speed that must be distance over time. That being the case, the dimensions of this proportionality constant must be distance over time squared. This is a formality that allows us to make use of algebra and write a single mathematical statement containing oranges on one side of the equality sign and apples on the other or, as in this case, speed on one side and time on the other. It will often be the case, as we will see shortly, that the proportionality constants in such equations have definite physical meaning. Before looking in more detail at the scaling and dimensioning functions of the proportionality constant, let us translate the original equation into a verbatim rendition of the ordinary language version of the postulate.

$$(\Delta v)_1 = a(\Delta t)_1 \tag{1}$$

For some other time interval equation (1) becomes

$$(\Delta v)_2 = a(\Delta t)_2 \tag{1'}$$

As equals divided by equals are still equal, we can divide each side of (1) by the corresponding part of (1′). That is

$$\frac{(\Delta v_1)}{(\Delta v_2)} = \frac{a(\Delta t_1)}{a(\Delta t_2)}$$

As the symbol "a" occurs in both the numerator and denominator of the right side of this last equation, it can be eliminated and the equation becomes:

$$\frac{\Delta v_1}{\Delta v_2} = \frac{\Delta t_1}{\Delta t_2} \tag{1''}$$

Mathematically when we *compare* the increase in speed during one time interval to the increase in speed for another time interval, we must take such ratios. The proportionality constant plays no role in the final statement. To say that one variable is proportional to another is equal to saying that one variable is equal to the product of a constant and the other variable.

Let us now examine the dimensioning role of the constant. As each variable in the mathematical statement used to describe physical situations means something, we identify the variables as representing things like time, length, force, work, mass, etc. Each concept has a particular meaning within physical science and, with each, there is associated a dimension. In order to manipulate the dimensions, the same mathematical laws that apply to variables are applied to the dimensions. In other words, the logic relating the dimensions is precisely the same as the logic relating the variables. Of course, as with any logical system, we must begin with postulates and undefined terms. The undefined terms for the dimensional system

Table 1

Dimension	*CGS System*	*English System*	*MKS System*
length *[L]*	centimeter (cm)	foot(ft)	meter (m)
mass *[M]*	gram(g)	pound (mass)	kilogram (kg)
time *[T]*	second (sec)	second (sec)	second (sec)

we will be using are "length," "time" and "mass." These are the common, undefined primitives of physics. Although I can tell you how to measure a length, or a time, or a mass by a formula, the concepts themselves are undefined. Either you know or you don't know what I mean when I use a phrase like "time passes." The dimensions of all other quantities in physics can be expressed in terms of these fundamental quantities.* In any physical equation, just as the numbers must be balanced, so must the dimensions or units. Generally, when dealing with dimensions, the "quantities" are abbreviated and placed in square brackets. Thus the dimensions of "length" are $[L]$, of time $[T]$, and of mass $[M]$. Table 1 lists several systems of units. One is not better than the other. They are different, stemming from different historical roots; each has advantages and disadvantages. More and more in the face of international technological cooperation, nations have adopted one of these systems: the MKS system.† Very few countries still adhere to the English system. In fact, America is one of the handful which has yet to switch.

The standard of length in the MKS metric system, the meter, was originally defined as one ten-millionth of the distance along the line between the north pole and the equator, passing through Paris, France (It was the French who invented the metric system). That length was inscribed on a bar of very stable metal (Platinum-Iridium) as the distance between two scratches. The bar was stored in a vault in France at a constant temperature. It became the primary standard. Secondary standards could then be made by direct comparison and sent to the bureaus of standards of other governments. These secondary standards could be used to make tertiary standards and given to manufacturers of measuring devices. The manufacturers, by direct comparison with the tertiary standards could then make quaternary standards which would be used directly in the manufacture of devices such as tape measures, meter sticks, etc. (The primary standard of length is no longer kept on a bar of metal. Rather, it is defined as a certain multiple of the wavelength of a particular line in the spectrum of the element Krypton.) The standard Kilogram is a cylinder of metal which is kept in a vault in Sevres, France. A second was once defined in terms of the rotation of the earth: one part in eighty-six thousand, four hundredth of the time required for the sun to make two

* Some systems of units make use of a fourth fundamental, undefined, quantity, "electric charge." If electric charge is not introduced as a primitive, it must be defined in terms of mass, length and time.

† The MKS system is actually only a core part of an extremely comprehensive system of units known as SI which has recently been adopted by the majority of countries in the world. There has always been a good deal of heat generated about which system is the best system. Some insight in that controversy can be found in the following: U.S. Metric Study Interim Report: *A History of the Metric System Controversy in the United States NBS SP 345-10* (Washington: Government Printing Office, 1971).

successive crossings of the meridian passing through Greenwich, England. This, however, was found not to be precise.

Currently, the second is defined as a certain number of vibrations of a particular electromagnetic disturbance which can be measured in the element Cesium.

Let us return to equation (1). Recall that it is the mathematical expression of the postulate with which Galileo began his analysis: the speed increases in proportion to the time. Since speed is defined as the time rate of change of position, the dimensions of speed must be length *[L]* over time *[T]*. The dimension of time—on the right side of equation (1)—is obviously time *[T]*. Therefore, we can rewrite equation (1) in terms of its dimensions as follows:

$$[L]/[T] = a[T]$$

We have not put the units of "*a*" in the equation because that is precisely the factor that now has to be determined. This equation can be solved for "*a*" by dividing both sides by *[T]*.

$$a = \frac{[L]/[T]}{[T]} = \frac{[L]}{[T]^2} \qquad (1''')$$

In other words, the units of "*a*" are length over time over time (that is the time rate of change of *speed,* that is, length over time, squared). But the time rate of change of speed is simply what we mean by "acceleration." Therefore, the proportionality constant in equation (1) plays the physical role of acceleration. In this case, the acceleration is a constant. (Of course, things do not *have* to accelerate in that way. As in the case of a car moving forward it might accelerate with different rates of acceleration.) So, for Galileo to have postulated that the speed increased in direct proportion to the time was the same as to have postulated that in free fall, the acceleration of an object is constant. In fact, since no object has been specified, the postulate states that in free fall, all objects accelerate at the *same* constant rate, regardless of shape, color, size, texture, etc. The only determinant of speed at any moment is the elapsed time.

Dimensioning the constant has helped us to gain insight into its physical meaning. Of course, this was a very simple case. In any event, just as the magnitude of the numbers on either side of the equal sign in an algebraic statement must be the same, so must the dimensions. Using the dimensions of "*a*" as determined in (1‴) we see that the dimensions of (1) are indeed the dimensions of speed on both sides of the equation. This check is a useful technique for insuring that a mistake in logic has not been made. For example, we said earlier that the logical outcome of the analysis which begins with the assumption that speed is proportional to time elapsed is that the distance traversed should be proportional to the square of the time elapsed, that is:

$$x = \tfrac{1}{2}a[\Delta t]^2 \tag{2}$$

As a check to see that no mistake has been made in the logic, we can check the dimensions of equation (2) to be certain that they balance. The dimensions of distance are $[L]$, the dimensions of acceleration are $[L]/[T]^2$ and, since the dimensions of time are $[T]$, the dimensions of time squared must be $[T]^2$. Equation (2) becomes:

$$[L] = \frac{[L]}{[T]^2} \cdot [T]^2 = [L]$$

This agreement gives us confidence that the analysis was not logically wrong. In order to see that this is not trivial, if you begin with the postulate that the speed of an object increases in proportion to the distance, the dimensions of the proportionality constant thus obtained are not the units of length over time squared. You can determine what they are yourself and show, by inserting the appropriate dimensions in the conclusion, that the conclusion of the argument cannot be that the distance is proportional to the square of the elapsed time.

We come now to the question of the magnitude of the proportionality constant. If you again examine equation (1) the size of "a" will scale the solution to the problem in that it will tell us how the speed is going to change during each second of time, and consequently, in equation (2) it will tell us how the distance will vary with changes in the time. Since we never measure speeds, the easiest way to determine the magnitude of the proportionality constant "a" is to use equation (2), observe the time required for objects to fall different distances and then solve the equation for "a." This, in fact, is done in practice. For Galileo the problem was very difficult because the motion of falling objects is very fast. He had to resort to special auxiliary arguments which we did not follow here. They are discussed in detail in Appendix 2. Today, the problem is not very difficult at all. Modern electric and electronic clocks make the problem easy to solve. At convenient heights, we stretch wires across the proposed path of a falling object. The wires are connected to clocks. The object (say a steel ball, Figure 6) is held by an electromagnet. When we turn the magnet off, the ball begins to fall. We use the same switch which turns the magnet off to turn the clocks on. As the ball falls it successively breaks each of the wires, thereby turning off the clock associated with that height. This gives us, within the limits of accuracy of our measurements in constructing the apparatus, a measure of the relationship between the distance traversed and the time elapsed. The value that one gets for the acceleration in free fall is subject to several local variations. Its value is very nearly 32 ft/sec^2 in the English system, 9.8 m/sec^2 in the MKS system, and 980 cm/sec^2 in the CGS system. In experiment after experiment, the distance fallen in one second is sixteen

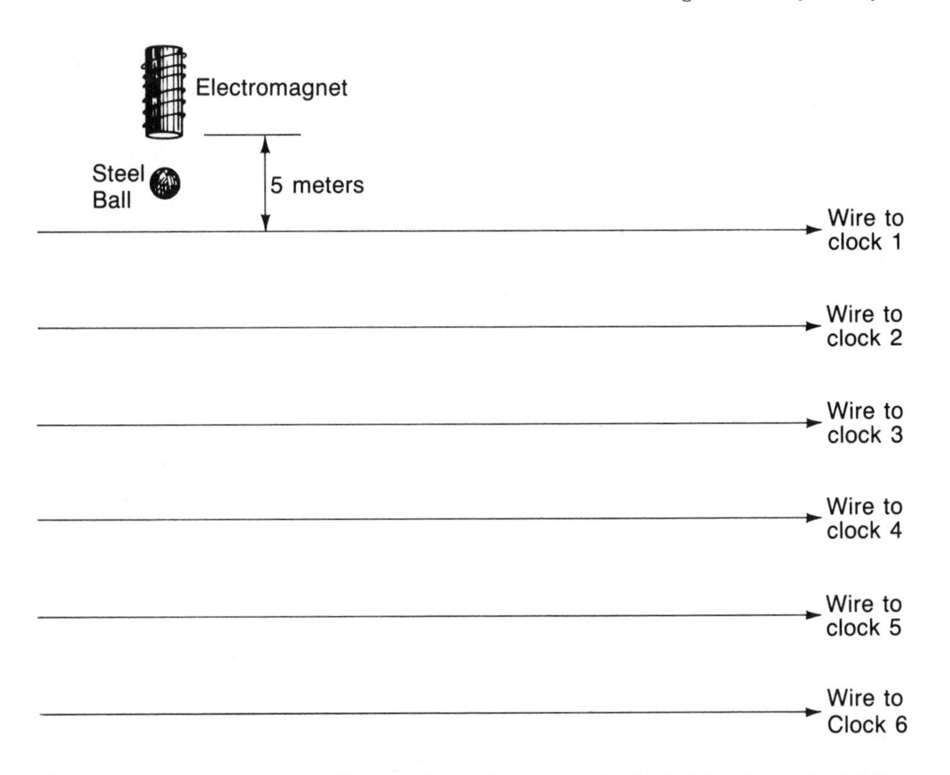

The wires have been placed a distance five meters apart. As the ball breaks each wire the clock associated with the wire is stopped. The time then can be read at the leisure of the experimenter. The errors which can be made are in the placement of the wires, the tension in each of the wires, and the internal workings of the clocks. In order to calculate "a", we use the formula $x = \frac{1}{2}a(\Delta t)^2$ and solve it for "a". $a = 2x/(\Delta t)^2$

Taking any two time in table 2 which is a set of data obtained in the experiment, squaring it, and dividing it into twice the total distance fallen will give a value for a. If now, all the values of "a" are added up and divided by the total number of "trials" we will get an experimental, average value of "a", $\bar{a}$.

Table 2

	t (seconds)	x (meters)
Clock 1	1.00	5.0
Clock 2	1.47	10.0
Clock 3	1.73	15.0
Clock 4	2.05	20.0
Clock 5	2.24	25.0
Clock 6	2.50	30.0

a_1 =	10.0 m/sec^2
a_2 =	9.26 m/sec^2
a_3 =	10.02 m/sec^2
a_4 =	9.52 m/sec^2
a_5 =	9.96 m/sec^2
a_6 =	9.60 m/sec^2

$$\bar{a} = \frac{a_1 + a_2 + \ldots + a_6}{6} = 9.73 \text{ m/sec}^2$$

Figure 1.6

feet (almost five meters); in two seconds, sixty-four feet (almost twenty meters); in three seconds, 144 feet (almost forty-five meters); etc. Recall that by postulating a constant acceleration that proportionality constant is divided by two in our conclusion. Table 2 represents the types of data that one might expect to obtain in such an experiment (forgetting, of course, that error will intrude). In the caption, we have calculated the value of "a" one would obtain from such data.

The analysis of free fall was only one part of Galileo's later studies of motion. He applied his conclusion to the problem of projectile motion. In that case, the postulates and the interaction of the postulates with the conclusions are more subtle, and the techniques for comparing those conclusions with experience more difficult. Galileo assumed that projectiles, for example, cannon balls, ballista, arrows, etc., were simply objects in free fall once they left the impelling implement, that is, the cannon barrel, the catapult, or bow. The role of the projecting device was to endow the object with an initial speed. Therefore, the analysis of the length of time it would require for the object to hit the ground should proceed precisely the same as it proceeds for an object in free fall, except for the effects of the initial speed.

Galileo had the incredible insight to realize that one could vastly simplify the analysis by taking advantage of the fact that the effects of gravity are only along the vertical. In Figure 7, there is a typical cannon barrel. There is no question that physically, the cannon ball will move along the line of the barrel. For the sake of analysis, it is convenient to divide the motion into two parts. For this purpose, as we show in Figure 8, we make use of trigonometry (Appendix 1) and decompose the motion of the cannon ball along the line in which gravity acts and along a line perpendicular to gravity, the horizontal. In that direction, as Galileo argued, the motion is uniform, unchanging (save for wind resistance and other frictional effects), and an object should cover the same distance in every equal interval of time.

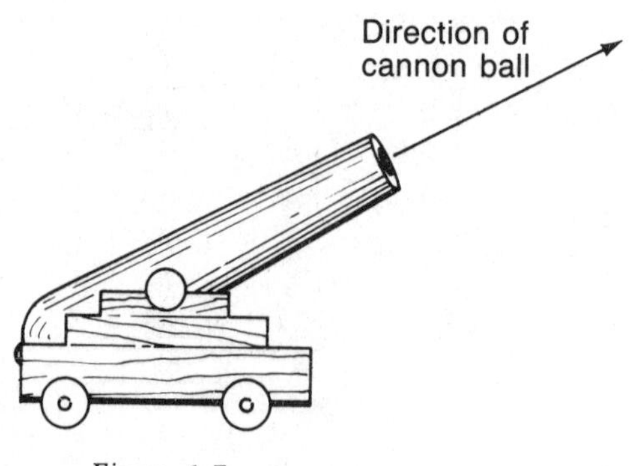

Figure 1.7

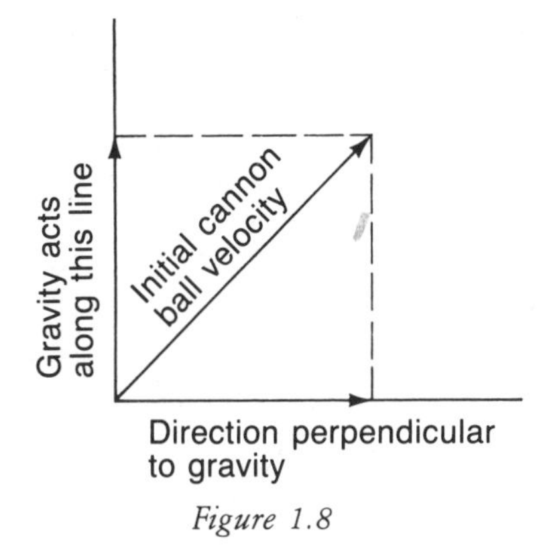

Figure 1.8

When both the magnitude and the direction of a quantity have significance, a special mathematical language is used. For example, in ordinary language, we often want to distinguish between motion (speed) and motion in a given direction (velocity). It makes a difference to the point of arrival if you are traveling fifty miles per hour north, or fifty miles per hour east. Your *speed* in both cases is fifty miles per hour. Your *velocity,* however, is different in the two cases. When only the magnitude is important, a number (or the variable representing a number) is called a *scalar.* When the magnitude and direction are both important, the number or variable is called a *vector.* Speed is a scalar; velocity is a vector. The mathematical rules for manipulating vectors are different than for scalars. For example, as illustrated in Figure 8, vector addition must be done geometrically. In this text vectors will be designated by a small arrow over the symbol, thus:

$$\vec{A}$$

Again, let me emphasize that *physically* the canon ball does not move in two directions at the same time. Physically, the cannon ball moves along the line of the cannon barrel. The analysis is one of convenience. In one direction the object acts like any falling body once it has left the projector, in the other direction the effects of the projector are to propel the object horizontally, equal distances in equal times. The details of this analysis are in Appendix 2. Some of the logical outcomes of this analysis are that the trajectory of projectiles should be parabolic, that is, the shape of the path followed is identified as a parabola, a curve having definite mathematical properties; the speed of the object when it hits the ground and the line of flight at that instant will be precisely the same as the speed and line of flight when the object left the projector (assuming the muzzle is on ground level), and the maximum range of the projectile occurs when the projector makes an angle of 45° with the horizontal. Figure 9 is an analysis of one trajectory.

Note that in each case we have examined, the connection between the

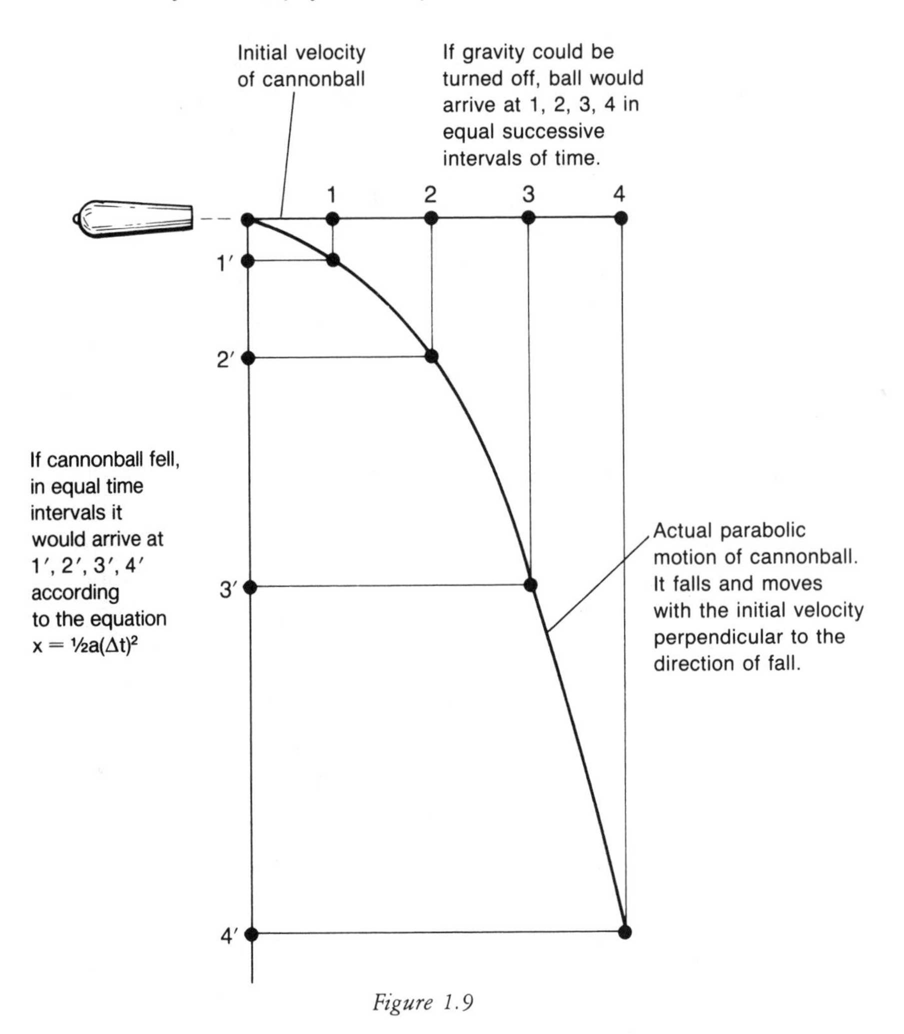
Initial velocity
of cannonball
If gravity could be
turned off, ball would
arrive at 1, 2, 3, 4 in
equal successive
intervals of time.
1
2
3
4
1′
2′
3′
4′
If cannonball fell,
in equal time
intervals it
would arrive at
1′, 2′, 3′, 4′
according
to the equation
x = ½a(Δt)²
Actual parabolic
motion of cannonball.
It falls and moves
with the initial velocity
perpendicular to the
direction of fall.

Figure 1.9

conclusions and the premises is a logical one. The measurements or observations have no logical relationship to the premises of the argument. It might be possible to arrive at the same conclusions from a different set of premises. For example, as I pointed out earlier in the case of free fall, even though we have become very sophisticated about how we measure distances and times, we, to this day, do not measure speeds directly. There is no way to measure a speed and there will never be a way to measure a speed directly. Thus, it will never be possible to prove (in the sense that most people refer to *proof* in natural science) that Galileo chose the assumption about free fall which is true, that is, the correct representation of the physical world.

We have used the well known case of Galileo's law of free fall as an example of the relationship between mathematics, logic and natural science, and it is a typical example. As will become clear in our discussion of the theory of relativity, the same relationships pertain between the various mathematical formulations of the theory and that which the theory claims about the nature of the physical world within its domain.

BELIEF AS A CULTURAL PHENOMENON

Our examination of the theory of relativity and how it was received will illustrate the theme that shifts in belief in science, as in all aspects of human activity, are cultural phenomena subject to intellectual fashion. There are similar shifts regarding the correct components of a curriculum in schools or the proper elements that should go into a work of art.

In making such an examination, it will be argued that, while experiment was an important ingredient in the development and propagation of the ideas in the theory of relativity, experiment did not play the usual role in the sanitized fairy tale version about ideas and how they are confirmed or falsified in science. We have already provided part of the argument for this idea. As we have seen, any experimental result can be explained by an appropriate set of postulates. Galileo chose the postulate of uniform acceleration in the case of free fall. He was convinced that when an object falls, the distance traversed is proportional to the square of the time: the logical outcome of his postulate. But it is not the only postulate that he might have chosen. In an earlier period of history, it would have been acceptable to invoke a postulate in which the desires of the falling object played a role, but such an argument was no longer acceptable by the time Galileo wrote.

Galileo did not explain how things fall by invoking motivations of the creator or the interactions of objects. His explanation began with the assumption that, whatever the reason, in free fall the acceleration is constant. When asked to explain why things fall the way they do, Galileo demurred. There is no way that one can prove that in free fall acceleration is a constant. On the other hand, there has never been another *mathematical* account of free fall which began with another set of assumptions and led to the conclusion that the distance traversed is proportional to the square of the time. We would agree this was the result of careful measurement relating the time of fall to the distance traversed.*

* Note again that this was not the case during Galileo's time. The technology for measuring short time intervals was very primitive and the motion of falling objects is very fast. Galileo's primary data was derived from observing metal spheres rolling down inclined planes. The first question then would be what is the relationship between motion on an inclined plane and free fall? In order to answer that question, Galileo turned to the motion of a pendulum. The details of these issues are discussed in Appendix 2.

No matter what the issue is, ultimately we have to agree about the data. "The needle points to six." Either it does or it doesn't. "The go light is on." Either it is or it isn't. If there were serious disagreements, we could come to an agreement about what happened, what the numbers were, what the sequence of phenomena are.

It is the impenetrable barrier between the results of experimental study and the postulates one uses to account logically for results which underlies the fact that belief in science is as much subject to fashion, social whims, fads, national and cultural differences as is belief in other fields of study.

The experimental results do not change with time. Given that the experiment is properly done, the result stands like a rock, forever. Of course, there may be technological improvements or new and ingenious techniques for arriving at answers to experimental questions. These developments are important. But what changes with time is the nature of the postulates that are acceptable for *explaining* experimental results. Regardless of the result, it can be explained by appeal to any philosophical viewpoint. Is the viewpoint culturally acceptable? is the question.

It was not experiment, nor superior predictive power, nor a physical observation that caused the rejection of the geocentric view in favor of the heliocentric view during the sixteenth and seventeenth centuries. In fact, the shifts in the outlook that individuals had of the world about social, psychological, philosophical, cosmological, and physical theories were so radical, that, *a priori,* it is not clear that one would go through such tortuous transformations in outlook in order to adopt the heliocentric view. With hindsight we can see that it was not a rational route. There were reasonable geocentric theories which explained observations as well as heliocentric theories. For reasons not yet clear, the *postulates* of all geocentric theories became unacceptable. There will always be a mystery about shifts in outlook, because the shift is irrational. It is disconcerting because the subject matter appears to be rational. But, as we can see as a result of our analysis of the relationship between formal logical systems and empirical observation, we are not talking about discovering the truth about the universe, but rather about individuals acting in concert. We are talking about social forces, culture, and belief. As is the case about other areas of contemplation, shifts in point of view in the social sphere are no more and no less rational than shifts in other spheres.

OBJECTIVITY AND THE SOCIAL INSTITUTIONS OF SCIENCE

If belief in science is subject to the same kinds of social forces as belief in other spheres of human activity, what of the commonly held views about the objectivity of scientific knowledge? Do we not rely on procedures like scientific method to insure that the results a scientist arrives at will be objective and independent of the biases of the scientist?

Scientists are human. Like all humans, they carry with them a set of predilections, biases, and preconceptions about the world which, for each person, results from a unique blend of familial, social, and individual interactions and backgrounds. Some may be due to heredity, some to environment. There will never be a way to sort it out, nor is it important to do so. Furthermore, as we have already seen, individual beliefs in science are culturally shaped and subject to cultural currents. These influences interact with each individual's predilections.

Nothing special happens to such a person when he or she is doing science. No mantle of objectivity descends when a white laboratory coat is put on; the mind doesn't become a blank slate before considering theoretical formulations; no special method of procedure insures that the persons doing scientific studies become neutral instruments of nature's way. In fact, there is no such thing as *a* scientific method for insuring objectivity. There are methods—approximately one for each scientist.

This must be the case if there can be no unique connection between the postulates about the nature of the physical world and the results of an experiment. It has been very common to train young people to believe in a *scientific method*. That method is usually encapsulated in an algorithm for keeping a notebook in the laboratories of science classes in elementary school and high school.

Everyone remembers those notebooks. Each experiment was described in a report with headings like "Purpose," "Equipment," "Theory," "Method," "Result," "Errors," and "Conclusions." In fact, no one works that way in science. Given our earlier analysis, it should be clear that a routine organizing scheme for reporting an experiment will not impart the objectivity that is usually associated with doing scientific work. In fact, there is no objectivity. People always have a stake in the outcome of their work; there is nothing wrong with that.

If there is no scientific method that will confer objectivity on the work of individual scientists, there *are* socially accepted canons of procedure that confer a kind of objectivity on the product of the work of individual scientists. Objectivity in science is not a function of individual effort; it is a result of the social institutions of science that evolved during the sixteenth and seventeenth centuries.

The most important aspect of the scientific revolution was not that discoveries were made about the world. As we have seen, those discoveries say more about changing belief than they do about nature. The most significant aspect of the scientific revolution was the emergence of scientific societies organized by individuals who knew that a social medium for sharing experiences and meaning was needed.

Before that time, people felt that they had a stake in keeping their work hidden from the eyes of others. For example, alchemists rarely said what they did in the laboratory. They wrote strange poems or mythical accounts in which were hidden the techniques they used. Leonardo da Vinci kept most of his significant work secret and much of it was not discovered in his notebooks until the beginning of this century.

In the sixteenth and seventeenth centuries, the fashion changed. There have been many attempts by historians and social scientists to account for this change in attitude, focusing on economic, technological and political forces.* Work that had hitherto been purposely private was now to be purposely public. If one were to participate in the activities of scientific societies, one had to agree to follow the canons of procedure laid down by these societies. These canons included public disclosure of techniques and data. This provided the possibility of replication of empirical and theoretical results. Peer review of analysis provided the possibility of examining, and sharing or rejecting individual biases and prejudices. In the seventeenth century, such formalized institutional canons in science were new. This was part of the track on which the driving engines of the scientific revolution were to ride.

Individual bias has not been eradicated, but the public posture protects the scientific culture from individual whim. It means keeping a notebook so that, if called upon, the techniques used in a result can be replicated. It is the social organization of science that protects individuals from themselves and from each other and which helps to insure theoretical and experimental results that will stand the test of time. It would be naïve, however, not to recognize that there are times when that social organization obstructs the road, protecting the *status quo* from techniques and ideas that some participants in the scientific enterprise find heretical or threatening.

SCIENCE AND TECHNOLOGY

There has been a good deal of confusion in western culture about science and technology since the time of Sir Francis Bacon. They are not interchangeable. For our purposes *science* is defined as an effort to *understand* the nature of the world of experience. *Technology* is a method of *manipulating* the world of experience. It should come as no surprise to the reader that one can become very good at

* The reader might consult the following: R. K. Merton, "Puritanism, Pietism, and Science," or "Science and Economy of Seventeenth Century England," in Barber and Hirsch (eds.), *The Sociology of Science* (New York, 1962) pp. 33–88. M. Purver, *The Royal Society: Concept and Creation* (Cambridge, Mass., 1967). M. C. Jacob, *The Newtonians and the English Revolution,* 1689–1720 (Ithaca, 1976). There are many other studies which might be consulted. The reader is directed to the excellent bibliographic essay in Jacob's book.

technical manipulation without understanding or caring how the machinery works. Consider the millions of expert drivers who care nothing about how an engine runs.

Science is almost never required for technological innovation. Consider the years of work in husbanding plants and animals before the creation of the science of genetics. Thomas Alva Edison understood little about the nature of electricity or sound, and Alexander Graham Bell was not knowledgable about electromagnetic theory. Both men had a genius for manipulating physical things and more important, for manipulating patent laws. Theoretical understanding is not required. It has been very popular in our society to argue that technological growth depends on the progress of science and scientific ideas. In fact, for every case where there seems to be a direct connection between a development in science and a technological breakthrough, one can probably point to twenty cases where it was technological breakthroughs that suggested to scientists questions about the way the world works. In the sense that we have defined science and technology, science does not progress but technology does.

Admittedly, there is a large grey area and it is sometimes difficult to separate science and technology, but it is no more difficult to sort out the contributions of science and technology to the creation of certain artifacts than it is to sort out the contributions of technology and art to the creation of certain other artifacts. There is, in fact, a one-to-one correspondence in the relationship of technology to science and the relationship of technology to art.

Just as it was the operation of the steam engine that suggested the science of thermodynamics, it was the creation of Corten Steel alloy that suggested the creation of many kinds of outdoor sculpture. The invention of epoxy paints has allowed for aesthetic expressions not earlier possible.

* * *

We turn now to the theory of relativity—the source of the ideas in the theory, how the theory was created, how it was received and assimilated—to illustrate these issues. The central ideas of the theory are simple. Anyone who can read can understand the scientific issues. I will use the development of that understanding to consider the fate of the ideas in the theory of relativity in emerging twentieth century culture within and without the scientific community.

·2·

The Rise and Fall of the Mechanical World View

WE WILL EXAMINE, IN THIS chapter, the development of Newtonian mechanics and how it related to the earlier work of Galileo as well as how it synthesized our understanding of the physical world, first by removing the distinction between phenomena on the earth and phenomena in the heavens. Later Newtonian mechanics underlay the view that all physical phenomena would be understood in mechanical terms. This view, referred to as the mechanical world view, incorporated the hope that Newton's laws would become the basis for explaining everything, not only the physics of motion, the physics of heat, electricity, magnetism, and light, and also chemistry, geology and biology, including the workings of the body, genetics, the working of the nervous system and the way the brain and the emotions function; *everything* was to be understood in mechanical terms. This program, which emerged at the end of the Enlightenment, was not successful in practice. It was during the attempts to modify it, near the end of the nineteenth century, by replacing mechanical interactions with electrical interactions as the basis for understanding the universe, that the theory of relativity emerged, simultaneously with growing skepticism about a unified account of the universe.

In Chapter 1 we reviewed the Aristotelian concept of motion and how that theory led to certain predictions about the behavior of objects in motion. We examined how that theory explained what was observed by positing a zeroth order abstraction: The motion of objects was explained by saying that what was observed to happen had actually happened. Things move only when forced to move. Left to their own devices things come to rest. Similarly, the observed motion of objects in the heavens had to be taken literally even though those motions were of an essentially different character from motions on the earth.

Galileo's analysis of free fall did not actually attack the premises of Aristotle's concept of motion. Galileo attacked the description of motion in which the Aristotelians had placed so much stock. The Galilean description of two kinds of motion, the motions of freely falling bodies and projectiles, was based on the premises that the acceleration due to gravity is constant for all bodies and that motion in directions perpendicular to the line along which the acceleration due to gravity acts is uniform and unchanging.

This view contradicted the Aristotelian view that the natural state of objects was the resting state, but Galileo did not attempt to replace that view with a new theory. He presented evidence that he accepted, implicitly, much of the Aristotelian theory. This is not strange since he was trained in that tradition.

It was only later that the premises of the Aristotelian theory of motion were replaced formally by Newtonian views. It should be emphasized that this does not mean that intuitions about the way things move were suddenly and dramatically changed. To this day, an untrained person will employ Aristotelian concepts when called upon to explain motions that are unfamiliar. These concepts emerge directly from naive observation.

THE SEVENTEENTH CENTURY

Isaac Newton was born in 1642, the year that Galileo died. His early years were a time of great ferment in England. There was civil war, Charles I was tried, and beheaded on January 20, 1649. The Commonwealth, a republican form of government under the leadership of Cromwell, himself proclaimed Lord Protector of the Commonwealth (and succeeded by his son upon his death), held power until the restoration of the monarchy in 1660.

When Charles II died in 1685 and the Roman Catholic James II succeeded him, the issue of religious freedom and the test of will between the crown and the anti-papists led Parliament to declare James' abdication. They offered the crown to William III and Mary. There ensued the Glorious Revolution and war with France.

Yet, it was in the midst of this political and social unrest that the scientific community was explicitly organized, not only in England, but in France and other parts of Europe. The Royal Society of London was founded in the year 1665. Its founding was the culmination of years of informal gatherings among leading natural philosophers. Early members included, not only Newton, but Edmund Halley, Robert Boyle, Robert Hooke, Christopher Wren, Samuel Pepys and Henry Oldenberg. Examination of the charter of the Society shows how closely related it was in concept to the Baconian notion, not of knowledge, but of *useful* knowledge and the wedding of knowledge to the idea of material progress. Bacon had died in 1629, but in the founding of the Royal Society we see his dream of the organization and discovery of knowledge that he had written about in *The New Organum* and *The New Atlantis.*

There were several notable changes in the behavior of natural philosophers during this period. Before the time of Galileo, it was generally accepted that scientists worked alone, kept their results to themselves, possibly for their own gain, but, if for no other reason, it was the custom. The Royal Society received a charter from King Charles II, and the members met regularly to share information, experiments, and theories. Demonstrations were organized on a regular basis to entertain and satisfy the curiosity of the membership, and the journal, *The Philosophical Transactions of the Royal Society of London,* contained not only summaries of the meetings, but reports of the work of scientists who communicated by mail with the Society. That which had hitherto been private and secret was now open and public. Indeed, public disclosure quickly became an important criterion for judging the worth of a piece of work. And, looking over the pages of early issues of the *Transactions,* one cannot but be impressed by a sense of how wonderful the world seemed. Suddenly it was the norm to believe that man could have direct knowledge of the way the world worked and that the path to that knowledge was experiment and observation. Knowledge of the world was seen, perhaps for the first time, not only as possible to obtain, but as a necessary precursor to technological development and material progress.

Similar attitudes were developing at the same time in other parts of Europe and America. In France, Louis XIV not only authorized the establishment of the Academy of Sciences, but provided funds to pay professionals to carry out technical work in the service of the Crown under the aegis of the Academy. It was the beginning of the Enlightenment, a period which, near the end of the eighteenth century, was characterized by the philosopher Immanuel Kant as the period when "man dared to know."

Newton's synthesis of the physics of motion, or, as it is more commonly known, Newtonian mechanics, was a key element in shaping and directing the character of the Enlightenment. It became the mode of explanation sought by others, from John Locke in attempting to understand the laws governing the growth of the mind and in proclaiming that there are natural laws and natural rights as well, to August Comte who tried to understand the natural laws governing the growth and interaction of societies. These attitudes were summarized by Alexander Pope in his *Essay on Man:*

> All are but parts of one stupendous whole
> Whose body Nature is and God the soul; . . .

All Nature is but Art, unknown to thee;
All chance, direction, which thou canst not see;
All discord, harmony not understood
All partial evil, universal good:
And in spite of pride, in erring reason's spite,
One truth is clear, what ever is, is right.

Newton had not only discovered the laws governing the motion of objects, he had also discovered that God's universe was exact and functioned perfectly like a large mechanical clock. To study and know nature was to study and know God. As Pope wrote elsewhere,

Nature and Nature's laws lay hid in night:
God said, *Let·Newton be* and all was Light.

Isaac Newton. Portrait by J. Vanderbank in the National Gallery, London. Courtesy of the American Institute of Physics Niels Bohr Library, W. F. Meggers Collection.

According to Newton and those who shared his outlook, the discovery of the laws of nature came about by *reasoning* from experimental evidence to the general relationships which are responsible for experimental behavior. One had to avoid idle speculation about causes. One had to avoid speculation of the kind which endowed inanimate nature with motivation. This outlook, mirroring as it does the Baconian notion of the relationship between evidence and belief, was referred to as mechanical, or natural, philosophy. It does not correspond to the view presented in this book as the relationship between evidence and belief, but more than that, with hindsight we can see that it does not represent a description of the behavior of natural philosophers who subscribed to the mechanical philosophy in the seventeenth and eighteenth centuries.

The work that is described today as Newtonian mechanics was produced by Newton, on the urging of others, in the years 1685–1687. But the core of the reasoning had been accomplished about twenty years earlier when Newton was only twenty-three. Newton had been a university student and had returned to the family farm for two years while the plague raged at Cambridge. He has written that in that period he discovered:

> . . . first the binomial theorem, then the method of fluxions [the differential calculus] and began to think of gravity extending to the orb of the moon and having found out how to estimate the force with which a globe, revolving within a sphere presses the surface of the sphere, from Kepler's rule, I deduced that the forces which keep the planets in their orbs must be reciprocally as the squares of their distances from their centers; and thereby compared the force requisite to keep the moon in her orb with the force of gravity at the surface of the earth, and found them to answer pretty nearly. All this was in the two plague years of 1665 and 1666 for in those days I was in the prime of my age for invention and minded Mathematics and Philosophy more than at any time since.

To put it mildly, Newton was a productive person. His genius was, in part, the ability to concentrate on the problem at hand with a ferocity that is foreign to most of us. There is also ample evidence that for at least part of his life, Newton was not a very appealing person. There are many anecdotes, some no doubt apocryphal, about his abuse of others, his paranoia, his unusual behavior. While it is fascinating, it is a side note to our main thread.* We turn to consider the nature of

* There are many biographical studies of Newton. The newest, by R. S. Westfall, *Never At Rest* (Cambridge and New York, 1980) must be considered the standard against which all others should be measured. Earlier biographical studies range in character from idealized lionizations to attempts at *post hoc* psychoanalytic reconstructions and analyses of his presumed neuroses and paranoia. Compare L. T. More, *Isaac Newton* (New York, 1934; repr. New York, 1962) and Frank Manuel, *A Portrait of Isaac Newton* (Cambridge, Mass., 1968). For a beautiful, short, yet penetrating

the evidence with which Newton and his contemporaries worked and the relationship of that evidence to Newton's theory and the mechanical world view.

The core of Newtonian synthesis was to unite the physics of motion for phenomena near the surface of the earth with the physics of motion for phenomena in the heavens, and to remove, as explanatory premises, anthropomorphizations of inanimate objects, the intervention of spirits, and the creation of special factors solely for the purpose of explanation. The Aristotelian and medieval identification of lack of change (change being a generalization of the concept of motion) with perfection was to be forgotten. According to Newton, whatever state of motion an object was in, was hypothesized to persist in the absence of intervention of outside influences. Since no motion was ever observed to persist, the Newtonian analysis was not to account for why things moved, but why things changed their state of motion. Rather than trying to explain only why a resting object would start moving, the Newtonian analysis focussed on why a moving object would come to rest.

The concept of the persistence of motion is called "inertia." It is a concept that was first discussed by the French philosopher and contemporary of Galileo, René Descartes. And although Newton adopted the Cartesian concept of inertia, he was opposed to Descartes' attempts to account for different kinds of motion. In fact, Newton's work is laced with anti-Cartesian jibes and jests.

In Descartes' attempts to understand motion and the nature of the universe, he persisted in inventing special entities which played key roles in the explanatory pattern. For example, hypothesizing that the universe was filled with a primal fluid, Descartes speculated that both light and matter were created from the fluid under special circumstances. According to the Cartesian mechanical concept, all interactions of material objects were by contact. Thus, the motion of the planets in the solar system was explained by hypothesizing that the primal fluid contained vortices and that it was the planets, including the earth, which were literally dragged around by such a whirlpool. Figure 1 is an illustration of the phenomenon as depicted by Descartes.

In his analysis, Newton undertook to show explicitly that such a concept could not account for the observed behavior of the planets. It has often been suggested that the entire structure of Newton's analysis was conceived of as an attack on naïve Cartesian mechanisms. Descartes had published an influential work, *Principles of Philosophy,* in 1644 and in this work Descartes enunciated a mechanical point of view. As modern and crucial as the concept was to the development of

analysis, see the essay by the economist John Maynard Keynes, "Newton the Man," in J. R. Newman (ed.), *The World of Mathematics* (4 vols.; New York, 1956) vol. I, pp. 277–285. For a classic account of the relationship between natural philosophy and the Enlightenment, see J. H. Randall, Jr., *The Making of the Modern Mind* (Cambridge, Mass., 1926, 1940, 1956), part III.

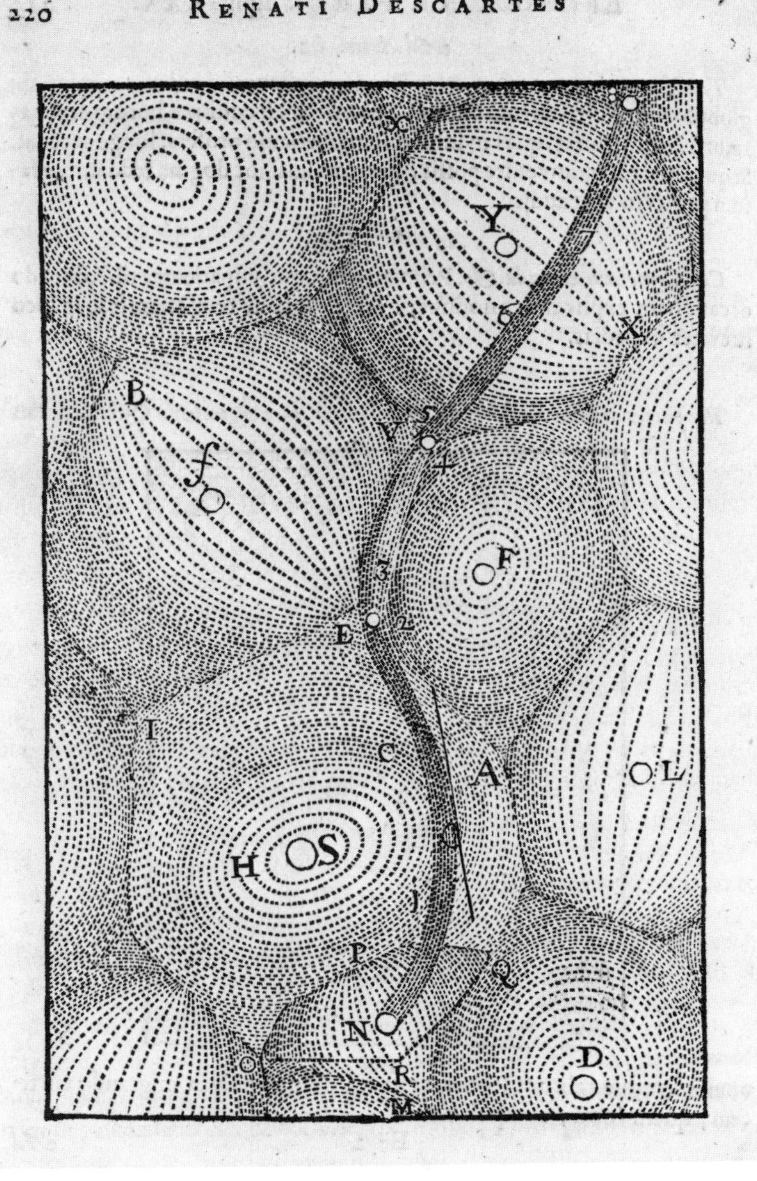

Descartes' depiction of the aether of space: Points Y, f, F, S, L, and D represent centers of different vortices (these centers are presumably stars) in this cross section of the ether. The vortices must rotate so that the motion of one does not impede the rotation of any of the others. The Sun is at S and the planets are dragged around by the motion of the aether vortex around S. The path 1,2,3,4,5,6 represents the motion of a comet being dragged along from vortex to vortex. (From Rene Descartes, *Epistolae* (Amsterdam, 1668).)

physical science, Descartes' schema, taken as a whole, could not be used for quantitative analysis. And, when Newton published his synthesis in 1687, the title was *Mathematical Principles of Natural Philosophy;* the words *mathematical* and *natural* appeared in red. This was considered to be the Newtonian answer to Descartes. Ironically, Newtonian physics was developed and took root first, and most firmly, in Cartesian France, where a school of mathematically inclined natural philosophers were well prepared to receive and understand it. In truth, the Newtonian synthesis was not only a synthesis of our understanding of the motion of the heavens with our understanding of the motion of objects near the earth, it was also a synthesis of the Cartesian concept of mechanism, with the Pythagorean, Platonic idea of the primary role of pure mathematics in understanding the world.

Newton's *Mathematical Principles of Natural Philosophy* is often referred to as *The Principia,* from a key word in the Latin title. Although at the time of Newton, algebra had become popular and familiar to natural scientists, he purposely recast his arguments into geometry, even those involving differential or integral calculus, subjects he had invented. *The Principia* is a formal, even formidable document. But so breathtaking was the accomplishment that to this day Newton's approach and treatment of mechanical problems occupies most of introductory physics sequences and is the spirit, if not the substance, of most advanced courses—especially those related to mechanics. We will sketch briefly the logic of Newton's accomplishment and its relationship to the evidence of his era. We also examine growing convictions in Europe and America during the Enlightenment about the mechanical world view. In Appendix 3, Newtonian mechanics has been developed more formally.

As with the Galilean approach to motion, detailed knowledge of the Appendix is not crucial to understanding the issues.

THE EVIDENCE

There are three or four kinds of phenomena which we will use as representative of the Newtonian synthesis. We have already analyzed the motion of freely falling bodies near the surface of the earth and we showed that such motion could be understood as resulting from uniform acceleration in the direction of motion. In the case of projectiles, the motion could be understood as the vector sum of motion which resulted from constant acceleration and uniform motion perpendicular to the direction of acceleration. Newton's mechanical system allows us to describe the conditions under which the acceleration will be constant and, furthermore, given the analysis, predicts that these conditions are satisfied near the surface of the earth.

We will also show that the Newtonian schema accounts, not only for motion, but for rest as well. In the Aristotelian scheme *rest* needed no explanation; it was the natural state of bodies. Not so with Newtonian mechanics. Rest, the absence of motion, must be accounted for.

There is a particular kind of motion for which Newton's system must account in the same terms that it accounts for all other motions. It is circular motion, and must be understood if one is to understand the motions of the planets. Therefore, we will examine circular motion as it is applied to the motion of the planets.

By Newton's era, understanding the motion of the planets had become synonymous with understanding Kepler's laws that supposedly characterized that motion. Johannes Kepler, a contemporary of Galileo's, shared convictions about heliocentricism, as well as many of his convictions about the importance of pure numbers and perfect shape in understanding the universe. Kepler had been obsessed with these ideas and his obsession drove him during a nightmarish twenty-five year search for the harmonious laws governing the motions of the planets.* His data were inherited from Tycho Brahe, a man who had devoted his life to careful and systematic measurements of the positions of heavenly bodies.† It was a propitious time. Before Tycho's time, such measurements had been made sporadically and were only accurate to about ten minutes of arc. Tycho's insistence on systematic measurements was accompanied by a virtuosic ability to make the most of available technology. For his day, Tycho's instruments were of Herculean proportions and his cold-eyed understanding of how to use them to minimize and assess the errors and uncertainties of measurement stood in sharp contrast to his explosive personality. Tycho had remained committed to a geocentric point of view. But Kepler used the data—primarily, the systematic daily observations of the motion of Mars relative to the stars as seen from the earth—to hew out general regularities about the motion of the planets on the assumption that the sun is the center of the solar system and the earth moves about the sun as one of the planets. These regularities are summarized as "Kepler's laws." There are three:

1. The shape of the orbits of the planets are ellipses with the sun at one focus of the ellipse.
2. A planet sweeps out equal areas in equal times.
3. The cube of the distance between the sun and the planet (the mean radius) is proportional to the square of the period of revolution (the time required to go

* An excellent and exciting study of Kepler and the nature of his work has been provided by Arthur Koestler in *The Sleepwalkers* (New York, 1959), Part IV, "The Watershed." This section has been separately published in The Science Study Series (Anchor Books) as *The Watershed.* See also N. R. Hanson, *Patterns of Discovery* (Cambridge, 1958), chapter 3.

† A brilliant historical novel based on the interaction between Kepler and Tycho was written by Max Brod, *Tycho Nahe Sein Gott* (Berlin, 1925). There is an English translation, *The Redemption of Tycho Brahe* (New York, 1927).

once through the orbit). Furthermore, the constant of proportionality is the same for all the planets. Mathematically, this can be written as

$$R^3 = KT^2$$

Kepler's laws are often referred to as empirical laws: the observational basis on which Newton proved that his system was applicable to motion of heavenly bodies. But it is misleading to consider these laws as empirical. First, the laws refer to concepts like the shape of the planetary orbits relative to the sun, or, to the time required to make one circuit around the sun, or, to the distance between the sun and the planet. Actually, the *observations* were made on the earth, not on the sun, and the observations were of the position of the planets relative to the stars as seen from the earth. In order to arrive at Kepler's laws, we would have to think how to transfer the measurements to the sun.

The critical test used by Newton to apply his system to Kepler's laws was the motion of the moon. In other words, in the earth-moon system, the earth is the central body occupying one focus of an ellipse, and the moon, in orbiting the earth, presumably obeys Kepler's laws. The laws are identical to those listed above except that the constant k, in the third law, would take on a different value since the central body is no longer the sun. Even when applied to the moon, the laws are not direct observational laws; as with the sun, the observations are of successive positions of the moon relative to the stars.

Another system to which Newton applied Kepler's laws was composed of Jupiter and its moons. Galileo was the first scientist to observe the satellites of Jupiter. But as always, the application of Kepler's laws to that system is achieved by comparing positions and times of the orbiting satellites. There is an enormous inferential chain between the observations and the relationships known as Kepler's laws.

Newton realized that Kepler's laws could not apply precisely to the various systems to which he was applying them. He showed that the conditions under which Kepler's laws would be accurate would be *ideal* when the central and the orbiting body were dimensionless points. In reality, this could not happen. Newton also showed that Kepler's laws are not perfectly obeyed by real systems to which he was to apply them because of the disturbing influence of the bodies upon each other. Thus, the moon only approximately obeyed Kepler's second law; that is, it did not sweep out equal areas in equal time intervals. The disturbing influence that the sun exerted on the motion of the moon is the reason. Newton also showed that, strictly speaking, Galileo's law of free fall was not accurate either; as one gets farther and farther from the center of the earth, the acceleration due to gravity decreases and that acceleration differs as one moves to different latitudes between the equator and the pole. Of course, when I say that Newton "showed" that Kepler's and Galileo's laws must only be approximate, I mean that he showed that using *his* theory, these laws, which are the logical outcomes of the application of Newton's premises, are true only under hypothetical conditions. In short, Galileo's laws or Kepler's laws that explain the motion of objects within their domain, became descriptive laws within Newton's theory. In *that* sense, Newton explained why Galileo's and Kepler's laws work.

Let us turn now to an examination of the structure of *The Principia* and its influence on the development of the mechanical world view, and the theory of measurement implicitly used by Newton in order to accomplish his ends.

Newton divided *The Principia* into three books. Book I is an examination of abstract dynamics, the motion of point masses under specified conditions. Book II considers the motion of objects in resistive media and Book III, subtitled "The System of the World," is an application of the first two books to the motion of the planets, the moon, comets, and tides. In Book I, Newton demonstrated how to understand Galileo's kinematical laws. The argument is prefaced by a series of definitions and axioms. Some of the details are provided in Appendix 3 which also contains a more detailed examination of the relationship of Newton's physics to planetary motion and Galileo's descriptive laws of falling bodies and projectiles.

We are not presenting an historical account of the development of Newton's physics. The history is a heuristic device for understanding the content and background of the physics and its relationship to cultural, social, and political issues.

NEWTON'S AXIOMS AND DEFINITIONS

Newton's treatise began with a series of definitions, followed by a discussion of absolute and relative time and space followed by three axioms or laws. These axioms are commonly referred to as Newton's laws. It is convenient to consider the definitions in the context of understanding the axioms. "Axiom" is used by Newton as we have been using the word "postulate." In other words, Newton's laws are postulates; statements set forth at the beginning of the argument without proof, and without the possibility of proof, but which are used to prove a series of further statements. In Chapter 1, we outlined the schema for the relationship between evidence and belief within natural sciences as follows:

If *A* then *B*.
I observe *B*.
Therefore *A*.

Newton's axioms are the "*A*" in the argument. One of the wondrous and bold features of Newton's physics is that the three laws are enough for all classical mechanics.*

Newton's laws are:

I. Every body continues in its state of rest, or of uniform motion in a right line, unless it is compelled to change that state by forces impressed upon it.

II. The change of motion is proportional to the motive force impressed; and is made in the direction of the right line in which that force is impressed.

III. To every action there is always opposed an equal reaction; or, the mutual actions of two bodies upon each other are always equal and directed to contrary parts.

Let us examine each of these in turn.

I. THE LAW OF INERTIA

Newton's first law is often called the law of inertia. Colloquially, it may be stated as follows: the state of motion possessed by a body persists unless a net outside force acts upon the body. This may be considered a criterion for whether or not there is a net force upon a body. Before stating the first law, Newton defined (definition 4) "impressed force" as that which alters the state of rest or uniform motion of a body. In the formal definition, the word "right" means straight. Therefore, not only must the magnitude of the motion remain the same, but its direction must remain the same, as well. Thus, contrary to the conceptions of Aristotle and almost all who followed him (including Galileo), Newton did not consider circular motion to be natural or inertial. If something moves in a circle, for example, a planet, there are net external forces upon it.

The term "motion" used by Newton in these laws has a technical, quantitative meaning. It was defined by Newton (definition 2) in two parts. The quantity of motion is defined as the product of the mass and the velocity of a body. It is a vector quantity and since the time of Newton has acquired the technical name: "momentum." The mathematical definition of the quantity of motion is

$$\vec{p} = m\vec{v} \tag{1}$$

The symbol $\vec{p}$ is traditionally reserved for momentum, $\vec{v}$ is the velocity of the object and m is the mass.

* In order to expand the explanatory power of his abstract system to the motion of planets, comets, the moon, the tides, falling bodies and projectiles, Newton required one additional postulate, the law of universal gravitation.

The mass was conceived of by Newton (definition 1) as "the quantity of matter." He also indicated that the mass of a body is proportional to the body's weight. As will be clear in the sequel, the former concept, "quantity of matter" refers to what is termed "inertial mass." The mass which is proportional to the weight of a body is termed "gravitational mass." It is a matter of experimental coincidence within Newtonian mechanics that the two masses are equal to each other in any body. (One of Albert Einstein's innovative propositions within the framework of the general theory of relativity was to postulate the equivalence of gravitational and inertial mass). In his definition of mass, Newton quantified it as the product of the density of the body and its volume:

$$m = \rho V \tag{2}$$

where ρ is the density and V the volume. But since Newton did not provide an independent definition of density and it is usually defined in terms of the mass and the volume, equation (2) is at best circular and at worst, meaningless. That Newton's logic was circular about the concept of mass was not realized until the last part of the nineteenth century when the physicist philosopher, Ernst Mach provided a reconstruction with the premise that all physical laws are merely economical summaries of experience and that they are linked directly to experience. As should be clear by now, Mach's biases are not shared here. But there is no question that, by providing an independent definition of mass and how to quantify it, he placed the foundations of Newtonian mechanics on a firmer footing. During the two centuries between Newton and Mach, the science of mechanics flourished; circular definitions, empty statements and all.

Newton's first law is frequently stated mathematically as follows:

$$\Sigma \vec{F} = 0 \tag{3}$$

$\vec{F}$ represents the force, a vector quantity that we will develop later and the symbol Σ represents the operation of summation. In words equation (3) might be stated: the sum of the forces, or the net force on a body is zero. Unstated in equation (3) is the important qualification that we are speaking of the situation in which the motion does not change. Since the quantity of motion is defined as $m\vec{v}$ and the mass m is a constant independent of the motion, to say that the quantity of motion does not change is equivalent to saying that the velocity of the object does not change. In using equation (3), one must apply it only to those situations in which motion is constant. The concept of inertia was not invented by Newton. The idea had been thought about by many scholars and Descartes has been credited with the first clear statement of the concept. On the other hand, Galileo held that circular motion was *inertial* and he had shown contempt for Kepler for suggesting that the orbits of the planets were not circular because, in Galileo's view, the motion would no longer be inertial.

II. THE SECOND LAW

In Newton's scheme, since both mass and impressed force had been independently defined earlier, the content of the second law was to quantify the magnitude of net forces by their effects on motion. Or, given a known force, the second law specifies the nature of an object's motion. Later, after Mach's critique of the foundations of Newton's logic, the second law became a defining equation for force. These philosophical differences are important, yet they do not affect the actual use of the second law.

From the analysis that followed, Newton's second law can be interpreted thus: By "the change in the motion" Newton intended the "time rate of change of the quantity of motion." It is *that* change that is to be set, proportional to the impressed force on the object. Since the quantity of motion is given by $m\vec{v}$ the time-rate of change of the momentum is $\Delta(m\vec{v})/\Delta t$. But as we have already seen, the mass of an object is a constant, a number we associate with the quantity of matter (or later, after Mach, a number arrived at by independent specification). Since the mass of the body will not change over time, $\Delta(m\vec{v})/\Delta t$ can be rewritten as $m(\Delta\vec{v}/\Delta t)$. But $\Delta\vec{v}/\Delta t$, the time-rate of change of the velocity of an object, is nothing more than the definition of acceleration, $\vec{a}$.

Quantitatively, Newton's second law is often written:

$$\Sigma\vec{F} = m\vec{a} \tag{4}$$

The same as

$$\Sigma\vec{F} = \Delta(m\vec{v})\Delta t$$

Since

$$\Delta(m\vec{v})/\Delta t = m(\Delta\vec{v}/\Delta t) = m\vec{a}\,*$$

There are two points to be emphasized. First, according to Newton's axiom the force is ***proportional*** to the change in motion. In developing equation (4), I followed custom in setting the proportionally constant equal to one. Note also that force is a vector quantity. This means, as Newton pointed out in a corrollary to his statement of the axioms, that forces add vectorially rather than algebraically. A

* For a challenge to this interpretation see E. J. Dijksterhuis, *The Mechanization of The World Picture* (Oxford, 1961) pp. 463–492.

colloquial statement of the second law might be that, when an external force is applied to an object, the object accelerates in the direction of the force and, the larger the force, the larger the acceleration. Earlier I stated that Newton's scheme would explain Galileo's law of free fall. We have seen that Galileo postulated constant acceleration.

But according to (4)

$$\vec{a} = \vec{F}/m$$

(We have dropped the summation sign before the $\vec{F}$, as it is understood that $\vec{F}$ represents the net force on the object.) Since $\vec{a}$ is constant in free fall, the ratio of the force to mass must be the same for all falling bodies. The details of that analysis are developed in Appendix 3.

It is important to note that, despite philosophical difficulties about the meaning of the second law regarding the nature of the universe or the nature of physics, there is no ambiguity about its *use*. Given a mass for an object, specification of the acceleration, on the one hand, determines the nature and direction of the net force on the object at any given time. On the other hand, specification of the net force at any time determines the magnitude and direction of the acceleration.

Let us take one very simple example in order to see how the second law is used in practice. An object is given a push across a table and it is noted that the object slows down and comes to rest. Therefore, there must be an acceleration of some magnitude (probably of changing magnitude because, as the object slows down, its rate of slowing down decreases) directed in the opposite direction from the original velocity. If that is correct, there must be a net force exerted *on* the object in a direction opposite from its original motion. (Since the acceleration of the object decreases as the object slows down, the magnitude of the force, but not its direction, must also decrease.) The question becomes: What supplies the force?

When we examine the physical situation, we find only one answer: The force is supplied by the interaction of the object and the table. It is often called a frictional force and it is usually explained as proportional to the speed of the object. (The constant of proportionality is referred to as the coefficient of kinetic friction.) We are not ascribing intelligence to the table and we do not imply that there is a sensory communication between object and table to inform each other that there is relative motion between them. The analysis is quite rational. We have said nothing about the *source* of the frictional force and, within Newtonian mechanics, we cannot and need not say anything about it. If you play Newton's game, you must use Newton's rules, specified by the axioms. Those rules demand the use of the concept "force." The historical development of the concept of force that led to Newton's formulation of the second law is interesting and historians of science continue to argue about it. Its roots go back to analyses of the motion of objects by scholars of the Middle Ages.*

III. THE THIRD LAW

Undoubtedly, the third law was the most ingenious of Newton's contributions to the axiomatization of mechanics. When we understand the application of the third law the nature and power of Newton's abstract analysis becomes clear.

The third law is often referred to colloquially as the "action-reaction" law, but one must be cautious about using the phrase. The law states that when two objects interact, if object A exerts a force *on* object B, object B exerts an equal and opposite force *on* object A. Note that, although the action-reaction pairs are equal and opposite, they are not applied to the same body and thus cannot be used to equilibrate each other. Whereas it is true that, if a horse pulls a cart with a certain force, the cart will pull back on the horse with an equal and opposite force, these two forces are not in equilibrium because they are not applied to the same body.

The setting forth of the third law by Newton represents keen insights and sensitivities on his part. For example, when you place your finger on a flat surface and push, what do you feel? Do you feel yourself pushing? One might argue that your sense of pushing is located in the muscles which are responsible for exerting the force on the table. And what do you feel in your finger tip? That, perhaps, is the reaction force of the table on your finger. However, in playing Newton's axiomatic game, it is important not to be trapped into anthropomorphization of the concept of force or the notion of action and reaction. Consider our earlier example: If, as an object moves across a table, the table exerts a force on the object in a direction opposed to the motion, the moving object must exert an equal and opposite force on the table; that is, a force in the direction that the object is moving is applied *by* the object *to* the table. Why doesn't the table move? That is the wrong question. The table doesn't move. Therefore, there must be other forces exerted on the table (for example, by the floor, through the legs) so that the *net* force on the table is zero, thus satisfying Newton's first law.

THE APPLICATION AND CONSEQUENCES OF NEWTON'S LAWS

We have indicated, to some extent, the manner in which Newton's laws may be applied to physical situations. In general, there are two types of mechanical

* See E. J. Dijkesterhuis, *The Mechanization of the World Picture,* R. S. Westfall, *Force in Newton's Physics,* (London, 1971). See also Max Jammer's study, *Concepts of Force: A Study in the Foundations of Dynamics* (New York, 1962)

problems: problems of statics and problems of dynamics. Static situations are those in which there is no net acceleration and, in most cases, the objects are at rest. Almost all such problems are solved by application of Newton's first law in conjunction with the third law. In some instances, the objects can be considered point masses, but in other cases—the construction of a large bridge, for example—such simplifications cannot be abstracted. But for simple cases such as the analysis of the situation shown in Figure 2, Newton's model is extraordinarily powerful. The practical question often is how big a support wire *AB* must be used to support the sign hanging on the wall, and attached at *C*.

From experience one learns that the solution is to make a "free body diagram" around a crucial point; for example, the point *B* where the support wire is attached to the sign. The sign has now become a ***point mass***. We required that it be at rest. Therefore, according to Newton's first law, the sum of the forces upon it must be zero. The forces are the weight of the sign, $\vec{F}_2$, the reaction force exerted by the wall on the sign, $\vec{F}_3$, and the tension, $\vec{F}_1$, exerted by the support wire on the sign. It is but a short step to resolve the force $\vec{F}_1$ into components in a Cartesian coordinate space in which the axes are along the lines of $\vec{F}_2$ and $\vec{F}_3$, and require that the sum of the forces in directions x and y be zero. In other words, the point *B* is at rest. It is an exercise for the reader to show that knowing the weight of the sign one can calculate the force $\vec{F}_1$. Once the magnitude of $\vec{F}_1$ is known, one can choose an appropriately sized support wire capable of transmitting a force of that magnitude (the transmission of a force by such a wire is often referred to as "tension") without breaking. Whereas this is a simple example of a statics

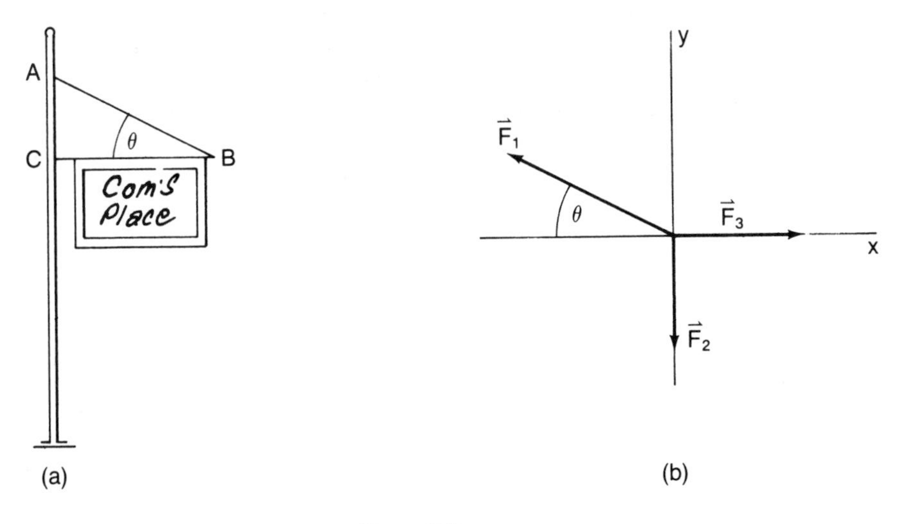

Figure 2.2

problem, it is typical of the puzzles presented to engineers and architects who are confronted with problems of stability and strength. The techniques may become very sophisticated but the skill of solving such puzzles, once mastered, will give a person a sense of power and competence.

The consequences of Newton's laws, however, far transcend the application of the laws. For example, as I have shown in Appendix 3, the first law together with the third law, implies the conservation of momentum. That law may be stated as follows:

In the absence of external forces, the momentum of any system is conserved in all interactions.

For the phrase, "the absence of external forces," an alternative phrase, "in any isolated system" is used. The trick becomes to specify such a system. For example, consider a cart carrying a bomb, shown in Figure 3a. The only significant object for consideration is the bomb. Let us assume that it has a momentum given by $M\vec{v}$ where M is its mass and $\vec{v}$ is the uniform velocity with which it is moving. After the bomb explodes (Figure 3b), according to the conservation of momentum, no matter how many fragments the bomb breaks into, the sum of the momenta of all of the fragments must still be $M\vec{v}$. We have depicted the bomb breaking into four fragments and,

$$m_1\vec{v}_1 + m_2\vec{v}_2 + m_3\vec{v}_3 + m_4\vec{v}_4 = M\vec{v}$$

Of course, had the bomb been at rest before the explosion, the sum of the momenta of the fragments would have been zero. The bomb is the system.

Let us abstract the problem still further. Suppose we had restricted ourselves to one dimension. The momenta in the forward direction would have to be equal to the momenta in the backward direction. Given that momentum is a vector quantity: it is a small step to see how useful it will be to analyze these problems in Cartesian space. Consider, for example, the problem of a rocket being accelerated from rest. The gases which come out of the back of the rocket presumably have

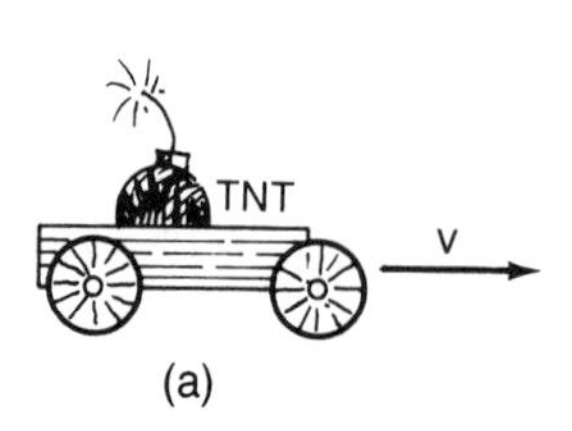

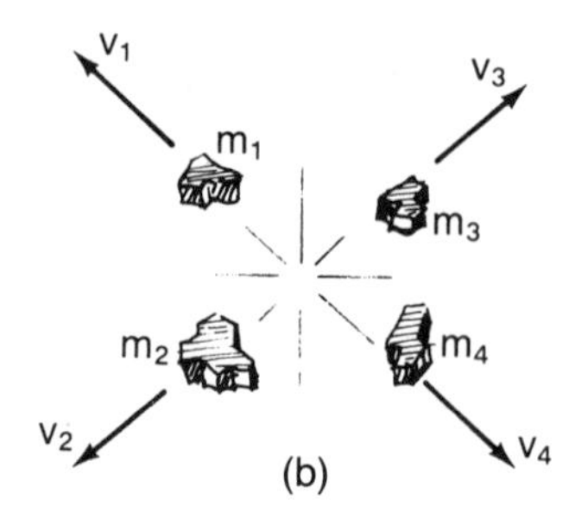

Figure 2.3

some mass and they come out with considerable velocity. If we consider the system being analyzed to be rocket and gasses, then at any time the momentum of the entire system must be zero. Hence the rocket must acquire momentum in the direction opposite to the direction in which the exhaust gasses are moving. In other words, the rocket will accelerate. The reader who wishes to pursue such analyses further is directed to Appendix 3.

Another consequence of Newton's laws is the conservation of kinetic energy. If m is the mass of a body and v is the speed of the body, the compound quantity $\frac{1}{2}mv^2$ takes on a particular significance called kinetic or moving energy. Originally, mv^2 was referred to as *Vis-Viva,* "living force." In Newton's time, there was a considerable controversy about the relative significance of the quantity we have called the momentum and the quantity then called *Vis-Viva* and related to what is currently called kinetic energy. Both are important. Under special circumstances in which there are no losses due to friction, to deformation of the objects which interact, or any other dissipative losses, the kinetic energy is conserved. For example, in Figure 4a two billiard balls approach each other for a collision. Assuming that there are no losses in the form of heat or deformations, the kinetic energy before the collision must equal the kinetic energy after the collision. If we specify the masses of the objects in Figure 4 and if we know the initial speeds of each of the balls, application of the conservation of momentum and conservation of kinetic energy allows us to predict the outcome:

$$\begin{aligned} m_1v_{1i} + m_2v_{2i} &= m_1v_{1f} + m_2v_{2f} \\ m_1v_{1i}^2 + m_2v_{2i}^2 &= m_1v_{1f}^2 + m_2v_{2f}^2 \end{aligned} \tag{5}$$

where the i and f subscripts refer to "initial" and "final." The first equation in (5) states that the momenta of the two balls before will be equal to the momenta after the collision. Similarly, the second equation in (5) states the same conservation regarding the kinetic energy. (Note that one cannot derive the second equation in (5) from the first by squaring the first.) It is simple to show that, as long as we do not consider the balls capable of spinning, and we restrict them to hitting on center, should the balls have the same mass and one of them be struck by the other while the second ball is at rest, the second ball will move off with a velocity equal to the velocity of the original ball, which will have come to rest.

One of the important consequences of the mechanical world view is the gradual expansion and delineation of the concept of energy. This is closely associated with

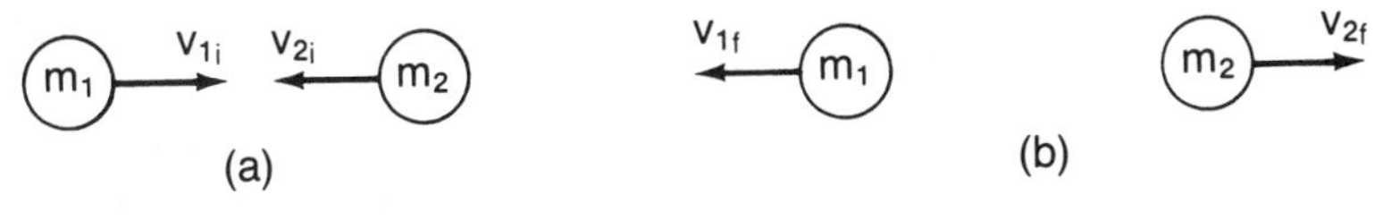

Figure 2.4

the concept of universal gravitation and with the gradual reinterpretation of the concept of heat. Earlier, heat was considered a material substance. By the middle of the nineteenth century it was believed to be a manifestation of the motions and mechanical interactions of the submicroscopic particles, the atoms, which make up material objects.

We can only sketch these developments here. More details are provided in Appendix 3 and Appendix 4. In addition to his three axioms of motion, Newton introduced the concept of universal gravitation. It is an axiom with no more status in nature than his other three axioms and laws. It was an ingenious invention which allowed, with one stroke, the unification of the motions of heavenly bodies and motions near the surface of the earth. Newton hypothesized that every object in the universe attracts every other object cojointly as the product of their masses, and inversely in proportion to the square of the distance between them. This attraction, called a gravitational attraction or force, is given by

$$\vec{F}_{12} = G\frac{m_1 m_2}{\vec{R}_{12}^2} \qquad (6)$$

$\vec{F}_{12}$ is the gravitational force of attraction between bodies having mass m_1 and m_2 and $\vec{R}_{12}$ is the distance between the two bodies.*

A measure of the boldness of Newton's hypothesis is his proposal that the constant G is a universal constant applying to the interaction of all objects throughout the universe. In order to explain, quantitatively, the orbiting of a planet around the sun or the moon around the earth, and the fall of an apple, Newton showed that for the case of spherical symmetry, one could treat an object as if all of its mass were located at a point at the center of the object. In order to make the argument, he required the calculus, a subject which he is credited with inventing.† The reason for requiring such an assumption is illustrated in Figure 5. On the left are two bodies, M and m. They attract each other, according to Newton, as in equation (6). What is the distance between M and m? One might think that one could break M up into tiny pieces, calculate the gravitational attraction between each piece and m and sum (vectorially) the resulting forces. But how tiny is tiny? Should we not do the same thing for the mass, *m*?

In fact, the invention of the calculus allowed Newton to show that for the case of a sphere, one could treat the problem as though all of the mass were located at a

* See J. M. Keynes, *op. cit.*, who discusses the priority fight between Newton and Leibnitz concerning the calculus.

† The subscript "12" is shorthand for the element "one-two," in this case the interaction of mass m_1 and m_2. Thus, $\vec{F}_{12}$ means the force exerted by m_1 on m_2. $\vec{R}_{12}$ is the distance between m_1 and m_2. Should one prefer to speak of the force exerted by m_2 on m_1, the standard notation is $\vec{F}_{21}$.

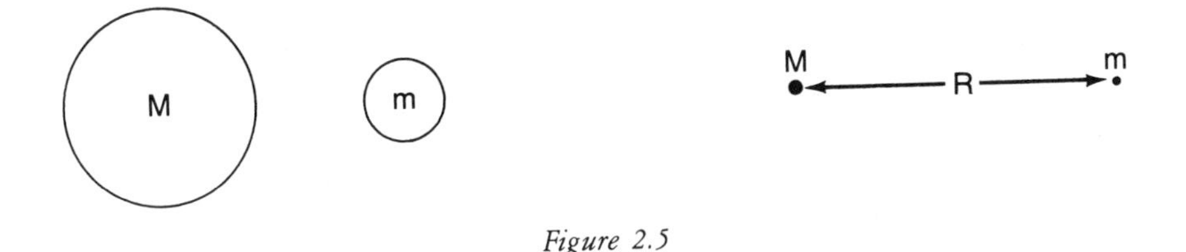

Figure 2.5

point at the center. Thus, the situation on the left of Figure 5 is reduced to the situation on the right.

Let us think about an object falling near the surface of the earth. According to the Newtonian conception the force that the earth exerts on the body is given by equation (6). At the same time, according to the second axiom, the force on the object will be proportional to the object's acceleration and according to the third axiom not only will the earth be exerting a force on the falling object, the falling object will be exerting a force on the earth.

It is at this point that the identity of gravitational mass with inertial mass becomes crucial. We can equate the forces as follows:

$$\vec{F} = G\frac{mM}{\vec{R}^2} = m\vec{a}; \quad \vec{a} = GM/\vec{R}^2 \qquad (7)$$

where $\vec{R}$ is the distance between the object and the *center* of the earth, m is the mass of the object, and M is the mass of the earth. The situation is depicted in Figure 6. As the center of the earth is approximately 4,000 miles from the surface, the value

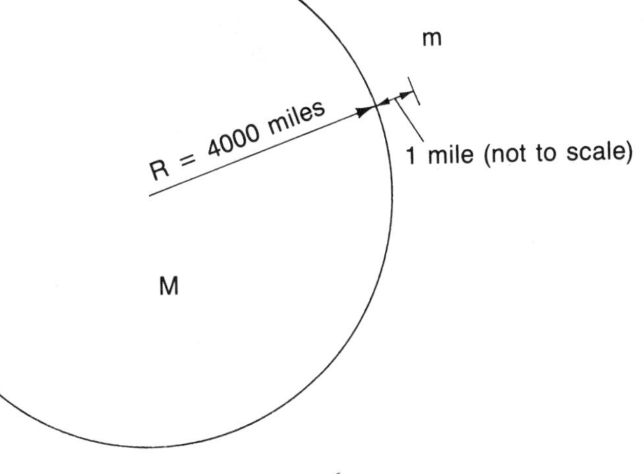

Figure 2.6

of R hardly changes for a considerable distance above the surface of the earth. Suppose an object drops from a height of one mile to the surface of the earth. The difference between 4000 and 4001 miles is less than .03 percent and, for all intents and purposes, we can consider R to be a constant. The terms on the right side of equation (7) governing the magnitude of the acceleration are, therefore, essentially constant. Newton has explained that all objects near the surface of the earth accelerate at the same rate by saying that the acceleration depends only on the mass of the earth and the distance to the center of the earth.

Consider now the kinetic energy of a falling body. As, near the surface of the earth, the body is accelerating at a constant rate, the square of the object's speed will be increasing in proportion to the height to which the object is raised. Since the square of the speed is proportional to the object's kinetic energy, a measure of the kinetic energy a body will acquire is just the height above the surface of the earth that the body will be raised. Technically, we can define a new quantity, *the work,* which is the product of the force exerted on a body and the distance the body moves under the action of that force. Lifting a body from the surface of the earth, or from a height near the surface of the earth to another height near the surface of the earth, requires a certain amount of *work.* In doing that *work,* we must exert a force equal to the force of the gravitational attraction between the body and the earth in order to lift the object to a certain height. Once we release the object, the object acquires kinetic energy. In a sense we have given the object the potential for acquiring that kinetic energy by doing work. As long as the force we are dealing with is an inverse square force such as the gravitational force, the amount of work done on the object in lifting it to a certain height is equal to the kinetic energy acquired, once the object is released and allowed to gravitate (that is, fall) back to its starting point. If we call the work done in lifting the object the *gravitational potential energy,* then at any time, for any closed system, barring frictional or other dissipative losses, the amount of energy in the system must be constant.

I will now discuss the roller coaster shown in Figure 7. All roller coasters are built so that the first hill is the highest of the ride. The cars are dragged up the hill by an electric motor, and after that, they are on their own, and at any time, assuming that there are no frictional losses, the amount of energy is constant. The energy is gravitational and as we can see from the figure there is a constant trade-off between speed (kinetic energy) near the bottom of the hills and height (potential energy) near the tops of the hills. This concept of potential energy can be expanded to all sorts of systems: for example, in compressing a spring, we endow the spring with the ability to do work. We have given the spring potential energy. Or consider the combustion of gasoline in an internal combustion engine. Through natural processes, crude oil has collected in pockets within the earth. The formation of that oil represents the accumulation of chemical potential energy that

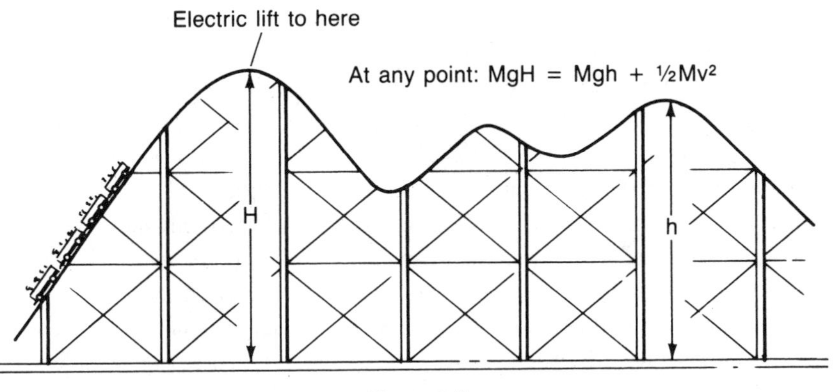

Figure 2.7

is converted into the kinetic energy of motion once the oil is refined and exploded. The concept of kinetic energy is abstract, yet powerful in a practical sense.

The idea of potential energy developed alongside a new realm of developments in mathematical analysis, often referred to as "potential theory." Rather than dealing with forces, which are vectors, and their effects on masses directly, which can be difficult, the potential energy a body acquires under the action of those forces can be calculated. Since energy is a scalar quantity, very often an analysis which had been difficult becomes relatively easy. Examples of this are given in Appendix 3.

One of the important developments of the nineteenth century regarding gravitational and electric potential theory was the development of the idea of the *field*. Given the existence of a mass, for example, the sun, one can calculate point by point what the force *would* be on an object due to interaction with the sun, no matter where in the universe the object was located relative to the sun. Similarly, one could calculate the potential energy between the object and the sun. This allows us to introduce the notion of a *gravitational field*. We can calculate, point by point, the potential energy on a unit mass at any point in space. We can then predict how an object of any other mass would move as a result of interaction with the sun were it introduced into our system. Potential fields are more useful than force fields because the *net potential* field at any point in space is the algebraic sum of the effects of all the masses in the universe at that point. The net force field at any point in space is the vector sum of the force fields of all the individual masses.

The concept of energy became generally understood during the first half of the nineteenth century. By the last quarter of the century, the concept had been expanded to include the subject of thermodynamics, or heat. At the beginning of the century, heat was considered a material substance, a fluid called *caloric*. By the middle of the century, the concept was superfluous as the notion took hold that heat was a manifestation of the mechanical interactions of the atoms and molecules that make up all matter. As I have analyzed in Appendix 4, after making certain assumptions about the nature of the fundamental particles, how they move and how they interact, Newton's laws are applied to the ensemble. If one then hypothesizes that the temperature of the object is proportional to the average kinetic energy of the collection of particles, the laws of thermodynamics begin to emerge. Thus, one reduced the science of thermodynamics to the science of mechanics. Moreover, if heat is considered a form of mechanical energy, those losses we have referred to in discussing the conservation of kinetic and potential energy, are *not* losses, but part of the system. The total energy of the system is conserved. This idea was so prevalent during the middle of the nineteenth century, that one historian of science has shown that there were eleven claimants to the discovery of the law of the conservation of energy.*

This law, which is also referred to as the first law of thermodynamics, states that the new flow of energy across the boundaries of any system is equal to the change of energy within the system. Obviously then, for an isolated system in which no energy flows in or out (analogous to the isolated system we referred to in speaking of the conservation of momentum), the energy is conserved. Rather than losses or additions, there are only conversions between potential, chemical, mechanical, electrical, magnetic, light, sound and heat energies. The force of the mechanical world view was to attempt to understand all such energies, not only heat energy, as manifestations of mechanical interactions.

Recall that we began our discussion of the expansion of the concept of energy by considering the fall of an object near the surface of the earth under the influence of gravitational force. Newton did not participate in the development and expansion of the concept of energy. It occurred after the publication of his work and indeed, mostly after he died. However, it is important to point out that the original Newtonian analysis applied to objects falling near the surface of the earth as well as to an object falling a great distance from that surface. The moon, for example, is such an object. Newton's conception was that, just as the apple is falling under the influence of gravitational attraction, the moon is also falling under the same influence. But we know that the path of the moon is, to a first approximation, a circle. Newton's contemporary and rival, Christian Huygens, a Dutch natural philosopher and a leading member of the French Academy of Sciences had shown

* T. S. Kuhn, "Energy Conservation as an Example of Simultaneous Discovery," in M. Clagett (ed.), *Critical Problems in the History of Science* (Madison, 1959), pp. 321–356. Kuhn's essay has been widely reprinted. See, for example, T. S. Kuhn, *The Essential Tension: Selected Studies in Scientific Tradition and Change* (Chicago, 1977), pp. 66–104.

that, for anything moving at a uniform rate in a circle, given the concept of inertia, there is a net force on the object (pulling it off its inertial course) toward the center of the circle. See Figure 8. The analysis, which is detailed in Appendix 3, is straightforward kinematics.

Suppose an object moves at a constant speed in a circle. How will it be accelerated? The answer, given the concept of inertia, is that the acceleration will be pointed constantly toward the center of the circle and be of magnitude, $v^2/\vec{R}$ where v is the speed of the object and $\vec{R}$ is the radius of the orbit. But, according to Newton, if an object accelerates, there is a force given by the product of the mass of the object and the acceleration. Therefore for an object moving in a circle, a force given by

$$\vec{F} = mv^2/\vec{R} \tag{8}$$

will be exerted toward the center of the circle. How that force is supplied depends on the physical situation. For a person slinging a weight on the end of a rope, the force toward the center of the circle, the centripetal or *central seeking force* is supplied by the person through the tension of the rope. Should the rope be cut, the weight will fly off tangentially. For a car turning in a circle, the road will supply the force to the tires. Should the road be unable to supply enough force, the car moves off tangent to the curve. In order to prevent this, curves are banked inward so that part of the weight of the car itself provides some of the centripetal force, as shown in Figure 9. For the loop in a roller coaster, the track must provide the force near the bottom and at other points, the force is supplied by the weight of the coaster car and its occupants. For the moon, it is the gravitational attraction of the earth on the moon that supplies the centripetal force. The concept then, developed in Appendix 3, is that the orbiting body is always falling but because of its inertial speed (how that inertial speed is acquired is not specified) which is always

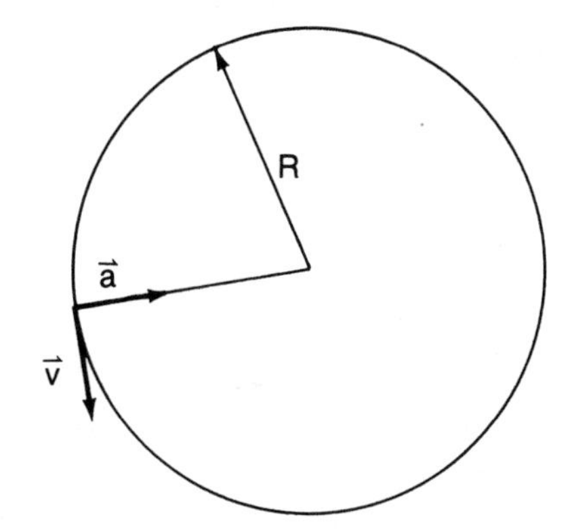

Figure 2.8

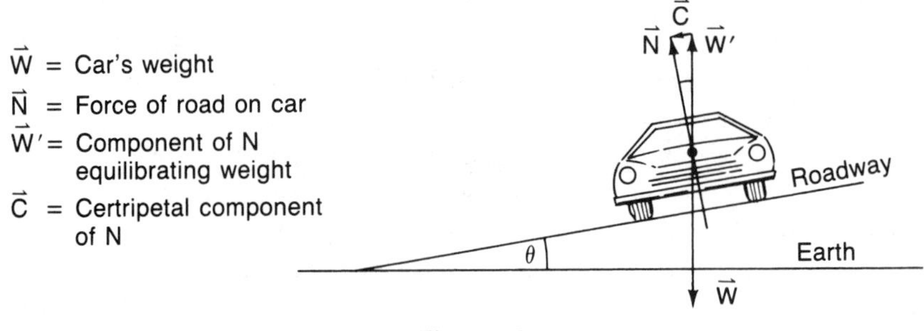

Figure 2.9

perpendicular to the direction of the gravitational acceleration, it is forced to move in a near circular orbit. (As pointed out in Appendix 3, Newton showed that the orbit is one of the conic sections: a circle, an ellipse, a hyperbola or a parabola. The exact shape of the orbit depends on the initial speed and direction of the object. Thus, the only requirement needed to orbit a satellite around the earth is that the satellite be given the proper initial velocity. Failing that, the satellite will either fall back to earth along a parabolic path, or be lost forever along a hyperbolic path).

In all cases the object in orbit requires not only the central seeking, gravitational force upon it, but also an inertial speed, perpendicular to the direction of that force. Such a speed is usually supplied to the satellites made by engineers by the orientation of the rocket that sends the satellite into orbit. The engine of the car, of course, provides the tangential speed to that vehicle as it makes a turn. The source of the tangential speed of the moon, the planets around the sun or the moons of Jupiter, or for the myriad of naturally occurring orbiting systems of the bodies in the heavens remains a mystery.

From Newton's time, there was no question that the operation of the universe is mechanical. In fact, after Newton the analysis of the energy of motion on or near the surface of the earth was applied to the motions of the heavens as well. If one took the entire universe as the system to theorize about, one could say that, at any time, the energy and momentum were a constant.

To take such a posture about the laws of physics near the surface of the earth and the farthest reaches of the universe is independent of beliefs that one might have about the meaning of life, the nature of the universe, a supreme being, or the relationships that might exist between these concepts. In Newton's time, almost all natural philosophers were explicitly religious and held beliefs about a supreme being from the Judeo-Christian tradition. To understand the universe was a way of knowing the creator and his purposes. Today, far fewer natural scientists subscribe to these views, but that says less about science than it does about the culture within which scientific and religious social institutions are now immersed. Our culture is less religious in the traditional sense than earlier cultures were. That has virtually nothing to do with science. I think that it has a great deal to do with our ever growing power and ability to manipulate the natural world. And as I pointed out earlier, manipulating the world does not require nor entail understanding it.

The Newtonian system of mechanics, created by Newton and commented upon by his eighteenth and nineteenth century successors, became a very elaborate system. Newton's laws were conceived to apply to the farthest reaches of the universe. All mechanical motions, from the translation of objects on the surface of the earth, to the translation of the solar system through the milky way, from the rotation of tops to the rotation of stars, from the somersaulting of gymnasts to the apparently chaotic motions of large assemblages of tiny grains of pollen suspended in water, from the motion of waves on the surface of the oceans, to the surging of blood in the veins of living things, from the motion of land masses on the surface of the earth, to the unfolding of flowers in the presence of warmth and light, all of these things and more were conceived of as being within their purview. The dream of the mechanical world view was not only to understand all motions which were *obviously* mechanical, but to reinterpret all other processes which were *not* obviously mechanical, in mechanical terms. One would understand in mechanical terms the mechanism which *caused* the flower to unfold. All chemistry could be reduced to the mechanical motions of atoms and molecules interacting with each other. All of psychology, how the brain worked, how moods changed, could be understood in mechanical terms. Biology, including why species breed true and the functions of various organs, would be understood in mechanical terms. In short, a person subscribing in a religious manner to this view, had the dream that everything would be understood ultimately in terms of Newton's laws of motion and the law of universal gravitation.

In discussing the fate of the mechanical world view, we will use only a few relationships as examples. In fact, we will restrict our considerations to those concepts which we have analyzed in the text and in Appendices 2, 3, and 4. In outline, the scope of our considerations will be as follows:

Premises:	Newton's Laws
	Law of Universal Gravitation
Conclusions:	Galilean Kinematics
	Kepler's Laws
	Conservation of Linear Momentum
	Conservation of Energy

The question is not whether or not these concepts agree with experience. Measurement itself contains a large theoretical component. During the eighteenth and nineteenth centuries, more and more phenomena were being understood with Newtonian premises. This is not more of a measure of the truth of the Newtonian premises than the widespread use of Euclidean geometry was a measure of the truth of Euclidean axioms. The success of the Newtonian system depended in part on matters which neither Newton nor his contemporaries, nor, for that matter, his successors made explicit. These matters are part of Newton's theory of measurement, an implicit theory which must precede a physical theory. As it stands before Newton's mechanics, it will be referred to as a "meta-theory."

NEWTON'S THEORY OF MEASUREMENT

In Chapter 3, the point of view will be developed that Einstein's special theory of relativity, unlike Newtonian dynamics, is not a theory about anything substantive; it is a theory of measurement. In that sense, the Einstein theory is a meta-theory. It is only after we realize that Einstein's theory is of measurement that we may raise questions about Newton's theory of measurement. Theories of measurement were not new in 1905 when Einstein first began to publish and theories of measurement have been an integral part of theorizing since early antiquity. As we will see, Newton had a theory of measurement and the special theory of relativity makes one modification in one of the assumptions in Newton's theory of measurement.

One of the reasons that few scholars pay attention to the fact that Newton had a theory of measurement is that the theory was never made explicit. It is embodied in a few paragraphs near the beginning of *Principia,* in a *Scholium* between the definitions and the statement of the three laws. It is a widely known passage which has been commented on by scientists and historians of science since Newton wrote it. ("*Scholium*" means only an explanatory comment on the text.) As the term "theory of measurement" is used, it means that the theory is used to understand how the fundamental primitives in the system, that is, mass, length and time, are measured and how one relates measurements made in different frames of reference.

In this *Scholium* Newton noted that he did not believe that the concepts of time and space, place and motion were understood by everyone. He thought that, for many people, these concepts were not defined independently but with regard to the relationships that they have to objects.

Newton intended to distinguish these concepts into the following categories: absolute and relative, true and apparent, mathematical and common. Let us examine Newton's definitions of space and time.

> I. Absolute, true, and mathematical time, of itself, and from its own nature, flows equably without relation to anything external, and by another name is called duration; relative, apparent, and common time, is

> some sensible and external (whether accurate or unequable) measure of duration by the means of motion which is commonly used instead of true time; such as an hour, a day, a month, a year.*

"Duration" is mathematical, true absolute time. There is no way to explain the concept of this time flowing "equably without relation to anything external." Either you understand what Newton is talking about or you don't. But as Newton points out later, one cannot measure this absolute time. Rather, what we actually measure is relative or common, or apparent time. We use clocks to keep track of such time. And, in fact, there is no guarantee that the measures of relative time have anything at all to do with the passage of absolute time. Newton next defined absolute space:

> II. Absolute space, in its own nature, without relation to anything external, remains always similar and immovable. Relative space is some moveable dimension or measure of the absolute spaces; which our senses determine by its position to bodies and which is commonly taken for immovable space. . . . Absolute and relative space are the same in figure and magnitude; but they do not remain always numerically the same. For if the earth for instance, moves, a space of our air, which relatively and in respect to the earth remains always the same, will at one time be one part of the absolute space into which air passes; at another time it will be another part of the same, and so, absolutely understood, it will be continually changed.†

Just as with absolute time, it is very difficult to talk about absolute space. Newton posits an ever present spatial universe within which matter is placed and relative to which matter moves. We have no way of measuring that space. Our sense of space is defined by the material bodies with which we have experience. As these are movable, they define, not absolute space, but relative space.

After defining "place" in terms of absolute and relative space, Newton then turned to the question of absolute and relative motion:

> Absolute motion is the translation of a body from one absolute place to another; and relative motion is the translation from one relative place to another. Thus, in a ship under sail, the relative place of a body is that part

---

* Isaac Newton, *Mathematical Principles of Natural Philosophy* tr. F. Cajori (Berkeley and Los Angeles, 1962), Vol I, p. 6. A. Koyre, I. B. Cohen, and A. Whitman have published an edition of the *Principia* which compares all published editions and Newton's handwritten manuscripts and notes. See A. Koyre, I. B. Cohen and A. Whitman, eds., *Isaac Newton's Philosophiae naturalis principia mathematica, the Third Edition* (1726) *with Variant Readings* (2 Vols.; Cambridge, Mass. and Cambridge, England, 1972).

† *Ibid.*

> of the ship which the body possesses; or that part of the cavity which the body fills, and which, therefore, moves together with the ship: and relative rest is the continuance of the body in the same part of the ship, or of its cavity. But real, absolute rest is the continuance of the body in the same part of that immovable space, in which the ship itself, its cavity, and all that it contains is moved. Wherefore if the earth is really at rest, the body which relatively rests in the ship, will really and absolutely move with the same velocity which the ship has on the earth. But, if the earth also moves, the true and absolute motion of the body will arise, partly from the true motion of the earth, in immovable space, partly from the relative motion of the ship on the earth; and if the body moves also relatively in the ship, its true motion will arise, partly from the true motion of the earth, in immovable space, and partly from the relative motions as well of the ship on the earth, as of the body in the ship; and from these relative motions will arise the relative motion of the body on the earth.*

This may appear to be very complex—perhaps unnecessarily so. Bear in mind that Newton was writing at the end of the seventeenth century and he is saying that if we observe motion from the surface of the earth and measure it, we are measuring a relative, not an absolute motion. Suppose, for example, the speed of the earth through absolute space were exactly equal to and oppositely directed to the speed of an object moving on the earth. The absolute speed of the object would be zero. It would always be at the same position in *absolute* space even though it appeared in motion relative to the earth. Newton gave his own example to illustrate his distinction between absolute and relative motion:

> As if that part of the earth, where the ship is, was truly moved toward the east [in absolute motion toward the east] with a[n absolute] velocity of 10010 parts; while the ship itself, with a fresh gale, and full sails, is carried toward the west, with a velocity expressed by 10 of those parts; but a sailor walks in the ship toward the east, with 1 part of the said velocity; then the sailor will be moved truly in immovable space towards the east, with a velocity of 10001 parts, and relatively on earth toward the west with a velocity of 9 of those parts.

This is easy enough to understand and, we will develop this example because it contains Newton's theory of measurement. Before proceeding to delineate that theory, let us follow Newton's discussion a little further. Almost immediately after giving the above example, Newton pointed out that

> . . . Because the parts of [absolute] space cannot be seen, or distinguished from one another by our senses, therefore, in their stead we use

* *Ibid.*

> sensible measures of them. For, from the positions and distances of things from any body considered as immovable, we define all places; and then with respect to such places, we estimate all motions, considering bodies as transferred from some of those places into others. And so, instead of absolute places and motions, we use relative ones; and that without any inconvenience in common affairs; but in philosophical disquisitions, we ought to abstract from our senses, and consider things themselves, direct from what are only sensible measures of them. For it may be that there is no body really at rest, to which the places of motion of others may be referred.*

Newton's comment is very important. He is saying that in measuring the speed of objects we are measuring them relative to an arbitrarily chosen point defined by a material object. For example, in measuring the speed of a train, we might use the train station from which the train started. We have no way of knowing if the station is in motion relative to absolute space since there is no way to perceive it. Nevertheless, nothing prevents us from realizing that the train may or may not be in motion absolutely.

Newton was careful to point out that there is one kind of motion, rotational motion, that is always detectable. We will not follow him in this discussion except to relate the following example. When an object rotates there is a tendency for its parts to recede from the center of rotation. Newton considered a bucket filled with water and hung from a rope. Let us suppose we wind the bucket around and around and then let it go. At first, the water in the bucket will not rotate with the bucket and the surface of the water will be flat. As the motion of the bucket continues, that motion is communicated to the water. Once the water is in motion, there is a tendency for the water to recede from the center of rotation. The surface of the water does not remain smooth, but becomes concave and the water at the edge begins to climb up along the walls of the bucket. Now picture an observer sitting on the rim of the bucket. While he or she might want to argue at first that the bucket was not in motion but that the world around him or her was spinning in the opposite direction, he or she could observe later that the surface of the water was concave and that would be a proof that it was the bucket in motion. Since, regardless of the relative motion of the observer, there would be absolute agreement that the surface of the water has a concave shape, there would be absolute agreement that the bucket was spinning.

In most of our discussion, we will not consider such rotations. They will fall outside the realm of discourse of the special theory of relativity. Let us return to Newton's example of the ship moving in the water. It was remarked earlier that

* *Ibid.*

this example contained Newton's theory of measurement within it. As will become obvious, *Newton assumed, implicitly, that information will be communicated from one place (the ship), to another place (the shore) in no time whatsoever.* In other words, there was an infinite signal speed. The special theory of relativity modifies that implicit assumption embodied in Newton's theory of measurement. Since the theory of measurement is prior to Newton's dynamical theory, the propagation of Einstein's modification through Newton's dynamical theory significantly changed the dynamical theory.

One cannot fault Newton, because not only did he not realize such an assumption was implicit in his work, but for the next three hundred years no one else did either. As Einstein pointed out in his "Autobiographical Notes," it is not surprising that Newton had gone the route that he did. After all, it led to sensible results:

> . . . Newton, forgive me; you found the only way which, in your age, was just about possible for a man of highest thought and creative power. The concepts which you created are even today still guiding our thinking in physics, although we now know that they will have to be replaced by others farther removed from the sphere of immediate experience, if we aim at profounder understanding of relationships.*

Very often one finds stated in elementary textbook treatments of the relationship between the special theory of relativity and Newton's theory that they reduce, one to the other, for speeds much less than the speed of light. Such a comparison results from confusing formal quantitative predictions of theories with their meanings. In fact, the analogy is like comparing apples and oranges; or more accurately, apples and their seeds. Newton's theory is a dynamical theory of mechanics which implicitly includes, in the *Scholium* we have examined, a kinematical theory of measurement compatible to the Newtonian understanding of the relationship between sense data and belief about temporal and spatial variables. In fact, it is only by understanding the epistemological status of Einstein's special theory of relativity that the *logical* role of Newton's discussion of absolute space and time becomes clear. For, as we will see, Einstein's special theory of relativity is a theory of measurement which changes only one of Newton's basic premises of measurement. When Einstein's theory of measurement is applied to Newton's dynamical axioms (Newton's three laws) we have non-Newtonian or relativistic mechanics. That mechanics is considerably different from Newtonian mechanics. The differences are the results of the propagation of the one significant

* A. Einstein, "Autobiographical Notes," in P. A. Schillp, *Albert Einstein: Philosopher-Scientist,* (New York, 1949).

difference between Newton's and Einstein's premises. Because that premise is at the start of the theory, it makes an important difference, not only in the outcome of quantitative predictions, but as we will see in Chapter 3, in the meaning of the terms of the theory. Newton assumed implicitly that it is possible to transmit information from one place to another instantaneously. Einstein assumed explicitly that this is not possible, and that there is a finite maximum speed with which information can be transmitted from place to place.

Having examined qualitatively the difference between the theory of relativity and the comparable Newtonian theory of measurement, let us return to the Newtonian theory. Again it should be kept in mind that for three hundred years from the time of Newton until Einstein reanalyzed the problem, it was not noticed that Newton had assumed the possibility of an infinite signal speed.

Newton's theory of measurement is conveniently summarized by a set of coordinate transformation equations now known as the Galilean transformation equations. They formalize the general relationship which, by example, Newton had discussed when he described the relative and absolute motions of a ship and the people on board the ship.

But, whereas they formalize those relationships, the Galilean transformation equations are straightforward, simple and intuitive for us. They were an important element in the propagation of the mechanical world view.

THE GALILEAN TRANSFORMATION EQUATIONS

Physically, the Galilean transformation equations represent the relationship between spatial and temporal measurements that are made by observers who are in uniform motion relative to each other. We have given an example of the use of the transformation equations at the end of Appendix 2. From the point of view of one observer, an object is in freefall. From the point of view of another observer moving relative to the first observer, the same object is not falling along a vertical straight line but is following a parabolic path.

As Newton pointed out in his analysis of relative and absolute space, all measurements that we make are relative. We choose a reference point from which to make the measurement. For example, when we say that a landmark is four yards away, we mean it is four yards from where we are. Measured from a point six yards beyond us, the landmark is ten yards away. We will call the point to which all measurements refer as the *origin* of a frame of reference. A frame of reference is shown in Figure 10; it is rectilinear. Three measuring rods, the axes, are oriented at right angles to each other and meet at point O, the origin. We have labeled the axes as x, y, and z, but any letters could have been used. Specifying three coordinate values will specify the location of an object in space. Locating any two numbers will specify a plane in which the object is located.

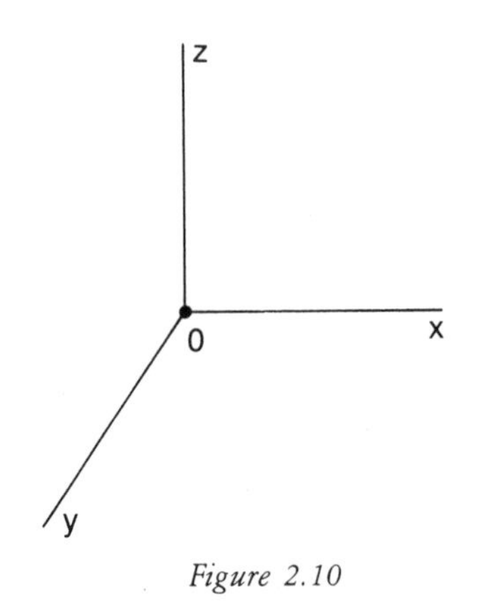

Figure 2.10

For many problems and situations, it is sufficient to use one or two dimensions. Nor need the system be rectilinear. A two-dimensional coordinate system is depicted in Figure 11a. For example, the point $P(x, y)$ could have been located in a polar coordinate system as in Figure 11b rather than a rectilinear one. In the polar coordinate system, rather than specifying two lengths at right angles to each other, we might specify the distance from the origin, r, and the angle from the horizontal, θ. $P(x, y)$ now becomes $P(r, \theta)$. The point P is the same point in both 11a and 11b. Only the *description* is different. The equations relating the two descriptions are called *transformation equations.**

The Galilean transformation equations relate spatial and temporal measurements made in one frame of reference to another, moving uniformly with regard to the first. For example, one frame of reference might be the surface of the earth. Another might be the ship to which Newton referred. Still a third would be the sailor walking on the ship. It is important to understand that if the person is not walking but standing on the ship, his or her frame of reference is identical with the ship's frame of reference.

We have depicted two abstract frames in Figure 12. The construction of the frame is arbitrary. Following custom, in Figure 12, the two frames are organized so that the direction of one axis is parallel to the direction of relative motion between the frames. If we use primes to designate one frame of reference, then the direction of relative motion between the frames is along the X and X' axes. In Figure 13, the problem has been simplified by removing one of the spatial coordinates, the y axis.

* For the situation in Figures 11a and 11b, the relationships between x, y, r and θ are given by

$$x = r\cos\theta$$
$$y = r\sin\theta$$

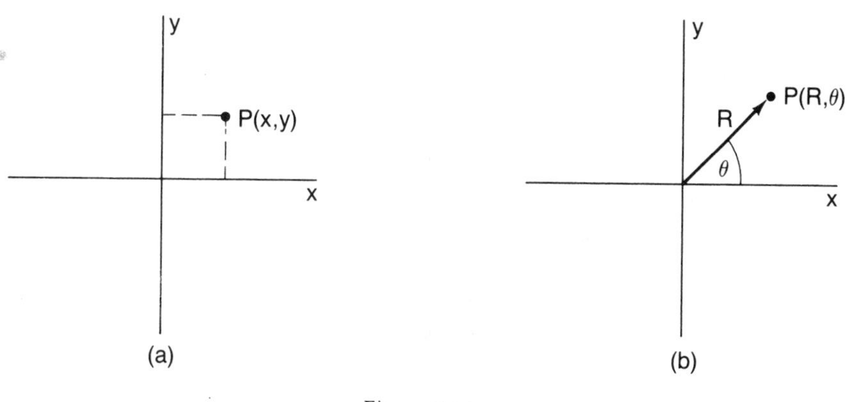

Figure 2.11

Suppose that in both frames of reference we have several identical clocks capable of running at precisely the same rate. They are spread out through the space wherever we need them in both frames of reference, capable of precise synchronization. (A technique for synchronizing clocks will be discussed later.) For simplicity we set all of the clocks in both frames of reference to zero, and we start them when the origins O and O' of the two coordinate systems coincide. At that instant, descriptions of the coordinates of all objects as seen from both frames of reference will be identical. After that instant, they will disagree with regard to only one coordinate: the distance an object is from the origin along the X and X' coordinate. In fact, relative to the origin O, the observer in O' will report an X' coordinate which will be less than the coordinate X measured in O by an amount equal to the product of the time elapsed from the moment when the origins were

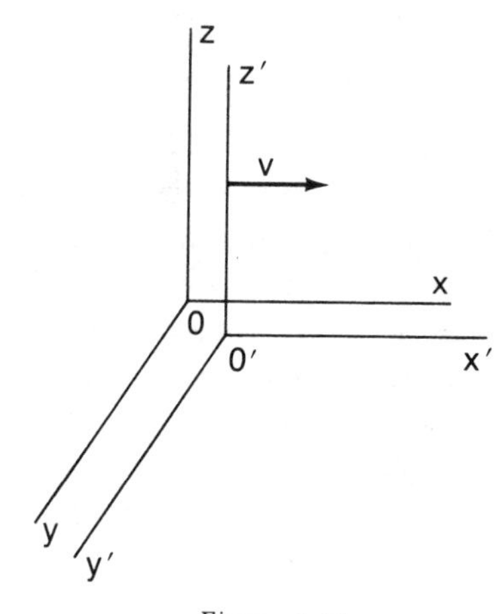

Figure 2.12

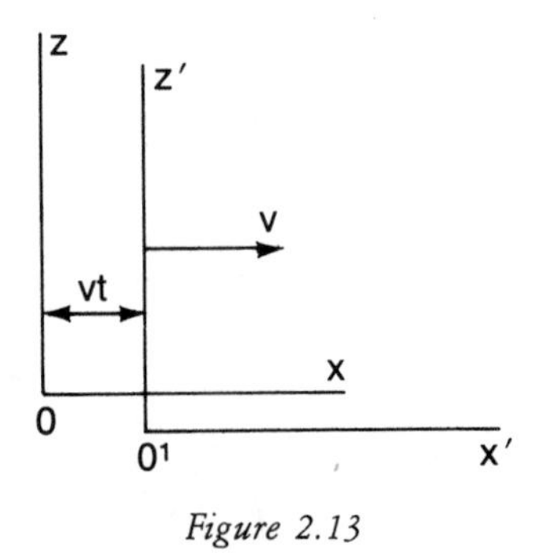

Figure 2.13

synchronized and the relative speed of the two frames of reference. Along the Y and Z coordinates, there will be agreement between the two frames of reference, and of course, they will agree on the time coordinate. It is these results which can be formalized as a set of transformation equations, the Galilean transformations, relating spatial and temporal coordinates in any frames of reference moving with respect to each other.

They are commonly written as

$$\begin{aligned} x' &= x - vt & x &= x' + vt' \\ y' &= y & y &= y' \\ z' &= z & z &= z' \\ t' &= t & t &= t' \end{aligned} \tag{9a}$$

where v symbolizes the relative speed of the two frames of reference. As the frame of reference that we label with the prime is arbitrary, it is common to write the x coordinate transformation as

$$x' = x \pm vt \tag{9b}$$

The Galilean transformation equations in (9a) and (9b) express in a formal way all of the content of Newton's discussion of the measurement of spatial and temporal coordinates in the *Scholium* with which this discussion began. The equations and the description which preceded them are intuitive in that they make sense, they correspond to our experience and expectations and they lead directly to other results that are equally intuitive: the Galilean velocity addition law. For the simple case of an object moving parallel to the X-X' axes in the two frames of reference, the velocity of such an object is the velocity of the object measured in the other frame of reference plus relative velocities of those frames of reference. This can be derived directly from the transformation equations and stated formally as

$$w = u \pm v \tag{10}$$

where u is the velocity measured in one of the frames of reference, w is the velocity measured in the second frame of reference and v is the relative speed of the two frames of reference. (This result has been derived in Appendix 3).

Let us take a simple case that is analogous to one Newton might have used except that in his day riding in trains and cars would not have been part of a person's experience. (Riding a stage coach is qualitatively different. It is unlikely that a stage coach moved at a uniform rate without jostles and bumps.)

Suppose a person is on a train, and we are standing by the embankment. We observe that the train is moving at a speed of 30 km/hr relative to us. The person walks from the back to the front of the train at a rate of 3 km/hr measured by observers on the train. The velocity addition law, based on the Galilean transformation equations, predicts that the velocity measured from the embankment will be 33 km/hr. This is what we expect intuitively. This experiment is feasible and has been performed many times in other contexts. The agreement with the prediction is also well documented.

It was only after Einstein's theory of relativity was published that a more sophisticated examination of the measuring process in these cases was suggested. For example, consider the following events: Event #1 — a person on the train opens the caboose door (The person is walking forward in the train); Event #2 — the person on the train arrives at the locomotive door. From the point of view of that person, both events take place at *his or her* original place, and in order to measure the time elapsed between the two events, the person requires only one clock.

Now let us consider the same events from our point of view. The observation that the person was at the caboose door was made by an observer opposite the occurrence along the embankment. That observer noted the time. By the time the person on the train had arrived at the locomotive door, the frame of reference we have called "the train" had moved a long distance down the track. When the person on the train arrives at the locomotive door, that event (Event #2) would have been observed by another person opposite the event who would also note the time. The time interval between the events in the embankment frame of reference is determined by the difference in the readings between the two clocks. The speed of the person on the train would be the ratio of the distance between those clocks and the time elapsed on the two clocks.

Similarly, the speed of the person to be measured by observers on the train will require two clocks, at the caboose door and the locomotive door. We have now described the same set of events from three points of view or three frames of reference. From the person's point of view he or she has gone nowhere. By definition, the person has remained at his or her origin; time has elapsed on his or her clock as the caboose receded and the locomotive approached. (We know, were one of us the person who was walking, that we were, in fact, walking. Walking requires work and movement on our part; nevertheless, almost everyone who has been on a train has had the experience of being unable to tell whether our train or the other train was moving.)

From the point of view of the observers on the train, the person's speed was measured by two clocks which are a known distance apart: the person's speed equals 3 km/hour. From the embankment, the person's speed was also measured with two clocks a known distance apart to find that the person's speed equals 33 km/hour. That there should be disagreements is not strange. We understand that they arise because the measurements were made from different inertial frames of reference. But note that in only one frame of reference were the measurements made with only one clock. In all other frames of reference more than one clock had to be used. Measurements of more than one event made with a single clock are referred to as *proper* measurements of time. A proper time measurement is made in only one inertial frame of reference. All other frames of reference, using more than one clock, make *nonproper* time measurements. (The terms "proper" and "non-proper" are used in a technical sense.)

Before proceeding, let us examine an alternative manner of depicting the physical and mathematical description summarized by the Galilean transformation equations. As it can always be arranged that the relative motion between inertial frames of reference is along the X-X' axis, let us eliminate *both* the Y and Z axes since there will never be measurement disagreements in those directions. Recall that the coordinate axes are abstractions, freely chosen to represent physical situations. The coordinates need not be perpendicular to each other in a frame nor do all the coordinates need be spatial. In Figure 14a we have shown a frame of reference in which time is measured along one axis and distance is measured along the second axis.

In Figure 14b a second frame of reference, the primed frame, is shown in which

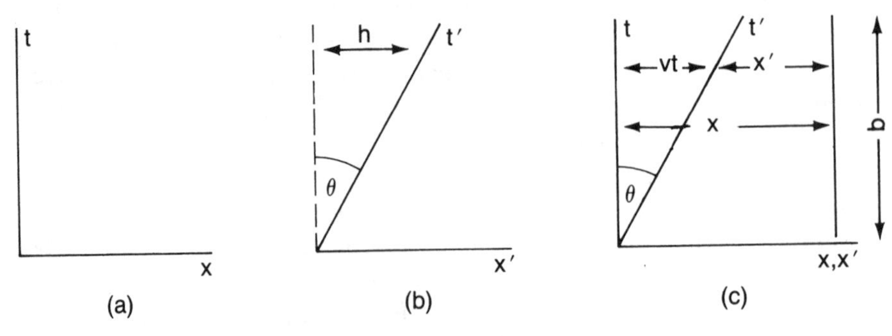

Figure 2.14

the time axis t' makes an angle θ with the t axis of 14a such that the distance h, the distance between the origins at any time t or t' equals the product of the relative speed between coordinates and the time elapsed.

$$tan\ \theta = h/t = h/t'$$
$$\tan \theta = v$$

In Figure 14c the two frames of reference have been joined. The line b represents an object which is at rest in the unprimed frame of reference. Its spatial coordinate does not change. It appears to be moving in a negative x' direction given by $x' = x - vt = x - \tan \theta$. In order to determine the spatial and temporal coordinate of any point in the x-t or x-t' spaces, as for the point c, auxillary lines are constructed *parallel* to one axis, intersecting the other axis.

After studying Figure 14c, you will see that there is something odd about it. The tangent of the angle represents the relative speed between the frames of reference. In Figure 15, we have constructed the line h perpendicular to the t axis. From the laws of Euclidean geometry, t should not be equal to t', yet that must be the case if the figure is to represent the physical situation of two observers in uniform motion relative to each other. Therefore, the unit of length in the unprimed frame of reference representing the time, t, must be different from the unit of length in the primed frame of reference representing the time t'.

This pictorial representation of spatial and temporal coordinates will be useful later. But depicting these relationships in unfamiliar ways helps us to keep in mind that we are not dealing with actual physical situations but with abstract representations of actual physical situations. In fact, later, not only will the relationship between the description of physical events be depicted by abstract figures, the geometrical rules will not satisfy the postulates of Euclidean geometry. This does *not* mean that either space is, or is not, Euclidean. It only means that the *representations* are not Euclidean. Whether space itself is Euclidean is a separate, physical question.

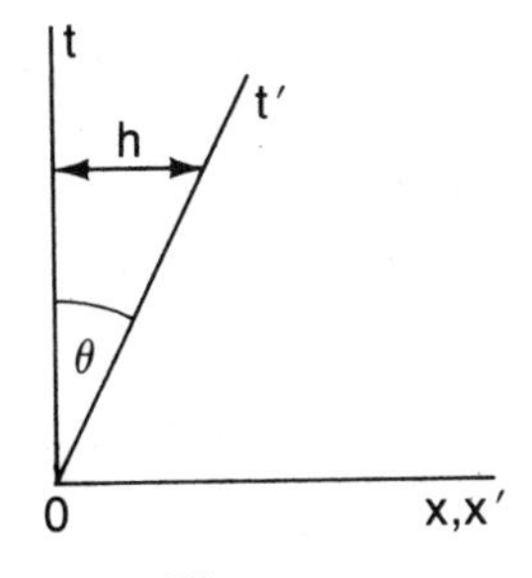

Figure 2.15

THE CLASSICAL PRINCIPLE OF RELATIVITY

We have pointed out that one of the major features of Newtonian mechanics is to unite the physics of motion of the heavens with the physics of motion near the surface of the earth. There is a separate, perhaps equally powerful principle that developed from Newtonian mechanics: the classical principle of relativity or the principle of relativity for mechanics.

The principle may be stated as follows:

> The laws of mechanics have the same form in all inertial frames of reference

The principle does not say that the results of measurement are the same, only that the laws have the same form.

Newton's laws apply to all inertial frames of reference since these are the premises from which we begin and the choice of those premises are ours to make. The principle of relativity makes a much more consequential, deeper claim. It says that, if certain laws of mechanics apply to a sequence of events in one inertial frame of reference, the laws will apply to *all* inertial frames of reference for those events. That doesn't sound *relative.* In fact, it sounds *absolute,* and, it is. The term "relativity" refers to the fact that, whereas observers will agree about the form of the law, they may disagree about the value that each of the measured quantities has.

This principle of relativity is a result of the Galilean transformation equations. Consider a situation in one frame of reference where the conservation of momentum applies (for example, the collision of two billiard balls). If, for this situation, the law applies in one frame of reference, it will apply in all frames of reference moving at a uniform rate. Observers in different frames of reference would disagree about the value of the total momentum for the system and the value of the momentum for each component of the system. All observers would agree that whatever the value, that value is conserved in the interaction. The quantitative derivation of this result from the Galilean transformation equations is demonstrated in Appendix 3.

A special term is used to denote that the laws of mechanics have the same form in all inertial frames of reference. It is said that the laws of physics are *covariant.* In other words, the quantities in the laws vary, each so that they leave the form of the law unvaried. An alternative statement of the principle of relativity for mechanics that summarizes this discussion is:

> The laws of mechanics are covariant under a Galilean transformation.

One can make a much stronger statement for the premises of mechanics—Newton's laws. These laws are not only covariant; they are *invariant.* This means that not only will the laws have the same form, but that in all inertial frames of reference, observers will agree that the numbers assigned to the different quantities are the same.

First, mass is an invariant in classical mechanics. A number is assigned to each object which, in Newton's day, represented the quantity of matter and which, after Mach, was a characteristic of how the object interacted with other objects. That number did not depend on the state of motion of the object; it was a *property* of the object. Furthermore, much as we are able to use the Galilean transformation equations to predict the velocity addition law that would pertain between inertial frames of reference, so, too, we can use it to predict how observers in inertial frames of reference will compare in their measurements of accelerations for objects. Since we are restricting ourselves to non-accelerating frames of reference (that is what is meant by inertial frames of reference), the calculation is exceedingly simple. As demonstrated in Appendix 3, not only will all inertial frames of reference agree that an object is accelerating, they will agree about the magnitude of the acceleration. The acceleration will also be an invariant. If in one frame of reference, the mass is measured by m and the acceleration is measured by $\vec{a}$, in a second frame of reference, a primed frame of reference, the mass, m' must equal m and the acceleration $\vec{a}'$ must equal $\vec{a}$. Consequently, in one frame of reference since one assumes as a premise $\vec{F} = m\vec{a}$ and in the second frame of reference one assumes that $\vec{F}' = m'\vec{a}'$, it follows that $\vec{F}' = \vec{F}$. Not only will all inertial frames of reference agree that forces are being applied in a given situation, they will agree on the magnitude of the force. Another way of stating this is as follows:

> Newton's laws of motion (the axioms of mechanics) are invariant under a Galilean transformation.

Despite the principle of relativity, the covariance of the laws of mechanics and the invariance of Newton's laws, those who followed Newton did not deny the existence or importance of absolute motion or absolute space and time. According to Newton these quantities exist whether or not we have direct knowledge of them. In a sense they were the sensorium of God, who, in the frame of reference of absolute rest, could state with absolute certainty whether or not a thing was *really* moving. One way of knowing God was to discover the absolutely resting frame of reference and the way of doing that would be to discover the motion of the earth relative to the absolutely resting frame.

Earlier I spoke of the mutual influence between Newtonian mechanics and Enlightenment philosophy. The feeling was widespread in western culture that Newton had discovered God's laws for the universe and that ultimately all physical phenomena would be understood in Newtonian terms. We have referred to this commitment as the mechanical world view. Were such a program successful, the principle of relativity for mechanics would necessarily apply to all phenomena. For this reason, the principle of relativity for mechanics is often referred to as "the restricted principle of relativity." As shown in Appendix 4, in the nineteenth century, the quantitative development of a molecular understanding of matter and the application of Newton's laws to the interactions of the molecular particles led to a reduction of the subject of thermodynamics to mechanics.* The immediate consequence was the applicability of the restricted principle of relativity to thermodynamics processes.

THE SEARCH FOR THE ABSOLUTE FRAME OF REFERENCE

A line of thought connected to the mechanical world view, which developed rapidly during the first part of the nineteenth century, gave natural philosophers the hope that they would be able to detect the absolute motion of the earth through space. This line of thought was closely connected with the wave theory of light. To say that light phenomena were wave phenomena means that the propagation of light from place to place must be understood as the propagation of a disturbance within a medium of propagation. Waves are not substantial objects like chairs. For example, water waves on the surface of the water are the propagation of a disturbance perpendicular to the direction of the disturbance (Figure 16) and, of course, take place in water. Sound waves are a disturbance, parallel to the direction of the propagation and take place in air.

In general, two kinds of disturbances are possible: those in the direction of propagation are longitudinal waves; those perpendicular to the direction of propagation are transverse disturbances. And, as shown in Figure 17, the wave can be described by three quantities: the wavelength, the frequency, and the speed of propagation. When we speak of a wave being propagated, as opposed to the propagation of a momentary disturbance, we refer to a regularly repeating disturbance that establishes a pattern of distortion. For example, regarding sound waves, a longitudinal disturbance in air, the density of the air is one of the variables that is distorted. Or consider the propagation of transverse water waves on the surface of water. (I am not speaking here of breakers, but rather waves on the surface when the depth of water is sufficient that the bottom is itself not a disturbing influence.) The disturbance which is being propagated in such a case is the height above or below the normal surface to which the water has been pushed or pulled.

---

* This is not to imply that the reduction was entirely successful or accepted by everyone. However, examination of the details would take us much too far afield. For a summary of the situation, see C. C. Gillispie, *The Edge of Objectivity* (Princeton, 1960), pp. 479–488.

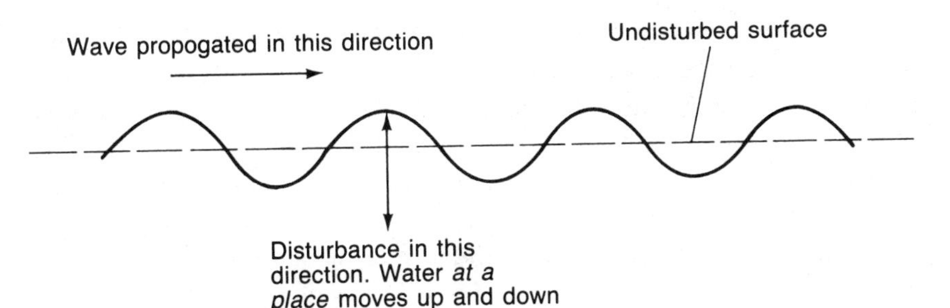

Figure 2.16

The distance between points representing a repetition of the cycle is called the *wavelength*. The number of wavelengths that propagate past a point in a unit of time is called *the frequency* and is often reported in terms of "cycles per second." The *speed* of the wave refers to the speed with which the disturbance is being propagated. Note that nothing material is being propagated and in transverse water waves, for example, the water merely moves up and down as the disturbance propagates along the surface of the water. Since water always conforms to the shape of the container, transverse waves are not possible below the surface. In longitudinal waves, the material supporting the disturbance undergoes a back and forth perturbation but again it goes nowhere.

Describing and understanding wave phenomena are perhaps the most difficult analytical problems in natural science because to speak abstractly about such phenomena, one must keep track of two kinds of variables at the same time: spatial propagation and temporal propagation.

During Newton's lifetime, there was considerable controversy about the nature of light. Newton had opted for a particle theory whereas his contemporary and

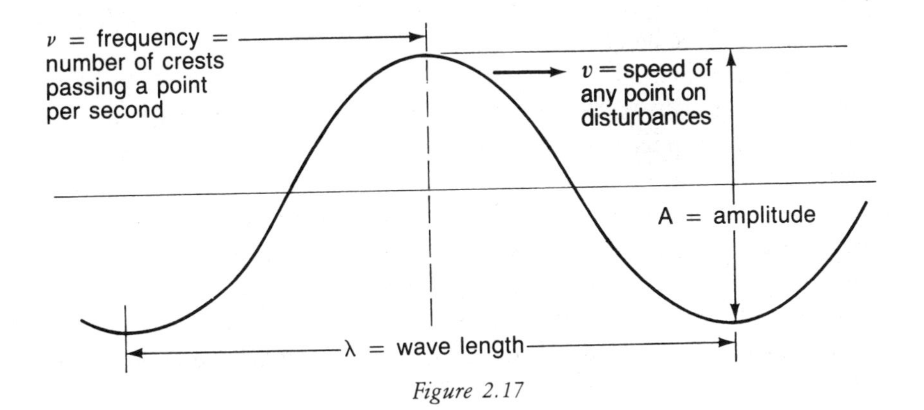

Figure 2.17

rival, Christian Huygens, had opted for a wave theory in which the light represented a disturbance in a given medium. There was no evidence that Newton knew, of light bending around obstacles as would be the case were light a wave phenomenon. Newton's influence on this problem cannot be overemphasized. Although wave theory could elucidate problems which were explained by the particle theory, the particle theory of light was paramount in natural philosophy for the next hundred years. And, at the turn of the nineteenth century when Thomas Young, the chief proponent of a wave theory, aggressively argued his case, he tried to convince his contemporaries that Newton, too, would have believed new evidence and would have chosen the wave theory.

The wave theory predicted that the speed of light in water would be less than in air. Particle theories predicted that the speed of light would be greater in water than in air. This, then, is a crucial test of the two theories. The technology for measuring the speed of light was not developed sufficiently until mid-century for such a test, but, as is often the case, when the test was carried out in 1853 by Jean Foucault, the questions had long since been settled by other means. In fact, by the third decade of the nineteenth century, not only had the wave theory of light been firmly established, but because of phenomena associated with polarization, light was understood to be a transverse wave phenomenon. If light, or more precisely, if those sensations we identify as light, are to be understood as manifestations of wave-like behavior, then the phenomenon is to be associated with a supportive medium. The sun is approximately ninety-three million miles from the earth. Light waves which are being propagated from the sun to the earth must be propagated in *something*. It is not air since there is no air beyond the earth. Furthermore, light passes totally through glass enclosed *vacua*. Not only must there be a medium to support the waves but the medium will act like an elastic solid because transverse waves cannot be supported below the surface of a liquid or air-like medium. The medium was referred to as "luminiferous ether" and was believed to fill all of space. Was there a place in the universe to which light could not, in principle, be propagated?

When I write that the medium behaves like an elastic solid, I have in mind steel, rubber, or perhaps Jello. A cube of Jello, when touched, jostles and jiggles throughout its volume, transverse to the direction of propagation of the disturbance.

Elastic solids are, strictly speaking, in the province of mechanics. Therefore, one way of subsuming optical phenomena would be to solve the riddle of the elastic solid medium through which light disturbances are propagated—the luminiferous ether of space. There is no question that this ether has remarkable properties in order to transmit the extremely high frequency, short wavelength disturbances associated with light. (The wavelength of light, which is the physical correlate of the perceptual variable color, is of the order of magnitude of 5×10^{-5} cm.). With speeds of almost two hundred thousand miles per second, the medium exhibits a tensile strength unheard of in any known material. It is stronger than the hardest piano wire steel. On the other hand, planets, stars and galaxies move through it as though it does not exist.

To you and me, it may seem extraordinary that natural philosophers were willing to accept the possibility of the existence of such a substance. But it is not strange nor atypical. The luminiferous ether was in the tradition of a long line of such substances. Phlogiston, for example, had been hypothesized as an imponderable fluid responsible for converting ores into metals. Caloric had been an imponderable fluid that was supposed to have been the substance of heat. In some circles, it is now being argued that the recent proposal of subatomic particles such as quarks is nothing more than another line of fanciful entities that defy detection.* Clearly, if light is a wave phenomenon, *something* must be supporting the propagation of the disturbance.

Suppose a train is at rest with its headlight burning. The model for how the light energy is propagated through space is, in principle, very simple. The ether in the immediate vicinity of the light is perturbed and that perturbation spreads outward in an ever growing sphere. Even if the headlight is turned off, the perturbations continue until a material object is encountered. The velocity of propagation depends only on the nature and state of the ether and is independent of whether or not the source moves, for once the medium is perturbed, it has nothing to do with the source.†

Suppose that now the train is moving at a constant velocity $\vec{v}$ along the track. It does not matter how fast the train moves, the velocity of propagation of light from the headlight will be the same when measured from the train as it was when the headlight was at a standstill. Following custom, we will always designate that velocity with the symbol $\vec{c}$. And as long as the ether does not move, the velocity of light on the moving train will remain the same as measured from the moving medium. But should the receiver move, the measured velocity of light from the medium will change. In this case suppose that you are moving toward the headlight on the front of the train (Figure 18). According to the Galilean velocity addition law, the velocity of light from the headlight is $\vec{c} + \vec{v}$. Should we be

* For two clear delineations of the nature of the evidence and the logic being used in arguing for the existence of quarks, see R. F. Schwitters, "Fundamental Particles with Charm," *Scientific American*, (Oct.) 1977, *237*:56. R. R. Wilson, "The Next Generation of Particle Accelerators," *Scientific American* (Jan.) 1980, *242*:42.

† If the source is moving, other wave variables change; for example, the wavelength of the wave becomes shorter in the direction of motion of the source and longer in the opposite direction.

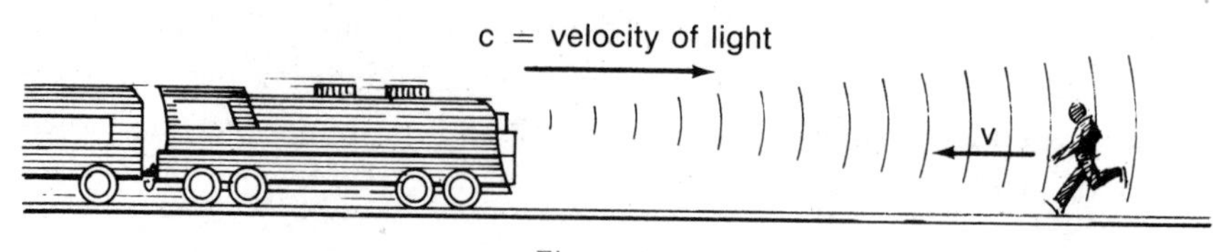

Figure 2.18

moving in the opposite direction, away from the headlight, the measure of the velocity of light is $\vec{c} - \vec{v}$.

Of course, we are assuming that the *medium* is at rest. Is that a reasonable assumption? There are three possibilities: 1) As an object moves through the ether, the ether is unaffected and remains at rest. 2) Instead of remaining at rest, the medium is dragged along at the same speed as the object and the relative speed between the object and the ether is zero. 3) The ether is partially dragged by the moving object so that if the object is moving at speed v, the ether is dragged along at the speed fv where f is a number between zero and one.

Assuming that the ether is fixed, if one could measure the speed of light in different directions or could measure the change, as a function of direction, of a physical variable that depends on the speed of light, we could use the variation to determine the speed and direction of the earth through the ether—that is, through absolute space.

These experiments are commonly called ether drift experiments. Many different kinds have been proposed and carried out although interpreting the result depends on belief about the nature of the interaction between matter and ether.

Thus far, the speed of light in empty space has been referred to. In other, transparent environments it is different depending on the wavelength of the light and the nature of the material in which the light travels. In water, for example, the speed of light is less fast than in empty space by a factor of 1 to 1.33; only three-fourths the speed in air. The speed of light in glass depends upon the type of glass. Typically, the ratio of the speed of light in a vacuum to the speed of light in glass is 1.5 to 1. That is, the speed of light in glass is only 0.67 of the speed in empty space.*

The ratio of the speed of light in a vacuum to the speed of light in glass or water or another transparent substance is called "the index of refraction" of the substance. As explained in Appendix 5, the change in speed of light as it enters a material substance is denoted by a change in the *direction* of propagation of light.

* Although the speed of light in air is slightly different from the speed of light in a vacuum, they are so close that we will take them to be the same.

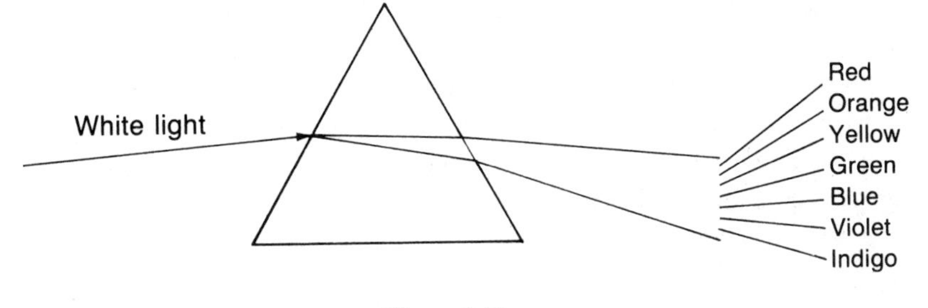

Figure 2.19

This property of transparent substances allows them to be fashioned into prisms and lenses.

Different wave lengths of light travel at different speeds in the same transparent substance. This is called *dispersion* because the direction of propagation changes with changes in speed. The physical variable wavelength is correlated with the perceptual variable *color;* thus, as white light enters a transparent medium under the right conditions, it is dispersed into bands of color from red to indigo—the rainbow. White light is a mixture of all the colors of the spectrum, and transparent media are said "to disperse white light into spectral colors" (Figure 19).*

In ether drift experiments the speed of light was not measured directly. Rather, a physical attribute of the instrument whose value depended on the speed of light was examined as the direction of the apparatus changed. If there were variations, a change in the speed of light as a function of the direction in which the light was traveling was indicated.

The problem of direct measurement of the speed of light going in different directions, and of comparing those speeds, can easily be demonstrated. Suppose that the light is coming from a star, low on the western horizon, early in the evening. Since we are moving in a direction toward the star (Figure 20), were we to

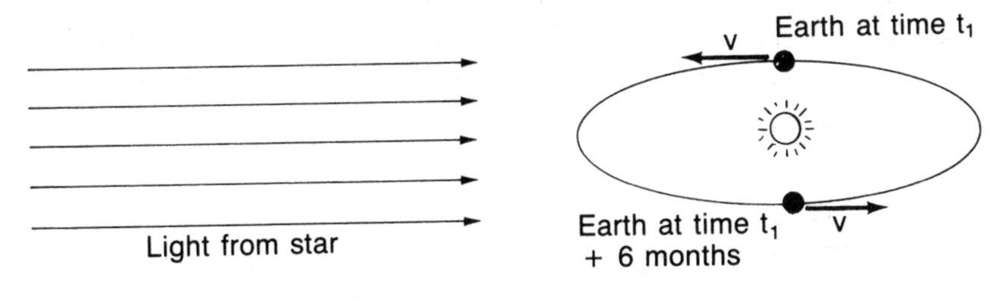

Figure 2.20

* The concept of white light as a mixture of the colors of the rainbow was first proposed by Isaac Newton. Do not confuse this discussion of spectral color with theories of color perception; that is, how color is perceived by human beings. They are different, although related, subjects.

measure the speed of light from the star, assuming the ether is at rest, it will be the sum of the speed of light in free space and the orbital speed of the earth. This assumes the application of the Galilean transformation equations. Six months later, we will be moving away from the star when it is low in the west. The speed that we measure for the light from the star will be the difference between the speed of light in free space and the orbital speed of the earth. The *difference* between the two values will be twice the orbital speed of the earth:

$$(c + v) - (c - v) = 2v$$

A moment's thought will convince you that we have oversimplified the problem by assuming that the net speed of the earth through absolute space is only the result of its orbital speed. We also have to take into account the motion of the entire solar system, the motion of the galaxy and the motion of any larger aggregate of which we are a part. Any significant discrepancy between the speed of light in free space and the speed of light measured from the star will be due to the *net* velocity of the earth through absolute space. It might be that by coincidence we have chosen two periods when the net velocity of the earth through absolute space is zero. Therefore, we will undertake to measure the speed of light from the star many times during the course of a year and we will find a maximum discrepancy from the speed of light in free space twice during that time interval. Those times are when the velocity of the earth in its orbit is parallel and anti-parallel to the direction of the sum of the solar system and galactic velocity.

We have also oversimplified the problem in another way. Measuring the speed of light is a very subtle and difficult experimental problem partly because the light moves very fast. There is no practical way to measure directly the speed of light from a star. There is not enough light to be used in just any apparatus. But by using a telescope the effects of the earth's motion on the speed of light from stars can be accurately inferred. The experiment described here was first done in 1810.

A telescope is a device for making objects appear closer. In one type of telescope, glass lenses are used to focus light from the star, depending on the lens's index of refraction. If c is the speed of light in empty space and c' is the speed of light in the glass, the index of refraction, n, is given by

$$n = c/c' \qquad (11)$$

Suppose (Figure 21) that the telescope is moving directly toward the star. According to the Galilean transformation equations, the speed of light in "empty" space is $c + v$. The speed of light in the glass is $c' + v$. The index of refraction will now be

$$n' = \frac{c + v}{c' + v} \qquad (12)$$

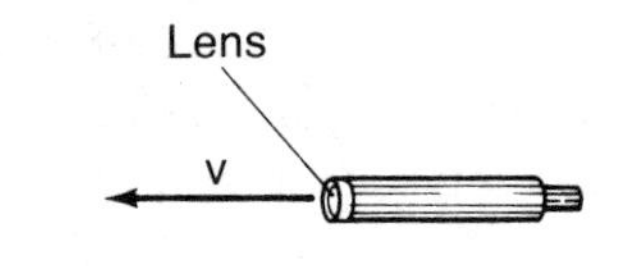

Figure 2.21

Compare equations (11) and (12). The value of n and n' are not the same. When a constant is *added* to the numerator and the denominator of a fraction, the value of the *ratio* changes.* If the index of refraction changes, the length of the telescope tube that is required for a sharp focus will change. This would be an indirect way of determining the change in speed of light from a star due to the motion of the earth.

The French physicist D. Arago is said to have undertaken this experiment in 1810. He located a star that was low on the horizon and focussed a telescope on it. The telescope was locked into position and the star was periodically observed during the following year. There was no change in the focus. In other words, the experiment could not determine the absolute motion of the earth through the ether.

That the focus of the telescope did not change over time means that in equations (11) and (12), $n = n'$. There appeared to be only one reasonable hypothesis to account for this remarkable result. As the earth moved through the ether, we assumed in equation (12) that the relative speed of earth and ether was v. In other words, the hypothesis was that the ether is absolutely fixed. Suppose that locally, as the earth moved through it, the ether was dragged along at just the right speed so that no matter what the angle between the velocity of the earth and the velocity of light from the star and no matter what the index of refraction of the material through which the light was moving, the index of refraction of the glass in the telescope did not change. In other words, we have to replace equation (12).

In 1818, the French physicist August Fresnel proposed a quantitative solution to the problem which found growing favor as the century progressed.† He suggested that the degree to which the ether is dragged depends on the index of refraction. Rather than the relative speed between the earth and the ether being v, Fresnel thought that the relative speed is

$$v(1 + 1/n^2)$$

---

* For example, take the fraction ⅔. If we add 1 to the numerator and the denominator, we transform the fraction to ¾.

† Along with Thomas Young, Fresnel was one of the major contributors to the revival of the wave theory of light in the nineteenth century.

where n is the index of refraction. The function $1 + 1/n^2$, became known as "the Fresnel dragging coefficient" and affects the velocity of light in transparent materials such as lens and prism to compensate exactly for the changes being sought to determine the motion of the earth through the ether. It was as though nature were conspiring to keep hidden the absolute motion of the earth through space.

The experiment we have described is sometimes referred to as a "first order experiment." The term "first order" refers to the expected effect of the motion of the earth through the ether depending on the ratio of the velocity of the earth to the velocity of the speed of light in the ether frame of reference, v/c. Later, we will encounter second order experiments where the effect will depend on $(v/c)^2$, the square of the ratio of the speed of the earth to the speed of light in the ether frame of reference. (Specific quantitative examples are given in Appendix 5.)

The proposal of the Fresnel dragging coefficient leads to finding even more astounding properties for the ether. Not only must it fill space and have the seemingly contradictory properties of behaving as the most dense elastic solid that we can imagine and at the same time being able to allow objects to move through it with ease, but it can also be picked up by matter and partially dragged along with it. One model suggested that the ether was compressed within matter; that is, like air entering the intakes of a jet engine, the ether is gathered up by the leading edge of moving matter, compressed, and ejected from the rear. Incredibly, the ether immediately comes to rest behind the object. Otherwise, one would find currents and eddies in the ether and these would manifest themselves as distortions of the images of distant objects such as stars and planets. Since no distortions are ever observed, this must mean that there are no local motions within the ether outside of matter.

There were many ether drift experiments performed during the nineteenth century. The details of some of them are examined in Appendix 5. In the Arago experiment, the variable that was examined was the index of refraction. Many other experiments relied on a different parameter related to "interference" phenomena. According to the wave theory, when light waves interact, the net effect can be either reinforcement of the waves (constructive interference) when crests and troughs from different waves coincide, or cancellation (destructive interference) when crests of one set of waves coincide with troughs of a second set, and vice versa. In intermediate states, partial reinforcement or partial cancellation is possible (See Figure 22). Such interactions between waves are referred to as phase relationships.

Relationships between waves can vary with time at a place or in space at a single instant. Obviously, it is a physically complex system which can be honed and simplified, experimentally, so that the experimenter is examining an interference

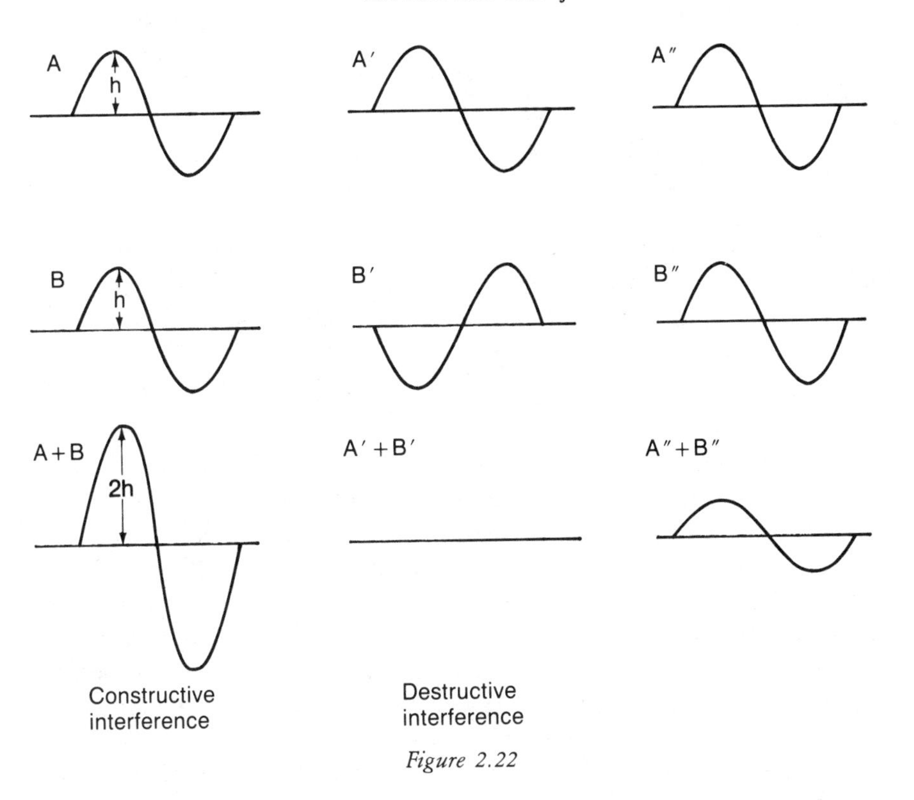

Figure 2.22

pattern created by two light beams which have followed different paths and through a series of mirrors and lenses have been brought back together in space. An interference or fringe pattern appears in Figure 23. Any change in the phase relationship between the interfering beams of light which give rise to these patterns will cause an apparent shift in the fringe pattern. For example, if the velocity of one or both of the beams change, the phase relationship will be affected and will manifest itself as an apparent fringe shift. An experimental relationship of this kind is always extremely sensitive; but, by the same token, it is extremely difficult to arrange.

There was remarkable confirmation of the partial drag hypothesis by Fresnel's countryman A. H. L. Fizeau using such a technique. In 1851, he undertook to compare the speed of light in water moving in the same direction as the flow of the water with the speed of light moving in the opposite direction. Fizeau's apparatus used interference techniques. The fringe shift observed in the interference pattern was correct according to the Fresnel dragging coefficient.

Despite confirmation and other successes of the partial drag hypothesis (See Appendix 5), the result was the frustration of those committed to the triumph of a

As the observer looks at a screen,
he or she sees bands of light and dark.

Figure 2.23

mechanical world view who sought a measure of the earth's absolute motion. Although the mechanical world view was not progressing, the electromagnetic synthesis was. That synthesis relied on the work of Clerk Maxwell, who recast the entire field of electricity and magnetism into a coherent physics of fields. In this scheme, charge was the source of the electromagnetic field as mass was the source of the gravitational field in Newtonian physics. The most startling outcome of Maxwell's synthesis was the reduction of the hitherto independent field of optics to electromagnetic theory. Light, according to Maxwell's theory, was an electromagnetic phenomenon, and the propagation of light was that of *electromagnetic* waves. Thus, three of the four ethers of the history of the eighteenth and nineteenth century, the electric, the magnetic and the optical ethers, were reduced to one—electromagnetic. And now, during the last quarter of the nineteenth century, the motivation to recast the electromagnetic and gravitational ethers into one coherent theory was high. If it were not possible to do so within the mechanical world view, perhaps the reason was that mass was not the fundamental entity. Perhaps the fundamental stuff of the universe was charge, and the proper program was to understand the nature of the physical universe, including mass and mechanical interactions, in terms of electric charge and electromagnetic interactions. It is this view and this program that current historians of physics label "the electromagnetic world view."

It would be a mistake to assume that there was unanimity among physicists of the last quarter of the nineteenth century about the detailed nature of the theories within this world view, and there was little recognition that they were attempting to forge something that we, looking back with one hundred years hindsight, have labeled "the electromagnetic world view." There were several different theories vying for appropriate expressions of the electromagnetic world view. The theories to which attention is paid here are those of the great Dutch physicist H. A. Lorentz. He was materially aided by the critical comments of Henri Poincaré.

H. A. Lorentz: engraving done in 1898 when Lorentz was 45. Reprinted from *H. A. Lorentz. Collected Papers.* The Hague: Martinus Nijhoff. Courtesy of the American Institute of Physics Niels Bohr Library.

The theory became known as the theory of electrons. It posited only two basic entities in the universe: electrons and ether. Matter was composed entirely of electrons and gross attributes such as elasticity, hardness, and ductility, were due to the manner in which the basic charges interacted with each other. The attribute of mass was explained, at least qualitatively, as electromagnetic inertia. The role of the ether was to transmit electromagnetic radiation between electrons. The electrical, magnetic, and thermal properties of matter were also explained by the interactions of the electrons with each other and with the ether.

This theory of Lorentz was not a kinematical theory. It was clearly a dynamical theory attempting to account for the nature of matter and how that matter would interact with radiation from its own parts or from other matter. There was a fair measure of success in deriving and explaining various quantities that had previously been defined, measured and assessed in older and more limited theories. The one area that resisted reduction to the Lorentz theory was commonly referred to as "the electrodynamics of moving bodies." In other words, the theory was not successful in predicting the behavior of materials, as understood within this theory, when they were observed from other inertial frames of reference. Lorentz struggled with this problem, at first alone and later with the advice and criticism of Poincaré. Specifically the theory could not predict the results of ether drift experiments while applying the Galilean transformation equations.

In 1892, in a monograph about the theory of electrons, Lorentz contented himself with invoking arbitrarily the notion of the Fresnel dragging coefficient to preclude the detection of first order effects of the relative motion of inertial frames of reference. Three years later, in a second monograph, he introduced a set of transformation equations to work between frames of reference. The transformation equations were a modification of Galilean transformation equations; Lorentz altered the way the time coordinate transformed between inertial frames of reference. For the record, these equations were:

$$\begin{aligned} x &= x' - vt' & x' &= x + vt \\ y &= y & y' &= y \\ z &= z' & z' &= z \\ t &= t' - \frac{vx}{c^2} & t' &= t + \frac{vx}{c^2} \end{aligned}$$

The meaning of the variables has not changed. A new variable, c, has been introduced; it represents the speed of light in free space. The transformation equation for time may appear to be a strange equation. To make it clearer we rewrite it:

$$t = t' - \frac{v}{c} \cdot \frac{x}{c}$$

In other words, the discrepancy is a function of the product of the ratio of the speed of the frame of reference to the speed of light and the time required for light to travel from the origin to a point. That is what the ratio of x/c is.

The first question that arises has to do with the viability of the entire Newtonian system, because that formalism assumed the Galilean transformation equations,

Albert A. Michelson in his laboratory. Courtesy of the American Institute of Physics Niels Bohr Library.

and implied a certain theory of measurement. To reject those transformation equations would be to reject that theory of measurement and to invalidate the conclusions (but not the analysis) associated with Newtonian physics. Lorentz was apparently well aware of this implication for he explicitly stated that these transformation equations are merely "an aid to calculation" and do not imply real or physical time, only something he called "local time." Physical time was still to be governed by the Galilean transformations. In this, Lorentz was preserving deep seated intuitions about the nature of time and its absolute character.

It is not strange to erect that kind of formalism to predict *physical* events while maintaining a cosmology and metaphysics whose spirit stood in contradiction to the predictive formalism. There was and remains a long tradition of such dichotomies in the history of science. For example, the metaphysical system behind Ptolemaic astronomy with its epicycles, equants, cranks and other complex mechanical structures was the simple Aristotelian concept of the centered earth surrounded by eight perfect circles. One was very careful never to get that cosmology and its implicit metaphysics confused with the predictive astronomy of Ptolemy and his later commentators. A similar dichotomy existed between Copernican cosmology which placed the sun at the center with planets, in perfect circular orbits, and the predictive system Copernicus used which also contained complex mechanical constructions. Currently there is a relatively simple cosmology associated with the quantum view of the atom. The complexities of the predictive system for organic chemistry stand in sharp contrast, and sometimes in sharp contradiction to that cosmology. The examples of this kind of dichotomy are endless once one becomes sensitized to the problem.

The Lorentz conception of an "aid to calculation" is very understandable. Heretofore, one accounted for the inability to measure the absolute character of the earth's motion in space in first order ether drift experiments by invoking the Fresnel dragging coefficient. As evidenced in Appendix 5, the calculations are very difficult. By using the modified transformation equations he proposed in 1895, Lorentz was able to bypass many of those computational difficulties. Whereas the transformation equations implied the Fresnel dragging coefficient, there was no other natural motivation for them nor was Lorentz able to provide any first principles, like those provided by Galileo and Newton for the derivation of the equations. Lorentz's justification was that the equations worked, and, in that sense, they were *ad hoc*.

Although Lorentz, in his 1895 monograph, had provided a set of transformation equations that worked for all first order ether drift experiments, the same document contained a remarkable last chapter entitled "Experimental Results Which Cannot Be Accounted for Without Further Ado." The experiments Lorentz described in this chapter were not first order ether drift experiments but second order. That is, they were physical arrangements for observing the interference of light patterns, or for observing the mechanical effects of electromagnetic interactions which depended, not on the ratio of the speed of the apparatus through space to the speed of light in free space, but rather on the square of that ratio. Since that ratio for the earth's orbital velocity is one part in ten thousand, the square of it is one part in one hundred million, a very small number indeed. That there were such experimental apparati speaks to the exciting progress of technology. It now allowed certain scientific questions which could not have been asked earlier. The most famous of these second order ether drift experiments was the Michelson-Morley experiment.

The apparatus splits a beam of light into two parts and directs these two beams along paths set at right angles to each other. The beams are then reflected back, rejoined, and the observer sees a pattern of interference fringes. Assume that one of the arms of the interferometer, as the instrument is called, is moving in the direction of the absolute motion of the earth. The other arm must be at right angles to that direction. If we rotate the arms so that they interchange their orientation relative to the motion of the earth through the ether, one will observe, *according to first order ether drift theory—that is, the Fresnel dragging hypothesis, or the transformation equations that Lorentz had just published,* a shift in the fringes due to second order effects. Michelson and later, Michelson and Morley had observed no such fringe shift when the apparatus was rotated although the experiment had been conducted over a six month period and the apparatus was capable of detecting a fringe shift whose magnitude was a hundredth of that being sought.

It is rather difficult to convey the sensitivity of the apparatus. Michelson constructed the first one in 1881 in a basement laboratory at the University of Berlin. The apparatus was so sensitive that one could detect the vibrations caused by people walking on the sidewalk outside the laboratory. Michelson was forced to move the apparatus to the countryside near Potsdam. It is also rather difficult to convey how astounding the results appeared to the physics community. Since 1881 the experiment has been performed dozens of times. It has been done in basements, on mountains, in glass enclosures; the apparatus has been built on a base of iron, of sandstone, of wood; the source of light has been varied from star light to sunlight, to moon light, to artificial light of various kinds. Rather than employing light waves, other electromagnetic radiation has been used, for instance, radar, and lately producers of coherent radiation such as masers and lasers. Even though such progress in the technology has produced ever increasing sensitivity, the result is always the same. There appears to be no detectable effect of the motion of the earth through absolute space.

The situation was more serious than was evident at first glance. Not only was there no second order effect of the motion of the earth through the ether, but the earlier model for the first order experiments would no longer serve. All first order experiments had been explained by assuming the partial drag hypothesis. The results of second order experiments suggested that the ether was being totally dragged by the earth so that there was no relative motion between the earth and the ether. It can't be both ways. It was a very serious dilemma.

During a ten year period between 1895 and 1904, Lorentz hewed out a solution which, while somewhat arbitrary and unmotivated, served to account for the phenomena. There were two stages to these developments. In the first stage, Lorentz suggested that the ether, rather than being totally dragged or even partially dragged, was unaffected by the motion of the earth. And, in addition, material objects were contracted in the direction of their motion by a factor related to the square of the ratio of the speed of the object through space to the speed of light. In symbols, if we call L_0 the length of the object when it is at rest with respect to an observer (that is, when the object and the observer are in the same frame of reference), and we call L the length of the object when it is moving in an inertial frame of reference relative to us with a velocity v in the direction of the length

$$L = L_0 \sqrt{1 - (v/c)^2}$$

In the Michelson-Morley experiment, the arm of the interferometer in the direction of motion would shrink enough to compensate for the otherwise expected time difference for the round trip of a light parallel and perpendicular to the direction of motion. Independent of Lorentz, the British physicist George F. Fitzgerald made the same proposal. It became known as the "Lorentz-Fitzgerald contraction."

Since the ether would now be absolutely fixed, one no longer had to think about the physics of dragging it along without creating other perturbations within it. More and more, the ether became the benchmark of absolute space for Lorentz. As to the epistemic status of the contraction, Lorentz left no doubt that he thought the effect a real one, and although he could supply no specific point by point independent quantitative theory for it, he suggested a qualitative dynamical theory to hypothesize that the effects of the motion of the electrons through the ether modified the forces between those electrons, causing the length of the object to change. Lorentz ignored the problem that, since the amount of the contraction depended on the relative velocities of frames of reference, different frames of reference would calculate different contractions for the same object. He also left no doubt that, by the absolute length of the object, he meant the length of the object when it was at rest with respect to the ether.

There was considerable dissatisfaction among physicists, Lorentz included, about these developments. They seemed too arbitrary, too *ad hoc*. In particular, Poincaré publicly urged Lorentz to generalize the theory and to perhaps give explicit recognition to a generalized principle of relativity which would apply, not only to mechanics but to all of physics. According to Poincaré, experiment demanded such a principle even though in the future a contrary experimental result might discredit it. Poincaré was referring to the negative results of all of the first and second order ether drift experiments. If there were an inductive generalization possible from these results, it seemed to be that first, there was no way to detect the absolute motion of the earth through space. In one way or another, nature seemed to conspire against the effort. Second, the laws used to describe mechanics, electricity, magnetism, optics, thermodynamics, and other branches of physics appeared to be the same in all inertial frames of reference. Whereas we might disagree about the numbers for one variable or another, we would agree about the way the variables related to each other; that is, to the *form* that the law had. And, the principle of relativity states that, in all inertial frames of reference, the laws of physics have the same form.

In 1904, Lorentz produced a new version of that part of the theory of electrons which dealt with measurements made in differing inertial frames of reference. At the start, Lorentz assumed a new set of transformation equations, which became known as "the Lorentz transformations." As was the case with his 1895 monograph, he did not rationalize or justify these transformations. He did not derive them. He seems to have worked backwards in order to determine what transformations would be necessary to satisfy the principle of relativity which Poincaré had been talking about, and, at the same time, predict such phenomena as the Lorentz-Fitzgerald contraction. He extended the notion of the contraction from macroscopic objects to the fundamental entities of his theory, the electrons, so that these carriers of the basic charge were now to be deformable, changing from spheres to elipses with the minor axis in the direction of motion.

The transformation equations are:

$$x' = \frac{x - vt}{\sqrt{1 - (v/c)^2}} \qquad x = \frac{x' + vt'}{\sqrt{1 - (v/c)^2}}$$

$$y' = y \qquad y = y'$$

$$z' = z \qquad z = z'$$

$$t' = \frac{t - vx/c^2}{\sqrt{1 - (v/c)^2}} \qquad t = \frac{t' + vx'/c^2}{\sqrt{1 - (v/c)^2}}$$

On the one hand, these look quite complicated, but on the other hand, we can see imbedded within them the old Galilean transformation equations and the first order modification which Lorentz had first suggested in 1895.

Lorentz's attitude toward these equations was precisely the same as the attitudes he had earlier expressed about the first order transformations. They were not to be considered physical; rather they were mathematical aids to calculation. From these transformation equations one could derive the length contraction and earlier results for the special case of first and second order ether drift experiments. Not only was length contraction derivable, but a relationship known as *time dilation* also resulted from the transformation equations. That is, the rate at which clocks ran in inertial frames of reference would depend on the relative speed of the frames.

Suppose we take identical clocks and put some of them in a frame of reference which moves at a constant rate with respect to us, and compare the clocks. We would find that the clocks moving relative to us were slower. From the transformation equations, the relationship would be

$$T = \frac{T_0}{\sqrt{1 - v^2/c^2}}$$

Lorentz swept these results aside as mathematical rather than physical. They did not make sense within the framework in which he was operating; the framework of Galilean-Newtonian notions of time and space. But the results were curious and sometimes useful byproducts of the major items of the analysis.

It will be recalled that the classical velocity addition law which stated that the speed of an object moving parallel to the direction of motion of two frames of reference moving relative to each other, was, as seen from one frame of reference, the sum of the relative velocities of the frames plus the speed of the object as measured in the other frame of reference. And that this intuitive result was derivable from the Galilean transformation equations.

One should expect a different velocity addition law from the Lorentz transformations and indeed, it is different.* The predicted relationship is as follows:

$$w = \frac{u + v}{1 + \dfrac{uv}{c^2}}$$

The symbols have the same meaning as before. This is not difficult to understand. First, without the denominator, $1 + uv/c^2$, the law reduces to the Galilean-Newtonian velocity addition law. When the *product* of the relative speed of the frames of reference and the speed of the object as measured in one of the frames is small compared to the speed of light, the denominator does, in fact, reduce the unity. The quantity uv/c^2 can be understood as the product of $\frac{u}{c} \cdot \frac{v}{c}$, that is, as the product of the fraction of the speed of light of the speed of the object as measured in one frame of reference and the fraction of the speed of light which

* It is important to bear in mind that in 1904, Lorentz did not himself derive the new velocity addition law.

the relative speed of the two frames of reference represents. If *either* of those ratios is small, the product will also be small and the result of this velocity addition law becomes *formally* close to the result of the classical velocity addition law.

There is an interesting implication of the Lorentz velocity addition law. Suppose that the speed we are measuring is the speed of light and that in one frame of reference we measure it as equal to c, the speed of light. What will the speed of that beam of light be if it is measured from a second inertial frame of reference moving with a speed v in the same direction as the light? In our formal relationship, the Lorentz velocity addition law, we substitute for u the speed of light, c, and the relationship becomes

$$w = \frac{c + v}{1 + \frac{vc}{c^2}} = \frac{c + v}{1 + \frac{v}{c}}$$

but that can be rewritten by rationalizing the terms in the denominator to $\frac{c + v}{c}$, inverting, and multiplying by the numerator as

$$\frac{c(c + v)}{(c + v)} = c$$

In other words, if one uses the Lorentz velocity addition law, it predicts that in all inertial frames of reference, the speed of light will have the same value, a result which corresponded to the known results of all first and second order ether drift experiments.

In the Lorentz theory, this is a *result* of his assumption of the transformation equations, and it would have had limited significance for him. Lorentz did not make explicit that the velocity of light would be the same in all inertial frames of reference. He did not seem to see that it was an implication of this new transformation equation. In fact, the title of his paper was "Electromagnetic Phenomena in a System Moving with Any Velocity Less than that of Light." So presumably there must have existed the possibility of frames of reference moving faster than the speed of light. The question of what the velocity of light would be did not arise in the Lorentz analysis.

To summarize, by 1904, Lorentz had produced a formal system, a subset of his theory of electrons, to account for all ether drift experiments by assuming a set of transformation equations commonly referred to as the Lorentz transformations. The theory's premise was the existence of electrons, the fundamental material entities of the universe, and an absolutely fixed ether, the benchmark of absolute space, responsible for mediating the propagation of radiation from electrons. Galilean-Newtonian premises concerning measurements of space and time were also accepted although the transformation equations formally contradicted those assumptions. Lorentz resolved these contradictions by claiming that the results of using his transformation were merely aids to calculation and had no physical significance. The invocation of such an explanation was usually reserved for results that contradicted our classical, intuitive notions of measurements of time.

The following year, Einstein published his first paper on the special theory of relativity. Whereas the results are *formally* the same as the Lorentz result,* the analysis, and hence the *meaning* of the formal terms, bears no resemblance to the Lorentz theory.

* Lorentz had made some errors. See Abraham Pais, *Subtle is the Lord . . . : The Science and the Life of Albert Einstein* (New York, 1982), p. 126.

·3·

Einstein's Special Theory of Relativity and Its Consequences

THE POINT OF VIEW THAT will be established in this chapter is that the special theory of relativity is a theory of measurement. The core of the theory can be understood as a modification of the Newtonian theory of measurement. It is important to emphasize that the Newtonian theory of measurement had been accepted long before Newton and was so much a part of the thinking of natural philosophers that Newton hardly discussed it and almost never referred to it. Yet the assumptions of that theory underlie the development of Newtonian mechanics and to an extent determine the meaning of the evidence that is gathered in support of the theory.

Once we have introduced and developed the point of view to be represented as that of Einstein's theory, it will become clear that in this interpretation, the theory says nothing about the nature of the world. It only speaks to how measurements are made when we begin to explore questions about that world. Therefore the special theory of relativity did not replace Newtonian mechanics. It replaced the Newtonian theory of measurement. The new theory of measurement can be applied to all the problems that were considered earlier, for example, momentum and its conservation, or energy and its conservation. Whereas the results are modified and the meaning of several terms will change, the basic premises of Newtonian mechanics are unaffected.

This is in sharp contrast to the import of the Lorentz theory, which was discussed at the end of the last chapter. That theory assumed the Newtonian theory of measurement, and in addition, Lorentz postulated the existence of fundamental entities: electrons. Lorentz attempted to account for all physical phenomena in terms of the interactions of electrons, the radiation from electrons and ether. Lorentz's theory was of matter and how matter interacted with ether.

In Chapter 2 it was pointed out that some of the conclusions of the Lorentz theory contradicted the basic premises of Newton's theory of measurement. For example, the Newtonian assumption of the possibility of transmitting information from one place to another instantly means that all clocks in all inertial frames of reference keep the same time. Furthermore, we cannot expect that the dimensions of objects will depend on the speed of the object relative to another frame of reference.

Yet these are two of the conclusions of the Lorentz theory. We are reminded of the conflict between Aristotelian cosmology and Ptolemaic astronomy. Generally when one spoke of the nature of the universe one referred to Aristotle, and to predict the positions of planets one referred to Ptolemy. Similarly, regarding the Lorentz theory, when one wanted to discuss the nature of time, one invoked the Newtonian theory of measurement as summarised in the Galilean transformation equations. All observers in inertial frames would agree on time intervals.

To predict the outcome of electromagnetic experiments in different inertial frames of reference one does not use the Galilean transformation equations. We use the Lorentz transformations and one result is to learn that clocks in different frames of reference do not keep the same time. The discrepancy between the clocks would depend on the relative speed of the frames of reference. Lorentz, in order to avoid the conflict between the Newtonian theory of measurement and his own predictive theory, referred to the times kept by different clocks as "local" time and "real" time. Real time was measured in the absolute frame of reference whose speed relative to the ether was zero. Local time had no reality save as "an aid to calculation."

THE SPECIAL THEORY OF RELATIVITY

The concept of simultaneity is central in Einstein's analysis. Einstein's conclusion that events that were judged to be simultaneous in one inertial frame of reference were not simultaneous in other inertial frames caused some disquietude to Lorentz and others. Let us turn now to the development of Einstein's special theory of relativity.

THE POSTULATES

Einstein's theory began by invoking as a starting point the principle of relativity: The laws of physics have the same form in all inertial frames of reference. In his first paper on the subject in 1905 he cited two reasons for this, both related to experiment. First he pointed that out the basic laws of electromagnetic theory, commonly referred to as Maxwell's equations, predicted certain asymmetries which did not seem to be "inherent in the phenomena." He noted that all attempts at measuring the absolute velocity of the earth had failed, suggesting that:

> . . . laws of electrodynamics and optics will be valid for all frames of reference for which the equations of mechanics hold good. We will raise this conjecture (the purport of which will hereafter be called the "Principle of Relativity") to the status of a postulate, and also introduce another postulate, which is only apparently irreconcilable with the former, namely, that light is always propagated in empty space with a definite velocity c which is independent of the state of motion of the emitting body.*

Einstein pointed out that in his analysis an ether would be "superfluous" since his analysis would not require the introduction of the concept of absolute space.

So in the introduction of his seminal paper on the special theory of relativity, Einstein indicated how experiment *suggests* to him the elevation of the principle of relativity to a postulate; in other words, he thought it necessary to assume the principle in order to construct the theory. It is not then subject to either direct confirmation or disconfirmation by experiment. This is an important and subtle point which is often missed by those who speak of the relationship between experiment and theory. Einstein has proposed the principle of relativity as a postulate to his theory because of the suggestive results of experiment. Experiment has not proved and cannot prove the principle of relativity to be true or false. No one has measured an effect on the speed of light by the motion of the apparatus relative to the light. Once Einstein made the principle of relativity a postulate, it could not be tested within the theory any more than Newton's laws of motion could be tested within Newtonian mechanics or Euclid's postulates within Euclidean geometry.

The second postulate states that the speed of light will be the same in vacuum in all inertial frames of reference regardless of the relative states of motion of the frames. Let us be clear about this postulate. Suppose that you and I are going to make measurements of the speed of light propagating from the sun to the earth. You, however, are on the earth and I am moving in a rocket ship on a line between the earth and the sun toward earth. I am moving at nine-tenths the speed of light. According to the second postulate, we would both measure the speed of the propagation of the light from the sun as c.

* A. Einstein, "On the Electrodynamics of Moving Bodies," in A. Einstein et al, *The Principle of Relativity* (New York, 1923), pp. 37–38.

On the one hand the postulate seems counterintuitive. As Einstein pointed out in the passage above, it appears to contradict the principle of relativity which heretofore had implied the Galilean transformation equations. They predict differences in the measured velocity of light in the two frames of reference.

On the other hand, in the ether drift experiments the speed of light was always the same. For that reason, some of the early proponents of the theory of relativity argued that the second postulate was really experimental, and that it was only a special case of the first postulate.

This argument runs as follows: according to the first postulate, the laws of physics have the same form in all frames of reference. That the speed of light is constant is a law of physics. Therefore, the speed of light is a constant in all inertial frames of reference. However, this argument fails because no experiment measures the speed of light in one direction. Since nothing can keep up with light, all measurements of its speed are round trip averages.

It is not clear from Einstein's paper which line of reasoning led him to the second postulate, or how it related to experiment or the first postulate. And although there are clues in the next section of his 1905 paper, in which he asks his readers what is meant when two events are simultaneous, he does not make the connections clear or explicit. He has, however, in other papers, given us his line of reasoning that helped him to build his theory on those two postulates.

Immediately after his statement of the two postulates, Einstein raised the question of the definition of *simultaneity.* He pointed out that there is no problem of understanding the simultaneity of two events which happen at the same place. Suppose, Einstein says, we are standing in the train station and the train arrives at seven o'clock. This statement means that the arrival of the train coincides with the small hand of my watch at the number seven. Furthermore, an observer on the train, B, moving by, who coincides with me when the train arrives at the station will also agree that the train's arrival is simultaneous with the time on his clock at that point. But what if we are talking about two events that happen at different points in space? What does it mean to say that they are simultaneous? What does it mean to say that by my watch, when I am standing in Times Square, it is midnight in Public Square in Cleveland, Ohio, and the Twentieth Century Limited has arrived there?

Before answering that question, it should be pointed out that this discussion was published in the most sophisticated and respected physics journal in the world at the time, *Annalen der Physik* (Annals of Physics). Finding this discussion in the midst of consideration of the most mathematically complex physical issues one can imagine, is tantamount to finding a lesson in reading for six-year-olds.

From the response to the paper, it seems to have been unclear to most readers why the discussion of simultaneity followed the statement of the two postulates.

One can cite only two men who understood Einstein and his motivation. One man was Max von Laue, an assistant to Max Planck in Berlin. In 1911, von Laue, who was to become a Nobel laureate for his work on x-ray diffraction, wrote the first monograph on the theory of relativity. The other man was J. J. Laub, now virtually unknown, who was a postdoctoral student under Wilhelm Wien at the Würtzburg University.*

Laub remarked that, in 1905, Wien requested that he give a seminar on Einstein's paper. Laub, who saw the issues simply, was puzzled because no one followed Einstein's reasoning. One reason might be that Einstein did not make a connection between the second postulate and his discussion of the definition of simultaneity which followed.

Nor did Einstein make explicit the connection, if any, his work had with the earlier work of Lorentz. Einstein made no secret of the fact that in physics, Lorentz was his idol. But specifically, by Einstein's account, he did not know of the publication of Lorentz's 1904 paper when he published the special theory of relativity in 1905. He had, however, studied and followed many of Lorentz's earlier works.

We can speculate about the influence that Lorentz's earlier contributions had on the development of Einstein's reasoning. In his "Autobiographical Notes" (p. 53) Einstein reminisced that it had taken him ten years to realize the need to elevate the principle of relativity to a postulate. The search had begun because of a paradox

> . . . upon which I had already hit at the age of sixteen. If I pursue a beam of light with the velocity c (velocity of light in a vacuum), I should observe such a beam of light as a spatially oscillatory electromagnetic field at rest. However, there seems to be no such thing, whether on the basis of experience or according to...[the fundamental electromagnetic] equations. From the very beginning it appeared to me intuitively clear that, judged from the standpoint of such an observer, everything would have to happen according to the same laws as for an observer who, relative to the earth, was at rest.

Actually, ten years earlier, when Einstein was sixteen, in 1895, Lorentz had published the final version of his first order theory. Einstein was a student at the Swiss Federal Polytechnic (ETH) in Zurich, and it is not likely that he studied that paper until several years later. When he did examine Lorentz's paper what might have struck him forcibly, attuned as he was to the peculiar behavior of light, was that no first or second order experiment had revealed the motion of the earth

* Subsequently, not being able to find a post in a German University, Laub emigrated to Argentina, where he took a post in physics at the University of La Plata. He became an Argentine citizen and a member of the Argentine diplomatic corps.

through the ether, and no experiment, regardless of the relative state of motion of the apparatus relative to the source of the light or the proposed propagating medium, had shown the velocity of light to be anything but *c*. Even more startling might have been that Lorentz had modified the Galilean transformations and proposed a transformation for the time coordinate which resulted in disagreements about the time coordinate in different frames of reference. Einstein writes further in his "Autobiographical Notes" (p. 53):

> One sees that in this paradox [the fact that no inertial observers can move with a beam of light] the germ of the special relativity theory is already contained. Today everyone knows, of course, that all attempts to clarify this paradox satisfactorily were condemned to failure as long as the axiom of the absolute character of time, viz., simultaneity, unrecognizedly was anchored in the unconscious. Clearly to recognize this axiom and its arbitrary character really implies already the solution to the problem.

Whether or not it was reading Lorentz's work which triggered the reasoning which led Einstein to the construction of the special theory of relativity we will never know, but it is clear that, not only in the first paper on relativity, but in his later ruminations about the theory, the importance of the definition of simultaneity loomed large in the creation of the theory.

It is not possible to say which steps Einstein followed to the publication of the special theory of relativity. We do, however, have a record of Einstein's early writings about the theory. He has left us very clear and unambiguous writings about his epistemological views. When one examines Einstein's philosophical musings on the nature of good theories and the relationship between evidence and belief, and compares them to the structure of his early papers we have further insights.

Throughout his life Einstein insisted that the postulates from which theoretical structures developed are not arrived at by inductions from experience, or logic. Einstein often referred to these postulates as "concepts" or "laws." He denied emphatically that they were the result of logical manipulation. For Einstein the postulates of a theory were arrived at by intuition which depended on "sympathetic understanding of experience." He sometimes referred to the postulates as "free inventions of the human intellect." Since Einstein considered *thinking* to be an incomprehensible mystery, those free inventions of human intellect represented by the postulates of theories could never be justified *a posteriori*.

As theories to account for sense experience were the free creation of the human intellect, one might think it impossible to choose between theories. Not according to Einstein. He frequently cited two criteria for selection among contradictory theories. The first criterion was that the theory could not be contradicted by experience. For Einstein, the bedrock of all knowledge was experience. This should not be interpreted to mean that a simple contradiction between experiment and theory would lead to rejection of an otherwise attractive theory. The contradictions would have to be deep and persistent. As we will see, several times Einstein held fast to the theory of relativity despite apparently contradictory experimental evidence.

Einstein well understood that there were an infinite number of theories that would satisfy the criterion that theory not contradict experience. A second criterion was required. Einstein often referred to it as the "inner perfection" as opposed to the "external harmony." This perfection is a function of the naturalness or logical simplicity of the premises of the theory. According to Einstein, it is the grand aim of all theory to make the postulates simple and to reduce them to the fewest numbers possible and still account for all of the evidence.

When one examines Einstein's early publications about the theory of relativity it becomes clear that Einstein kept faith with his vision. His writings for a professional audience began with only two postulates, the principle of relativity and the invariance of the speed of light. All deductions flowed from them. One of the reasons that other physicists found Einstein's reasoning difficult had little to do with mathematical obscurities. The mathematics Einstein used was well known to his peers. Rather, few of his peers in physics shared his conception of a good theory. He found attractive in the theory of relativity its parsimony about postulates, and that mystified his colleagues. Most would not have accepted Einstein's convictions that the premises of theories are not arrived at by an inductive process from experience. They found the background of his theory too lean, too barren and unconnected to experience. This was especially true of the second postulate, the invariance of the speed of light. Indeed, when one examines Einstein's introduction to this postulate in his scientific papers, it stands alone. It is not very physical. Einstein's early papers in science reflect his convictions about the nature of good theories.

On the other hand, Einstein also wrote accounts of the theory of relativity for lay audiences. In those writings, he apparently did not feel obliged to make the structure of the paper reflect his views on the nature of good theories. These accounts were much more expansive in examining the reasonableness of postulates, the sources of the postulates, and how they are related to experience. In the introduction to his most famous popularization of relativity, Einstein wrote:

> The author has spared himself no pains in his endeavour to present the main ideas in the simplest and most intelligible form, on the whole, *in the sequence and connection in which they actually originated.**

* A. Einstein, *Relativity, The Special and The General Theory,* p. v. Emphasis added.

Let us now explore and make explicit the sequence and connection among the first postulate, Einstein's ideas concerning simultaneity, and the second postulate. We follow Einstein's account in his fine, albeit parsimonious, popularizations.

SIMULTANEITY

We begin by asking if two events are simultaneous and how one can determine the simultaneity of such events. As long as the two events are at the same place there is no problem. Observers in all inertial frames of reference will agree on whether or not the events took place at the same time. The problem arises when we think about the temporal order of events which are not at the same location.

We have a very long, straight, train track. We initially use the track to lay out a straight, measured course. We lay out a distance *AB* (Figure 1) and half way between, at *M*, we erect two mirrors at 90° to each other and 45° to our line of sight along each direction of the track. *AM* is equal to *BM*. Given the arrangement, and standing at *M*, we can look down both directions of the track at the same time. Since this is a thought experiment, we are free to imagine almost any arrangement. Let us suppose that we can arrange to have lightning strike the track at *A* and at *B*. Standing at *M* we then have to decide if the lightning strikes occurred simultaneously. Later, we can walk down the track and look for the physical effects of the lightning (a charred mark on the track, for example) to be certain that it struck at *A* and at *B*. As the lightning comes down, the light spreads out in all directions, (Figure 2) and, standing at *M*, we see both strokes in our mirrors at the same time. Therefore, the observer says, "the events lightning struck *A* and lightning struck *B* were simultaneous." After a moment, I answer "But how do you know that the speed of light is the same in the directions *A* to *M* and *B* to *M*? After all, even though we agree that we saw the strokes at the same time, it might be that the speed of light is not the same in all directions, and that perhaps one of the strokes occurred before the other, and this was compensated for by the difference in the speed of light in the two directions."

This presents you with a quandary. As we realize, the speed of light is the fastest signal speed we know, and there is nothing that can keep up with it. In order to make a measurement of its speed, the best that we can do is to send the light out and return it with a mirror. We have only measured the average round trip speed,

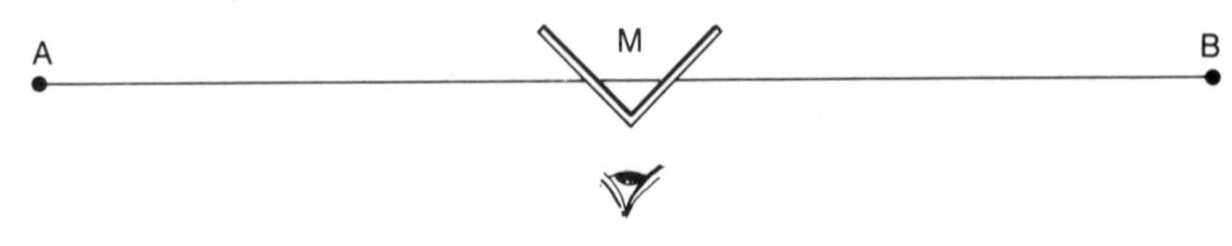

Figure 3.1

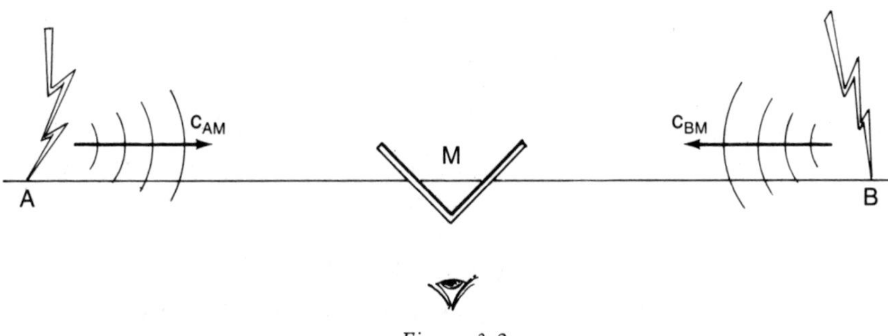

Figure 3.2

and, should the speed in one direction have been different from the speed in the opposite direction, we would not have known it.

On the other hand, you might propose that we use two clocks and two observers in each direction to measure the speed of light along the line A to M independently of the speed of light along the line B to M.

Suppose that we modify our thought experiment: we will take three identical clocks and place them at A, B and M. We will measure the time difference between the clocks at A and M and between the clocks at B and M to determine the time required for light to travel the distance AM and BM. It becomes simple to compute the speed in each direction. But before we begin, the clocks must be synchronized. This must be done *in place* since we have no guarantee that motion will not affect the rate of the clocks. Each clock is set as follows: the clock at M is set at zero; the clocks at A and B are set at a time t equal to the ratio of the distance $MA = MB$ to the speed of light, c.

$$t = \frac{MA}{c} = \frac{MB}{c} \tag{1}$$

Each of the clocks at A and B is equipped with identical photocell switch circuits, and when we start the clock at M, a flashbulb will go off. When the light reaches the clocks A and B, the photocell circuits will start the clocks, and all three clocks will then show the same time (Figure 3).

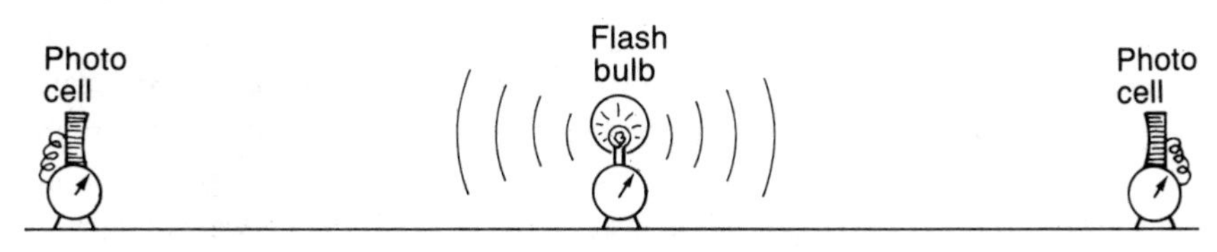

Figure 3.3

"Fantastic." I say, "However, we still have a problem. Everything depends on the speed of light from *M* to *A* being the same as the speed of light from *M* to *B*. Is it the same in all directions? That is what we were using these clocks to determine. If you will recall, a proper measurement of the time between the events light leaves *M*, light arrives at *B*, or light leaves *M*, light arrives at *A* must be made with one clock. There can be no such proper measurement of the time between those events because there is nothing that can move at the speed of light. Had you sent the signal to the two clocks using a stiff wire rod, I could have checked the speed with which the rod moved in both directions using sound signals. But I would have to raise the question of the uniformity of the speed of sound in all directions, and I would be forced to fall back on my ultimate signal speed, the speed of light, if I am going to use light to define simultaneity. There is no way that I seem to be able to determine if space is isotropic — that is, the same in all directions — with respect to the speed of light. Non-proper measurements are required — that is, we must use two clocks. In order to synchronize them, we must invoke the very thing we are trying to determine by using the clocks — the isotropy of space with respect to the speed of light."

For a while after this long speech, you look crushed. Finally, you say to me with considerable relief, "Regardless of what you have said, I maintain that the lightning strokes were simultaneous. Since nothing can keep up with light, and there is no way to determine if space is isotropic with respect to the speed of light, I am free to take any view that I want to, and I *stipulate* that the speed of light in this frame of reference is the same in all directions *in order that I may come to a meaningful definition of simultaneity.* I have to start somewhere."

And this is the approach that Einstein took, and it is the approach that anyone must take when beginning any theory. One begins with a set of definitions, perhaps operational measuring routines and certain stipulations about the nature of the problem with which one is working. To do otherwise would be to end with hopeless circularity. Einstein had realized that one cannot assume, as Galileo, Newton and even Lorentz had, an infinite signal speed. Given that the fastest signal speed is the speed of light, one must stipulate that space is isotropic with respect to it. Should we eventually discover a signal speed faster than light, for example, the propagation of gravitational effects, the situation would in principle be unchanged. We would continue to stipulate space as isotropic with respect to *that* signal speed.

We would not have to use the speed of light, nor do we have to stipulate isotropy. We could stipulate that the speed of the signal we are using is different in one direction from another. Suppose that we did that. And suppose that in deference to the fact that light is the fastest signal speed known we had chosen light in stipulating our criterion of simultaneity. When we saw the two lights from distant events at the same time, they could not represent simultaneous occurrences. We would probably have to make calculations for decisions about such information. Simultaneous means "at the same time." Therefore, we have stipulated that the speed of light is the same in all directions so that information from distant events may be evaluated relative to the question of their simultaneous occurrence in accordance with our intuitive and traditional notions of simultaneity. Having stipulated isotropy of space about the speed of light, not only can we make judgments about distant simultaneity, we can specify operations for synchronizing clocks and thus make nonproper measurements of the time interval between events that happen at different points in our frame of reference.

THE RELATIVITY OF SIMULTANEITY

There is nothing exceptional about our frame of reference. In fact, according to the principle of relativity, the same process is necessary in *all* inertial frames of reference. Let us look at another inertial frame of reference. We picture a long, fast train on the track along which we are to determine the simultaneity of distant events. Occupants of that frame of reference consider themselves to be at rest.

We communicate with the passengers and explain the experiment we have done about the judgment of simultaneous events. They agree that our logic is faultless, and they too undertake to perform the experiment. They have laid out the course $A'B'$ and have placed M'at the midpoint so that $A'M' = B'M'$ (Figure 4). We further arrange that when M' on the train is opposite to M on the embankment, the lightning strokes will come down and leave marks not only at A and B but also at A' and B' (Points on the embankment can be made infinitely close to corresponding points on the train when they are lined up). There are observers at M and M' in each frame of reference who are looking at the mirrors.

For the sake of the discussion, we assume the exalted position of looking down on the scene. We observe the train moving in the direction A to B and the embankment. But from the train, the passengers' perspective is that the embankment is moving from B' to A'.

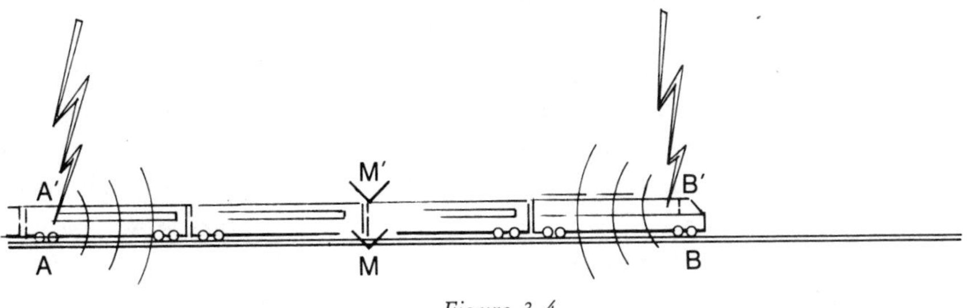

Figure 3.4

When M and M' are adjacent to each other, the lightning strikes the track and the train as planned. As before, the observers on the embankment see both flashes at the same time, and their judgment is that the flashes occurred simultaneously. But observers at M' come to a far different conclusion. They are moving toward B' and away from A', therefore, the light from the lightning stroke at B' will reach them before the light from the lightning stroke at A' from which they are rapidly receding. Having seen B' before A', they will conclude that the lightning strokes were not simultaneous. It is important to emphasize that from their viewpoint the speed of light is the same in all directions just as it is for the frame of reference of the embankment. That is, in the train's frame of reference, the speed of light in different directions is stipulated to be the same in all directions so that they too can make judgments of distant simultaneity. And within the frame of reference of the train, they can check to see that the lightning struck at A' and B'. Since the strokes occurred equidistant from M' and since space is isotropic with respect to the speed of light, the only conclusion possible to the passengers is that stroke B' occurred before stroke A'. The strokes were not simultaneous.

At best the observers in the two frames of reference can agree to disagree. But while disagreeing, the explanations given by the different inertial observers will be consistent.

From the point of view of the observer on the embankment the judgment of the observer on the train is the result of the train's moving toward B and away from A. On the other hand, the observer on the train has decided that the two events were not simultaneous. He or she observed that the embankment is in motion relative to his or her frame of reference away from B' and toward A'. Since the stroke at B' happened before the stroke at A', it is understandable that the observer on the embankment might judge them to be simultaneous. Since such an observer must be moving toward A and away from B the light from B' would require more time to reach the moving observer than it might were he or she at rest.

Picture a train moving in the opposite direction from the first train relative to the embankment. Observers in this train will also participate in the experiment. A moment's thought will convince us that they will judge that the lightning stroke at A'', corresponding to points A and A' occurred before the stroke at B'', corresponding to points B and B' (Figure 5). Now we have three judgments about the order of the events lightning stroke at A and lightning stroke at B. From the embankment the events were simultaneous. In one train B was seen as coming before A, and in a second train A was viewed as coming before B. If the principle of relativity is to be believed, in all frames of reference the laws of physics have the same form. In all frames of reference we must stipulate that space is isotropic with respect to the speed of light, that is, the speed of light is the same in all directions. These results can be made consistent using the following postulate: We assume as

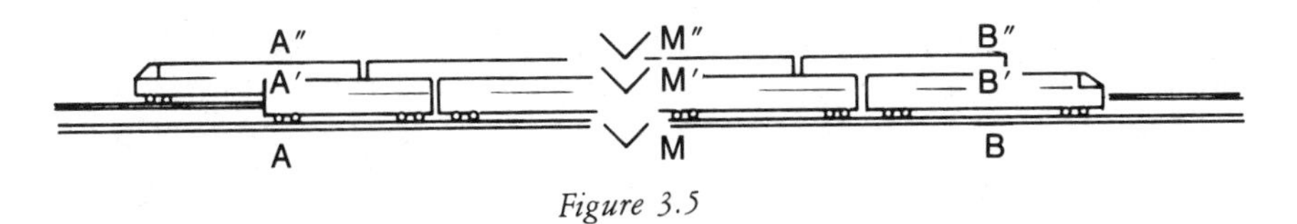

Figure 3.5

a postulate independent of the principle of relativity, that the speed of light must have the *same value* in all directions in all these frames of reference. This is one logical solution to the existence of inertial frames of reference in motion with respect to each other all assuming isotropy with regard to the speed of light. In all inertial frames of reference the speed of light will have the same value. A quantity whose value is the same in all inertial frames of reference is an *invariant*. We have arrived at the postulate, not as a special case of the principle of relativity and not as a result of an experiment. Indeed, there is no experiment that can help us since it is impossible to make a one-way measurement of the speed of light.

THE ORDER OF EVENTS AND CAUSALITY

Let us return to the thought experiment we have spent so much time on. It is unsettling to be forced to accept that the order in which the events lightning stroke at *A* and lightning stroke at *B* are judged is not an invariant. Are they actually simultaneous and the judgments in the other frames of reference misguided? No. There is no actuality about judgments of simultaneity. Einstein had realized that the moment one drops Newton's assumption of an infinite signal speed, judgments of distant simultaneity can no longer be invariant and absolute. But the disagreement about the order of the events cannot have physical consequences. The result cannot, for example, lead to the birth of a child before the birth of a parent, or the reversal of causally related events.

Let us explore this further because some of the early criticisms of the theory of relativity rested upon a misapplication of the postulates. That there are disagreements about the order of the lightning strokes has no effect on causality. In order for an event to cause another, the information that the first event occurred has to arrive at the point at which the second event is going to happen.

With light waves, a certain time is required for the light waves to leave one point and arrive at a second, thereby causing the second event. Recall that we proposed synchronizing clocks using light signals and photocell circuits. There is no frame of reference by which to judge the situation different from the arrival of the light signal coincident with starting the clock. Both events happen at the same place. Nor is there any frame of reference to judge that the clock started before sending the light signal. There is an important difference between causally connected events and the occurrence of the lightning strokes in the other thought experiments. The occurrence of the lightning strokes were independent of each other. One stroke is not the cause of the other's occurring. Another way of stating this is to say that the lightning strokes occur in a time interval less than the time required for light to get from one of them to the other. And light, after all, is the *fastest* signal speed known.

It will always turn out to be the case that disagreements about the order of events occur only in cases where there can be no causal relationship between the events. For those events which can be connected by light signals in a time less than the time between the events, the order in which the events occur will always be invariant and beyond disagreement between different inertial observers.

That Einstein enunciated the two postulates does not, *ipso facto,* mean that they are sufficient or necessary. It might, as Einstein suggested, *appear* to be the case that the two postulates are contradictory. The most direct way to decide these questions is to see where the two postulates lead. This is where logic enters the development of scientific theories: connecting premises to conclusions.

Einstein used the two postulates to *derive* a set of transformation equations relating spatial and temporal measurements in different inertial frames of reference. Rather than leading to contradictory results, the postulates led to a set of transformation equations which were identical to the Lorentz transformation equations. For convenience, we repeat them here:

$$x' = \frac{x - vt}{\sqrt{1 - v^2/c^2}} \qquad x = \frac{x' + vt'}{\sqrt{1 - v^2/c^2}}$$

$$y' = y \qquad y = y'$$

$$z' = z \qquad z = z'$$

$$t' = \frac{t - vx/c^2}{\sqrt{1 - v^2/c^2}} \qquad t = \frac{t' + vx'/c^2}{\sqrt{1 - v^2/c^2}}$$

The first implication of this result, as Einstein wrote, is that the resolution of the apparent paradox between the two postulates is the Lorentz transformation equations. The postulates are independent of each other, not contradictory, and both are necessary. They lead to a consistent result although it is different from that arrived at when one assumes the principle of relativity plus a second assumption that the signal speed is infinite. In that instance we arrive at the Galilean transformation equations.

THE RELATIONSHIP OF EINSTEIN'S THEORY TO LORENTZ'S THEORY

The immediate consequences of using the postulates which Einstein had enunciated, is that one derives the transformation equations that Lorentz had assumed.

Whatever is predicted in one theory about the relationship between variables in different frames of reference will be predicted in the second theory. In other words, there is no way to distinguish between the theories on the basis of prediction nor is there any experience that can distinguish between them regarding qualitative effects or accuracy.

This is not exceptional with competing scientific theories. This does not mean that there will emerge no criteria for selection. It means that the common belief that one theory succeeds another because of superior predictive power cannot be defended. It also means that there is often confusion about the relationship between the Lorentz and the Einstein theories.

Not everyone agrees that the possibilities of different interpretations of formal results or different theoretical accounts of experimental results are significant. Certain practitioners of empiricism hold that the only meaning which scientific theories have are proved in formal results. In other words, if two theories make identical predictions in all instances, as the Lorentz and the Einstein theories do, they are, in fact, the same theory. Whereas there are many shades of empiricism, in general most empiricists subscribe to the view that the interpretive aspects of the theories discussed here are metaphysics in that they are speculative about the nature of the world that they describe.

Many empiricists would also argue that such speculation, while perhaps interesting, has no meaning, at least for understanding the world. In the most radical branch of the empiricist school, known as logical positivism, the approach generally is to attempt to construct a logic connecting sense data (for example, the perceived behavior of the fringes in interferometer experiments) to theoretical generalizations by a series of logical statements. Some empiricists have used the following analogy: The combination of sense data and theoretical statements represents a net staked to the ground. The stakes are the observations and the perceptual data. One can run along parts of the net which represent combinations of statements until one encounters one of the stakes. In other words, all statements are ultimately tied logically to sense phenomena. This is referred to as a "causal nexus."

Although this empiricist view is not subscribed to here, we will follow the analysis further. From the viewpoint of this book, one of the problems with empiricism is that science is not a field of logic. In Chapter 1 I pointed out that all scientific arguments can be cast into the following idealized form regardless of how the information was gathered:

If A then B
I observe B
Therefore A

Let us illustrate this with the Lorentz and Einstein theories.

> If the Lorentz equations are true (If *A*)
> Then one should expect no fringe shift when the Michelson apparatus is rotated (Then *B*)
> No fringe shift is observed (I Observe *B*)
> Therefore the Lorentz equations are true (Therefore *A*)

We have already noted that there is a major logical fallacy in this conclusion. For example, consider the following statements:

> If the ether is dragged totally by the earth (If *A*)
> Then no fringe shift is to be expected when the Michelson apparatus is rotated (Then *B*)

The remainder of the argument would proceed as before. There is more than one way to get from *A* to *B*. There might be a *C*, *D*, *E*, or an infinite number of arguments to allow me to assert that I can expect to observe *B*, whatever *B* is. In order to make the argument complete and free of logical error, the first statement would have to be:

If and Only if *A* then *B*

But scientists have traditionally argued about what the proper *A*'s are in each instance.* This is one of the problems with adopting an empiricist position.

Einstein and Lorentz agreed about the correct transformation equations to use in relating variables from one frame of reference to the other. The premises which Einstein used in obtaining those equations were:

1. The Principle of Relativity
2. Isotropy of space in any frame of reference regarding the speed of light which implies the relativity of simultaneity.
3. The Invariance of the Speed of Light

Lorentz used only one premise:

1. The inability to detect the motion of objects in the ether, which implies the Lorentz transformations.

The empiricist might say that all of the conclusions and manipulations which allowed the two men to arrive at those equations are meaningless or irrelevant; the significant starting point for the theory of relativity is the transformation equations.

* Of course, such arguments are not restricted to science. It is a feature of any abstractive discipline. The common view is, however, that such disagreements are not a part of the enterprise of science.

One of the problems with this point of view is that it is difficult to excise those steps which Einstein took to the equations that Lorentz assumed. Furthermore, it develops that it is those steps, and not the formal parts of the theory, which give the Einstein view a different meaning from that of Lorentz. That meaning is heuristic in the sense of suggesting new pathways along which to move, whereas the Lorentz view finishes his theory of electrons, and is, in that sense, an end. But, just as it is impossible to prove the correctness of an explanatory theory in science or social science, so too, it is impossible to prove the correctness of a philosophical position. However, a philosophical position that excludes metaphysical statements contradicts its own precept, for to make such a demand is itself metaphysical.

We will now illustrate some of these points. Recall that we have two theories—the Lorentz theory and the Einstein theory—both of them accept the Lorentz transformations. Of the great quantity of phenomena that both theories predict about measurements of variables between inertial frames of reference, I will examine several in order to point up differences in substance and historical context.

TIME DILATION AND THE RELATIVITY OF SIMULTANEITY

We have discussed the Lorentz transformations which show that observers in one inertial frame of reference think that clocks in another inertial frame of reference run slow. This is true even if the clocks were originally built to identical specifications and even if, when they were at rest with respect to each other, they kept precisely the same time. "Time dilation" is the term applied to this phenomenon. What we mean by "the same time" is that the interval between clicks, or rotations of a gear wheel, or the swings of a pendulum, or the rhythm of a quartz oscillator are identical in all of the clocks. The fact that the time kept by these clocks in a frame of reference moving with respect to us is slow, means that the interval between clicks, or the interval between steps in the rotations of a gear wheel or the interval between successive swings of the pendulum on those clocks is longer than in identical clocks in our frame of reference.

As we have seen, Lorentz's interpretation of this phenomenon was to deny that it had a physical reality. The result was only a mathematical byproduct of the Lorentz transformations and had no physical meaning. He distinguished between "true time" and "local time." True time is the time used to measure physical events in the world.

The Einstein interpretation of the transformation equations and the derived time dilation relationship was different and led to different considerations although the equations are precisely the same. To make the distinctions clearer let us imagine the following situation: O and O' are two observers in different frames of reference moving inertially with respect to each other along the X-X' axis with a velocity v. Earlier, each frame of reference was provided with clocks from an identical set. For the sake of discussion, let us assume that each point in space in the primed and the unprimed frame of references (hereafter designated as Σ and Σ') is equipped with such a clock. Furthermore, observers in each frame of reference have undertaken synchronization operations, and the clocks have been synchronized. And we have already seen that in the Einstein analysis one must assume isotropy of space with respect to the speed of light in order to come to a meaningful definition of simultaneity, and that the observers will not think that the second group of observers have performed these operations correctly. That is, according to the Einstein theory, judgments of simultaneity are relative. Two distant events in Σ which O judges to be simultaneous will be judged as *not* being simultaneous in Σ'. Similarly, two spatially distant events which O' judges to be simultaneous in Σ' will be judged not simultaneous by O in Σ.

Notice, however, that not all things are relative. Whereas judgments of simultaneity must be relative, by postulate all inertial observers agree on the magnitude of the speed of light from any source at any time. The speed of light is an invariant.

We will see shortly that in the Einstein theory, time dilation is not a mathematical device. It is a genuine phenomenon because observers in different frames of reference will report that the clocks in frames of reference moving relative to them are slow. Given Einstein's analysis, this is not a statement about the physical nature of clocks, rather it is an artifact of how we measure. It can ultimately be traced to the postulates of relativity and from them to the stipulation that was made at the beginning of the analysis about isotropy of space with respect to the speed of light.

For the moment, let us concentrate on the fact that observers in different frames of reference will disagree about the simultaneity of two distant events. This disagreement was unacceptable to Lorentz and to a number of other physicists. As late as 1927, a year before his death, Lorentz wrote about simultaneity:

> You will remember our two observers A and B, using the different times t and t' each able to describe phenomena in exactly the same way, though what is simultaneous for one is not simultaneous for the other. The theory of relativity emphasizes the fact that one of these is exactly as good as the other. A physicist of the old school says, "I prefer the time that is measured by a clock that is stationary in the ether, and I consider this as the true time, though I admit that I cannot make out which of the two times is the right one, that of A or that of B." The relativist, however, maintains that there cannot be the least question of one time being better than the other.

Einstein and H. A. Lorentz. Courtesy of the American Institute of Physics Niels Bohr Library.

Of course this is a subject that we might discuss for a long time. Let me say only this: all our theories help us to form pictures, or images, of the world around us, and we try to do this in such a way that the phenomena may be coordinated as well as possible, and that we may see clearly the way in

> which they are connected. Now in forming these images we can use that notion of space and time that have always been familiar to us, and which I, for my part, consider as perfectly clear and, moreover, as distinct from one another. My notion of time is so definite that I clearly distinguish in my picture what is simultaneous and what is not.*

Lorentz denied the validity of the relativity of simultaneity and his confusion about Einstein's claims becomes clear. Einstein also used his theory to help form pictures, and he also used pictures, in the form of thought experiments, to form theories. It was his picture of the process of synchronizing clocks which were not at the same point in space that led to the concept of the relativity of simultaneity. Einstein undoubtedly would respond to Lorentz that he, too, could distinguish in his picture that which is simultaneous from that which is not. However, it does not follow that events which were simultaneous in one frame of reference would also be simultaneous in other frames of reference. Lorentz had confused an idea of whether or not two events are simultaneous with the notion of the absolute or relative character of simultaneity. He was a great physicist. That the distinction escaped him illustrates how difficult it can be to understand simple analyses which organize experience in ways different from those normally accepted.

Whereas the judgment of simultaneous events is relative to the frame of reference in the Einstein analysis, observers agree that the events are not causally related, that is, that the time between the events, be the events judged simultaneous or not, is less than the time required for light to travel from the spatial coordinates of one of the events to the spatial coordinates of the second event. As we have seen, the requirement that events be causally related translates into a statement that the time between the events be no less than the time required for light to travel the distance between the events.

This discussion of the differences in the Einstein and Lorentz views underscores how important the concept of "event" is in the Einstein analysis. The term was never part of Lorentz's working vocabulary.

Let us now return to the problem of time dilation from Einstein's point of view. The concept of "event" is crucial. It is true, as Lorentz would have said, that observers in each frame of reference would report that the clocks of the other frame of reference run slow. However, all observers would agree to the following: For any two distant events no more than one frame of reference would make a proper measurement of the time between those events. Recall that by a proper time measurement is meant a time measurement made with one clock. The word "proper" is used in a technical sense. For events that cannot be causally related, there will be no proper time measurement made since no inertial observer could

* H. A. Lorentz, *Problems of Modern Physics* (New York, 1927) pp. 220–221.

move from the point of one event to the point of the second event in the time between the events. But for events for which that is not true, there will be a proper measurement made in some frame of reference. Furthermore, the observers will agree about who made the proper measurement. That judgment is absolute and invariant. And all observers will agree that the proper time interval is the shortest, measured *between those two events*. That will also be absolute. For two *other* events, the proper measurement will be made in another frame of reference. That will be the shortest time interval between *those* two events of all measurements made by other inertial observers.

The key to this problem is the events that are considered. For any *two events* all things are *not* relative. Generally, the time dilation means that the rate at which clocks run is judged to be slow in other inertial frames of reference. But for any two specific events, only one frame of reference makes the proper measurement. All observers will agree about that value although they will not agree that they measured that value. The disagreement is not serious. It arises from asymmetry in the measurement process between frames of reference, from the premise that the speed of light is an invariant, and from the need to stipulate isotropy of space for the speed of light. Questions of the reality of the time dilation phenomena are not significant. The discrepancy is an artifact of the measurement process and says nothing about the nature of clocks or of reality.

Let us examine a specific example. Suppose we are traveling between earth and Mars at a speed close to that of light. We consider two frames of reference. The first is composed of the earth and the fixed stars and the second is the rocket ship. Only one frame of reference makes a proper time measurement for two events, rocket leaves earth, rocket arrives on Mars. That is the frame of reference of the rocket. Observers in the rocket (and we ignore the fact that if the rocket ship accelerates at the beginning and the end of the trip, it is not then an inertial frame of reference) require only one clock to record those two events. In order for us on earth to measure that time interval, we would have to compare the readings of previously synchronized clocks located here and on Mars for the events "rocket leaves earth, rocket arrives on Mars."

"Well," you might say, "how is it that in a similar case, we know how long it took for our Mariner space probes to arrive at Mars in 1976–77? We did not have clocks on Mars, and I don't recall anyone describing some synchronization routine." Actually, in all cases of space travel, thus far, the speeds are nowhere near the speed of light, and, hence the discrepancies between proper and various nonproper time measurements have not arisen.

There are examples of time dilation available to us, however. The notion of a clock is a general one. All biological processes associated with aging are a kind of clock. For example, hearts beating, cells dividing or growing are types of clocks.

Radioactive decay is another kind of clock. By using high energy accelerators, we are able to accelerate radioactive materials almost to the speed of light. These materials decay at a known rate, and in order to compare the radioactivity of different materials, "the half-life" is generally reported. This is a statistical measure which allows us to predict the behavior of large populations of radioactive particles of a certain species although it would be impossible to predict with certainty the behavior of an individual particle over a finite period of time. When radioactive species having a known half-life are accelerated to speeds close to the speed of light the half-life of the species seems inordinately long. In common parlance we would say that from our perspective, the clocks in the frame of reference of the particles are running slow. The proper time measurement of the events "particle has attained speed .999c" and "particle disintegrates" is, of course, a measurement made in the particle's frame of reference. As far as an observer on the particle is concerned, his or her clock is keeping the same time that it always has, and the entire cohort of particles that are in his or her frame of reference, have the same half-life that they always had. Strictly speaking, the half-lives reported in tables of radioactive materials are proper half-lives.

The importance of the concept of "event" in this discussion cannot be overemphasized. Without the concept of event one might argue that, according to the theory of relativity, all things are relative. According to the theory of relativity, all things are *not* relative. In this instance, not more than one frame of reference will make the proper time measurement between any two events, and all observers will agree on which frame of reference that was and what that time interval was.

The mathematical relationship between the proper time interval measurement and the nonproper time interval measurement between two events is derivable from the Lorentz transformation equations. I have derived it in Appendix 6. I have also shown in that Appendix how one might arrive at the result by considering thought experiments in which one frame of reference makes proper time measurements and a second frame of reference makes nonproper time measurements.

If T_0 is the proper time measurement and T is the nonproper time measurement between two events they are related to each other, as Lorentz predicted (Chapter 2):

$$T = \frac{T_0}{\sqrt{1 - v^2/c^2}}$$

LENGTH CONTRACTION

According to the Lorentz theory, observers in different frames of reference will report different results for the measurement of the length of the same object in the direction of motion. Lorentz explained this by pointing out that the object is squeezed up in the direction of motion as a result of electromagnetic interactions of the object and the ether through which it moves. In a series of lectures delivered in 1906 at Columbia University, published in 1909 and revised in 1915, Lorentz said:

> The hypothesis certainly looks rather startling at first sight, but we can scarcely escape from it, so long as we persist in regarding the ether as immovable. We may, I think, go so far as to say that, on this assumption, Michelson's experiment *proves* [sic] the changes of dimensions in question. . . .*

There is a curious circularity in this argument. Lorentz proposed the contraction to explain the results of the Michelson-Morley experiment, and then said that the Michelson experiments proved that the contraction occurs. Fitzgerald, who proposed the contraction independently of Lorentz, apparently did the same. According to Oliver Lodge, a colleague of Fitzgerald's, Fitzgerald came upon the idea while sitting in Lodge's study:

> The Fitzgerald-Lorentz hypothesis I have an affection for; I was present at its birth. Indeed, I assisted at its birth; for it was in my study . . . with Fitzgerald in an armchair and while I was enlarging on the difficulty of reconciling the then new Michelson experiment with [other ether drift experiments] . . . that he made his brilliant surmise: Perhaps the stone slab [the base of the interferometer] is affected by the motion." I rejoined that it was a 45 degree shear that was needed. To which he replied, "Well that's all right—a simple distortion." And very soon he said, "And I believe it occurs and that the Michelson experiment demonstrates it."
>
> And is such a hypothesis gratuitous? Not at all: in the light of the electrical theory of matter, such an effect ought to occur.†

That both Lorentz and Fitzgerald came upon the same notion at about the same time in roughly the same context is significant. The import will be explored later. Suffice it to say that ideas are in the air, and often several people will simultaneously make the same suggestions about natural phenomena and will construct the same kind of argument. In this case, the explanation has little logical force since the experimental evidence to explain the phenomena is also used as the basis for the principle.

The situation is even more serious and inconsistent. Whereas Lorentz dismissed the time dilation phenomenon as having no physical reality, the contraction was

* H. A. Lorentz, *The Theory of Electrons,* p. 196.

† O. Lodge, *Continuity: The Presidential Address to the British Association at Birmingham* (London, n.d.) See S. Goldberg, "Early Response. . . .," pp 336–337.

considered to be a genuine phenomenon. This is very strange because, according to the theory, the amount that the object is contracted depends on the frame of reference from which the observations are made.

What is the true length of the rod? Although it is inconsistent with his analysis of the reasons for the contraction, Lorentz almost certainly would have said that the true length of the rod was the length measured when the rod is at rest, that is, the length of the rod as measured in the ether frame of reference. That absolute length is identical with the length measured by an observer at rest with respect to the rod.

To make the comparison of Lorentz's analysis with Einstein's clearer, let us assume the situation that we described when we spoke of the problem of synchronization and time dilation. We imagined two frames of reference Σ and Σ' containing observers O and O', respectively. The observers in one frame of reference are in communication with the observers in the second frame of reference. The frames have a speed relative to each other of magnitude v, along the X-X' axis. The observers have agreed to construct identical rods having a length $L_0 = L_0'$ when at rest with the respective observers in each frame of reference (the subscript merely denotes "at rest.") According to the Lorentz theory, O' would say that the rod in Σ was contracted relative to the rod in Σ', whereas O would say that the rod in Σ' was contracted relative to the rod in Σ. The amount of the contract would be the same in each instance:

According to O

$$L' = L_0'\sqrt{1 - v^2/c^2}$$

and according to O'

$$L = L_0\sqrt{1 - v^2/c^2} \tag{3}$$

In the Lorentz analysis, which, in the first instance proposed the contraction as a technique for explaining the Michelson-Morley (and other second order) experiments, the situation appears to be symmetrical. From the point of view of observers O and O', rods in the other frame of reference are contracted by the same amount. And though Lorentz spoke of the "true" and "effective" dimensions of an object, his analysis of the situation dwelt on a dynamical explanation of the forces responsible for contracting the rod. It was only later that he derived the length contraction by assuming the transformation equations.

The Einstein analysis does not, in the first instance, depend either on assuming the contraction or the transformation equations. It depends on assuming isotropy of space about the speed of light and thereafter on the two postulates. With time dilation, and length contraction, all things are *not* relative, and *all* observers will agree that there are essential differences between the conclusions in different frames of reference.

In addition, observers will agree that the rod will be at rest in only one inertial frame of reference. We call the measurement of the length in that frame the "proper" measurement of length analogous to the "proper" measurement of time. I emphasize again that the terms *proper* and *nonproper* are not used in an ethical or judgmental sense, but technically. The measurement of a proper time interval is the time interval measured with one clock. That is possible in only one frame of reference for the time interval between two specific events. The proper measurement of length is the measurement made when the object is at rest with respect to those making the measurement. The observer making the proper time measurement will measure the shortest time between those events. Observers making the proper measurement of a length will report the length of the object as greater than all other reports of the length. All inertial observers in all frames of reference will agree about who makes the proper measurement of length and there will be general agreement about the magnitude of the proper length. The relationship between proper and nonproper length is given by equation (3).

Suppose observers in Σ' make a proper measurement of the length of the rod in that frame of reference. We designated that length as L'_0. Let us consider how that measurement might be made in other frames of reference. In Σ' we compare the ends of the rod when it is at rest with respect to a standard of measurement like a meter stick or tape measure. Should the rod be longer than any available standard for length measurement, we would lay the standard end over end until we had measured the length of the rod.

How would a length be measured in other frames of reference? The simplest technique would be for *one* observer, the observer O in Σ, for example, to start a stopwatch at the event, "first end of rod in Σ' opposite to O" and to stop the watch at the event "second end of the rod in Σ' opposite to O". The length would be the product of the relative speed of the two frames of reference and the measured time interval.

In general the length so measured would be:

$$L = v(\Delta t)$$

It is important to note that in only one frame of reference will the time interval for the two events "front end of rod in Σ' opposite to O" and "back end of rod in Σ' opposite to O," be a proper time measurement. Therefore, for *those two events,* the time interval measured by O will be the shortest obtained for those events by all frames of reference who measure them. In particular, when comparing the time elapsed between those two events for O and O' the result is $T' = T/\sqrt{1 - v^2/c^2}$, satisfying the general absolute result that the proper measurement of time will differ from the nonproper measurement by the usual factor of $\sqrt{1 - v^2/c^2}$. That being the case, the length of the rod reported by O will be less than the length of the rod recorded by O' by the same factor. To see this directly, consider the time interval T. According to O, that time interval must be equal to the ratio of the length of the rod to the relative speed between the two coordinate systems; that is, $T = L/v$. Similarly, for O', $T' = L'/v$. If we now take the ratio of these two equations, keeping in mind that observers in both frames of reference agree on the relative speed between the frames (This must be the case to satisfy the principle of relativity):

$$T'/T = L'/L$$

We have earlier expressed T' as a function of T, therefore:

$$L = L'\sqrt{1 - v^2/c^2}$$

This result has been derived in Appendix 6. There are several issues we must make clear before moving on. First, whereas Lorentz tried to understand the contraction in physical terms, the question of the reality of the contraction did not arise in the Einstein analysis. Rather, the contraction is an artifact of the way we measure and of the discrepancy between different inertial frames of reference about the time interval between any two events. Taking the analysis one step further, since the discrepancy between time intervals between specified events is a result of the postulates, we can argue that the length contraction is an artifact of the discrepancies that exist between inertial frames of reference about the synchronization of clocks.

What about other frames of reference? Suppose that we have several others moving inertially, relative to each other and to Σ and Σ'. For the time intervals measured for the specified events, front of rod opposite to O and back of rod opposite to O, none of the other frames of reference make a proper measurement. Observers in these other inertial frames of reference say that the time interval is longer than the time interval reported by observers in Σ although they do not agree with observers in Σ'. On the other hand, *these are not the events which any other inertial frame of reference would use to measure the length of the rod by the method we have specified.* Consider the frame of reference Σ'' for example. In that frame of reference, the length of the rod is determined by measuring the time interval between the events, "front end of rod is opposite O''" and "back end of rod is opposite O''." Since O'' is making a proper time measurement for *those* events (and only O'' is making such a measurement), the time interval will be the shortest of those reported by other frames of reference *for those two events.* If the relative speed between the frames of reference Σ and Σ'' is w, the discrepancy in the length between those frames of reference will be

$$L'' = L\sqrt{1 - w^2/c^2}$$

This makes the problem clear. The discrepancies are not a result of squeezing the rod; rather, a result of the way we measure. For in each case that we examine with techniques for measuring length, the proper time measurement for the events will be shorter than the time interval measured within the frame of reference making the proper length measurement. As a result, the proper length measurement of the rod or other object in the direction of relative motion between the frames of reference, will be longer than the length determined in other frames of reference moving with respect to the rod or another object.

The reader is invited to repeat the analysis, for the measurement of the length of the rod in Σ'' relative to Σ'. It is in Σ'' that the proper measurement of length for *that* rod will be made. All other frames of reference will make a nonproper measurement. Note that, whereas the situation is symmetrical, it is an oversimplification to say that all things are relative. As we have pointed out, with regard to the measurement of the length of a given object or the time interval between two particular events, all things are not relative, and there are no disagreements about who will make the proper and nonproper measurements. The discrepancies that appear in the measurements are the result of the different modes of measurement in different frames of reference.

Thus far we have concentrated on one technique for determining nonproper lengths. There are other techniques one might consider. For example, suppose that in Σ we place a revolver at each end of the rod (Figure 6). The revolvers will be triggered by photocells mounted adjacent to them, and when the revolvers fire, the bullets will be embedded in a rod in Σ'. The two frames of reference are moving relative to each other with a velocity v along the X-X' axis, and we can, in principle, bring the two frames as close together as possible, thereby minimizing

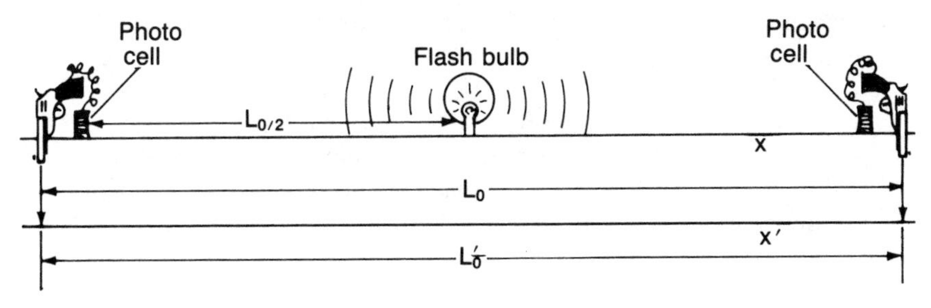

Figure 3.6

any chance that the bullets will be impeded in their travel between the frames of reference. In addition to the photocell-fired revolvers placed at the ends of L_0, we also place a flashbulb precisely in the center of the rod so that the distance from each revolver to the center is $L_0/2$. O fires the flashbulb, and the bullets are fired simultaneously and are embedded in the rod-like object in Σ'. After the bullets are fired, O' makes a proper length measurement of the distance between the bullet holes since the holes will be at rest with respect to his frames of reference.

Even though both O and O' are moving with respect to each other, it might be expected that they would agree as to the length of L_0 since they have made a proper measurement of its length.

But, this is not the case. O' claims that the rod is shorter than the measurement reported by O. According to O the bullets were fired simultaneously. That is, the light from the flashbulb traveled at the same rate over an equal distance in both directions, and the photocells responded at the same time. But O' concludes, even though the speed of light is the same in all directions, that one of the photocells approached the spreading envelope of light from the spent flashbulb, and the other photocell receded from the same envelope of light. Hence, O' judges that the guns were not fired simultaneously, and that the distance between the bullet holes in his frame of reference are farther apart than they would have been had the guns fired simultaneously.

It must be emphasized that these disagreements are not the result of an aberration in perception or psychological illusion. There is between the two frames of reference a fundamental disagreement about the simultaneity of the events, the firing of the revolvers in this case. The best that can be done is to agree to disagree and to understand the sources and ramifications of the disagreement. And so, rather than hypothesizing about a dynamical change being responsible for the contraction, the second technique for comparing lengths illustrates again that in the Einstein theory, the contraction is an artifact of how measurements are made in the two frames of reference. This second example also illustrates the role that disagreements in judgments of simultaneity play.

Lorentz never changed his mind about the interpretation of the mathematical formalism. Shortly before his death he wrote the following:

> I should like to emphasize the fact that the variations of length caused by a translation (i.e., a change of place or motion) are real phenomena, no less than, for instance, the variations that are produced by changes of temperature.*

Lorentz wrote further that, under suitable conditions if our observers O and O'

* H. A. Lorentz, *Problems of Modern Physics*, p. 95.

were to take photographs of one of the rods which are the examples, "it may very well be" that one of the images would be shorter than the other. Why the equivocation occurred to Lorentz is not clear and perhaps his conclusion seemed strange to him. More recent work on the problem of photographing an object moving at high speeds relative to an observer reveals that the object does not appear shorter in the photograph taken from an inertial frame of reference different from that of the object. This is because the length of time required for light from the different ends of the rod to reach the film are different, and will compensate the contraction being looked for exactly. However, the perspective of the moving object will shift because of the motion, so that although the two frames of reference are moving parallel to each other along the X-X' axis, the moving object appears to be rotating with respect to the observer.

The persistence with which Lorentz maintained his interpretation illustrates a point made by Max Planck in his *Scientific Autobiography.* According to Planck, new ideas do not gain favor by changing the minds of established individuals in the field. Rather, as the older members of a profession die, they are replaced by younger men who become familiar with the newer ideas, and eventually recognize the advantages of replacing the old with the new. T. S. Kuhn, in his influential *Structure of Scientific Revolutions,* suggested a similar thought when he proposed that one of the mechanisms for spreading new ideas efficiently is through their appearance in textbooks used to train new generations of scientists. We will return to this question in Part III. The problem to be addressed for now is how the process begins. As for the theory of relativity, individuals, Planck included, elaborated on the consequences of the theory. It was that elaboration that more and more revealed the heuristic power of Einstein's formulation and the lack of it in the Lorentz formulation. It led to the gradual appearance of the Einstein theory in textbook formulations.

Let us examine another physical relationship which further reveals that the theory of relativity is one of measurement. This examination will also emphasize the importance of the meanings that individual scientists attach to the formal relationship of a theory.

MASS

As we saw in Chapter 2, there was some difficulty with the role perceived for *mass* in the classical, Newtonian formulation. Two centuries after Newton, the concept was subject to intense analysis by Mach, who concluded that the Newtonian concept had been introduced by a circular method.

There is always a difficulty with the concept of mass. It has been hard to teach students in beginning physics what "mass" means. Elementary physics text-books in American secondary schools before 1945 are filled with arm waving about the concept, in addition to statements that mass represents "the amount of stuff in an object." This definition corresponds to Newton's definition of mass as proportional to the density of matter.

Since 1945 there have been several notable attempts to introduce the concept of mass with a Machian empirical approach in which bodies interact with each other in isolation. This complies with Newton's first and third laws, and thereby anticipates the conservation of momentum.

Whatever one believed about the concept of mass, in classical physics, it was *a* number which was constant, unchanging, and invariant with regard to different frames of reference. It only changed when the object itself changed, for example if the object began to disintegrate. Furthermore, whether or not one defined inertial mass by observing objects interact as Mach proposed, or in terms of how objects accelerated under the influence of forces, the results were the same.

Later there was to be another criticism of the Newtonian concept of mass, this one by Einstein. He pointed out that Newton and those after him had assumed that the mass of an object determined with pushes and pulls of mechanical contact forces, was the same as the mass of an object determined by the pulls from gravitational sources. There was no *a priori* reason that the mass of an object be the same in those two circumstances. In the Newtonian scheme the same number for the inertial and the gravitational mass was a coincidence. This observation was a starting point for the general, as opposed to the special, theory of relativity.

There is a different kind of question that one can ask about the concept of mass, which was asked during the nineteenth century: "What is the source of mass?" Mass has traditionally been a primitive in physics, and the primitives, such as the concepts of length and time, are not defined, but rather are the starting points for analysis. In recent practice, electric charge has been treated as such a primitive, but, at the beginning of the nineteenth century, under the aegis of the mechanical world view, it was believed that ultimately electric charge and magnetic poles would be understood as a manifestation of the mechanical action of mass. Also, in the mechanical world view, it was established that the light ether would be a solid. Later, when it was accepted that light was a manifestation of electromagnetic radiation, the light ether, the electrical ether and the magnetic ether were replaced by one electromagnetic ether. In the view of some of the proponents of the mechanical world view, ordinary matter was then derived from this ether and electric and magnetic poles resulted when the ether was subject to various forces, such as shears or twists. It is a remarkably Cartesian view of the world containing a primal material responsible for all matter and all radiation.

When the electromagnetic world view became prominent during the last decades of the nineteenth century, the emphasis changed. It was incumbent on proponents of that view to account for mass as manifestations of electromagnetic interaction. The fundamental entity in the electromagnetic world views was still ether. Now, however, charge was the manifestation of forces exerted on the ether, and mass was the result of the interaction of charge with radiation. Mass was often described as the result of the inertia of electric charge moving through external fields that it encountered, in addition to the electromagnetic inertia which resulted from the charge moving through its fields.

It is not a simple concept, especially for those who are not familiar with the elementary physics of electricity and magnetism. Let us illustrate what those committed to the electromagnetic view mean by electromagnetic inertia by using the example of the operation of a basic electric motor and generator.

The principle of the electric generator is based on the observation that, when a conductor of electricity is moved through a magnetic field, an electric field is created in the conductor and, should the conductor be part of a complete electrical circuit, there will be a flow of electricity in the conductor and through any electrical appliances included in the completed circuit. Actually, it is not necessary that the conductor move; all that is required is that there be relative motion of the conductor and magnetic field.* In practice, in generating electricity, a prime mover such as water or steam is frequently used to move the conductors in the magnetic field. The effect is maximized when the relative directions of the motion of the conductor and the sense of the magnetic field are perpendicular to each other. The created electric field is in a direction mutually perpendicular to the sense of the magnetic field and the direction of motion of the conductor.

The principle of the operation of an electric motor is similar. It is based on the observation that, when a current flows through a conductor in the presence of a magnetic field, a force is exerted on the conductor, and the direction of that force is mutually perpendicular to the direction of the current and the sense of the magnetic field.

You may notice that when the conductor is moving in the case of the electric motor, it is moving in a magnetic field, and hence it will act like a generator. Alternatively, when the current begins flowing through the conductor in the generator, it is a current carrying conductor moving in a magnetic field, and it will act like a motor. In what direction does the force on the current carrying conductor in the generator act? If it acted in the same way that the prime mover was moving the conductor, there would be no need for the prime mover; it would be in perpetual motion. The analysis shows that the force created in a generator by the

* This was precisely one of the bits of physical evidence Einstein used in the opening paragraphs of his first paper on special relativity as suggestive of the principle of relativity.

current carrying conductor moving through the magnetic field is in a direction opposite to the motion. It is this phenomenon which proponents of the electromagnetic world view label as electromagnetic inertia. Anyone who has tried to hand crank an electric generator to operate even the smallest appliance light bulb knows how much work is required. This work is to overcome the motor effect.

Since the motion of the electric motor is a function of current carrying conductor moving in a magnetic field, the electric current developed by the generator effect must be to oppose the original motion creating current. Otherwise we would again have perpetual motion.

Let us apply these ideas directly to the electron theory. Conductors of electricity, for example silver or copper metal, contain electrons which are free to move through the metal under the influence of electrical forces. Electrons are the fundamental entities of electric charge. When they move, they constitute an electric current. Should the electron, in its motion, be in the vicinity of a magnetic field, a force should be exerted on the electron in a direction opposed to its motion. This type of force is referred to as electromagnetic inertia of the electron. A precise calculation of the magnitude of the force, and hence the magnitude of the electromagnetic inertia, depends on the assumptions that are made about the nature of the electron.

Beginning in the year 1901, the German experimental physicist Walter Kaufmann undertook a series of experiments to measure the inertia and the mass of electrons moving at speeds which were significant proportions of the speed of light. The experiment was carried out before the days of particle accelerators, and the source of his electrons was a kernel of radium bromide—that is, he was using electrons that were ejected from a naturally occurring radioactive material. He measured the ratio of the charge of the electron to the mass as a function of the electron's speed, and found that the ratio decreased as the speed of the electrons increased. This is what one would expect in the electromagnetic world view. Assuming that the charge on the electron remained fixed, the decreasing value for the ratio of the charge on the electron to its mass (referred to as e/m) could only be interpreted as evidence for the increase in the mass. How much of the mass of the electron was due to this kind of electromagnetic inertia and how much was mechanical? For proponents of the electromagnetic world view, it was hoped that none of the mass would be mechanical; it would all be the result of electromagnetic inertia.

The problem was to construct a viable theory to account for this effect. It is extraordinarily difficult. In 1902, Max Abraham, a colleague of Kaufmann's at Göttingen University, undertook to construct such a theory. He assumed that the electron was an absolutely rigid sphere. If the electron were not rigid, then one would have to introduce other forces to hold the electron together under the influence of the repulsive forces due to the close proximity (The dimensions are of the order of magnitude of 10^{-14} cm.) of a relatively large amount of like charge. If one was going to be true to the electromagnetic world view, the introduction of non-electromagnetic forces would be a contradiction, and Abraham skirted the problem by stipulating that the electron be rigid. He then calculated the forces that a body would experience as it moved through its own and external electromagnetic fields. It is a difficult problem in the application of the fundamental laws of electricity, and it speaks for Abraham's technical virtuosity that he was able to come to a reasonable solution. The forces are different, parallel and perpendicular to the direction of motion of the electron. It is through the calculation of these forces (assuming that the magnitude of the charge does not change) that Abraham was able to tease out the mass of the electron. The electromagnetic mass thus calculated was not only a function of the speed of the electron, but also a function of the direction of motion.

The mass in the direction of motion of the electron relative to the ether was termed the *longitudinal mass*. The mass of the electron in the direction perpendicular to the direction of motion was termed the *transverse mass*. The only experimental situations which have ever been devised for measuring the mass of the electron have been those in which the forces are perpendicular to the direction of motion. The exact expressions that Abraham calculated are very complex and can be expressed as the product of a basic mass m_0 and an infinite series of powers of v/c. Since experimentally, the only mass that is measured is the transverse mass, we give only Abraham's approximation for the transverse mass. It is

$$m = m_0(1 + 2/5 \frac{v^2}{c^2} + \ . \ . \ .)$$

where the ellipsis . . . refers to higher order terms in v/c. Abraham also concluded that this analysis agreed substantially with Kaufmann's experimental result even though that result was uncertain. There was no material or mechanical mass; all of the mass of the electron was electromagnetic inertia. Such a result fitted well within the percepts of the electromagnetic world view. During the next several years, Kaufmann improved his experimental apparatus and measuring techniques and each time he performed the experiment, he claimed closer agreement with the prediction of Abraham.

In 1904, when Lorentz published his final version of the second order theory which explained ether drift experiments, his prediction for the change in the mass of the electron disagreed with the Abraham prediction in several ways. Lorentz assumed the Lorentz contraction, and therefore, his electron was deformable. Thus, to an observer moving relative to the electron, the electron's radius would be contracted in the direction of motion. The originally spherical electron would be an ellipsoid when in motion. This posed problems for Abraham, who recognized that under those circumstances, nonelectromagnetic forces would be required to hold the electron together. In fact, proponents of the Lorentz electron, such as Henri Poincaré, tried to solve this problem by hypothesizing that the ether itself exerted the correct pressure on the electron not only to hold that charge together in a relatively small region of space, but also to contract the electron. It was not a conceptually or quantitatively satisfying solution.

The Lorentz expression for the change in mass of the electron was derived by him using the assumption of the deformable electron and the Lorentz transformation. We can compare the result directly to the Abraham result by expanding the Lorentz equation for the change in mass into an infinite series in powers of v/c.

$$m = \frac{m_0}{\sqrt{1 - v^2/c^2}} = m_0 \left(1 + \tfrac{1}{2}\, v^2/c^2 + \ldots\right)^*$$

---

* That this approximation can be made in this situation arises from the following line of reasoning: By direct algebraic division or by a process of mathematical induction, it can be shown that $1/\sqrt{1 - v^2/c^2}$ can be expressed as the sum of the terms in an infinite series in powers of v/c. The first few terms of that series are

$$1 + \tfrac{1}{2}\, v^2/c^2 + \tfrac{3}{8}\, v^4/c^4 + \ldots$$

where the ellipsis refers to still higher order even terms in v/c. The physical situation is that the ratio of v/c is always a number less than one. (Nothing can go as fast as the speed of light). Therefore v^2/c^2 is going to be even smaller and for most physical situations the ratio of v^4/c^4 will be totally insignificant.

The approximation may be tested in the following way:
Assume

$$\frac{1}{\sqrt{1 - v^2/c^2}} = 1 + \tfrac{1}{2}\, v^2c^2$$

Is this correct? Square both sides of the equation.

$$\frac{1}{1 - v^2/c^2} = \left(1 + \tfrac{1}{2}\, v^2/c^2\right)^2$$
$$= 1 + v^2/c^2 + \tfrac{1}{4}\, v^4/c^4$$

Ignoring terms higher than the second order in v/c gives:

$$1/(1 - v^2/c^2) = 1 + v^2/c^2$$

By direct multiplication this gives

$$1 = (1 + v^2/c^2)\,(1 - v^2/c^2)$$
$$= 1 - v^4/c^4$$

which is true if we ignore terms higher than the second order in v/c. Notice that this also suggests a useful power series approximation for the expression $1/(1 - v^2/c^2)$.

If we compare this with the Abraham prediction, we see that they differ from each other by one part in ten in the second order term in v/c, not very different. In fact, Lorentz devoted himself, in his 1904 paper, to showing that his predictions were as adequate for Kaufmann's experiments as the predictions of Abraham.

In the year that followed, several significant things occurred. Abraham published a popular advanced textbook on electrical theory that was to become standard in German universities and was to go through many editions. A later edition is still in print. The book makes clear that the Abraham theory of the electron has limited applications. For almost all other aspects of electromagnetic theory of moving bodies, Abraham relied on the Lorentz conceptions: the transformation equations, the notion of the contraction of macroscopic objects and time dilation, etc. However, as to the electron itself, Abraham departed from the Lorentz view, and insisted on his own conception, the rigid electron. He also rejected the principle of relativity because it was at odds with the results of his theory of the electron, and, in general with the electromagnetic world view.

Another notable event in 1905 was the publication of Einstein's theory. As I have indicated, virtually no one realized the significance of Einstein's proposal. In the minds of many, since the predictions of Einstein and the predictions of Lorentz were the same, they were seen as aspects of the same theory. Even supporters of Einstein shared this confusion; for example, Max Planck referred to the Lorentz-Einstein theory. Hermann Minkowski, the man who is credited with generalizing Einstein's theory to four dimensions remarked that Einstein's work was a generalization of Lorentz's. With hindsight we are in a position to see that that remark reveals a lack of understanding of the important distinctions between Lorentz and Einstein. Nevertheless, the confusion was widespread.

Another notable occurrence in 1905 was Kaufmann's further refinement of his experiments on the relationship between the speed of the electron and its mass to distinguish between the Abraham theory and the theory he referred to as the Lorentz-Einstein theory. Interestingly enough, Kaufmann, who acknowledged that he was a partisan of the Abraham notion of the electromagnetic world view, reported that his experimental results did not support the prediction of Einstein and Lorentz. He wrote:

> The results of the measurements are not in accord with the Lorentz-Einsteinian basic assumptions.

Later, Kaufmann elaborated:

> . . . *if this* [*the Lorentz and Einstein prediction*] *is considered as refuted so also the attempt to base the entire body of physics, including electrodynamics and optics under the principle of relative motion can be labeled as failing.* . . . We must remain with the assumption that physical appear-

> ances depend on motion relative to a completely determined coordinate system that we designate as the *absolute resting ether.* If we have not as yet succeeded in detecting by optical or electrodynamic experiments any such influence of the motion through the ether, this does not exclude the possibility of such detection.*

This situation is different from others thus far described about the relationship of experiment to theory. We have shown that the Michelson-Morley experiment, while perhaps suggestive, does not give us information about the postulates of relativity. We can take some heart that it does not contradict the theory. On the other hand, we have also pointed out that no experiment can distinguish between the predictions of the Lorentz theory and those of the Einstein theory. In this case, however, we have two *different* predictions, one from the theories of Lorentz and Einstein, and the other from the Abraham theory. And according to Kaufmann's experiment, the predictions of the Abraham theory appeared to come closer to the experimental result than those of the theory of relativity.

As far as anyone could tell, Kaufmann's procedures had been unflawed. That being the case, one might be tempted to say that the Lorentz and the Einstein theories were false, and the Abraham theory was verified, or at least, corroborated. That should have ended that.

Obviously that did not end that, and there were three basic responses to the Kaufmann experiments. One was to call for the rejection of the "Lorentz-Einstein" theory on the grounds that its predictions did not correspond to experimental verification. Another group called for a suspension of judgment because it was only one experiment; further, there was an uncertainty about the experimental results; the smallest change in one of several experimental parameters could overturn the results. This group was composed of those who liked one or another aspect of the theory of relativity and were not willing to reject it without a more convincing demonstration. At first, this group had only one active spokesman, the physicist Max Planck. Most other physicists were indifferent. Whether they agreed with the theory of Einstein or the theory of Lorentz, or the "Lorentz-Einstein" theory, or on the contrary, whether or not they were partisans of the Abraham rigid electron theory, they did not appear to care about the experimental results. One of the major proponents of the Lorentz theory, Henri Poncaré, said that one experiment such as that done by Kaufmann could overturn all of electricity, magnetism and optics. Although he said that, there is no evidence that the Kaufmann experiment affected his *beliefs* about the worth of the Lorentz theory in any way. It brings to mind that Einstein once said: "If you want to find out

* W. Kaufmann, "Die Konstitution des Elektron," *Annalan der Physik,* 1906, *19*:487–553. The quotation (pp. 534–535) is my translation. See S. Goldberg, "Early Response. . ." p. 82.

anything from theoretical physicists about the methods they use, I advise you to stick closely to one principle: don't listen to their words, fix your attention on their deeds."*

The various shades of opinion of the meaning and importance of the Kaufmann experiments were brought together in confrontation during the 1906 meeting of the Society of German Scientists and Physicians. This was an annual affair of one of the central social organizations of science in Germany. It has its counterparts in the major western countries; for example, in Great Britain the organization is known as the British Association for the Advancement of Science (BAAS), in America, the American Association for the Advancement of Sciences (AAAS). The society is usually organized by field so that there are sections devoted to mathematics, mathematical physics, experimental physics, biology, chemistry, etc. By looking at the degree to which distinctions are made by such organizations, one can get an insight into the degree of abstraction a science has undergone and the narrowness with which individual members of the scientific community identify their colleagues. Such organizations not only organize meetings whereby members working on similar problems can share ideas through papers and commentaries, but they organize symposia about pertinent subjects and maintain standing committees to investigate major technical problems and to review progress in areas of scientific work. They frequently oversee the publication of one or several journals.

Very often, the proceedings of the meetings are reported in the journals. The proceedings of the meetings of the physics sections of the Society of German Scientists and Physicians, including the discussion between those delivering the papers and those making comments were printed in a journal known as *The Physical Journal* (*Physikalische Zeitschrift*).

The Kaufmann experiments were the subject of discussion at the seventy-eighth meeting of the Society, held in Stuttgart September 16–22, 1906. The vehicle for the discussion was a paper delivered by Max Planck on September 19. It had been about a year since Einstein had published his original paper on relativity. Planck had already defended the "Lorentz-Einstein theory" on the grounds that, although it did not appear to conform to Kaufmann's experimental results, it yielded beautifully simple solutions to complex problems in ways that Planck had found pleasing. He had, therefore, urged restraint in rejecting the theory. At the Stuttgart meetings, Planck undertook an independent analysis of the Kaufmann experiments, subjecting the experimental arrangement to critical scrutiny and independently reevaluating Kaufmann's computational techniques as well.

* A. Einstein, "On the Method of Theoretical Physics," in *Ideas and Opinions* (New York, 1954), p. 270.

Planck stated that he could see nothing wrong with the experiment, but he pointed out that the technique was delicate, subject to experimental uncertainty, and that it would require only very small changes in some data to overturn the result. He again urged restraint on the part of his colleagues.

In the discussion that followed, both Abraham and Kaufmann pressed Planck about what appeared to be a simple issue: the experiment gave results which, within experimental error, was in accord with Abraham's predictions and not with the predictions of Lorentz and Einstein. But Planck remained firm, repeating again the conclusions his analysis had led to, that to decide on the basis of one experiment would be premature.

The tenor of the argument changed dramatically. Abraham pointed out that more important than any issue which had thus far been discussed, only his, the Abraham theory, conformed to the electromagnetic world view. The Lorentz electron, he said, required non-electromagnetic forces to hold it together, and Einstein had said nothing about the nature of the electron, but had come to the result on the basis of the postulate of relativity. In making the distinction between Lorentz and Einstein, Abraham showed a sensitivity about theoretical distinctions that most of his colleagues did not have.

Planck's response to Abraham is important. He agreed that, of all the theories making predictions about the increase in the mass of objects in motion relative to the observer, only the Abraham theory was based on pure electromagnetic assumptions. But, Planck continued, that was after all an assumption which Abraham preferred, not dictated by a higher authority. Planck said that he preferred the assumptions of the theory of relativity.

Einstein took no active part in the debate. This was typical for him; he rarely responded to such disputes. In this case, there is some evidence that he knew nothing about it because several months later he was apparently informed about Planck's early contributions in defense of the "Lorentz-Einstein" theory by a letter from Johannes Stark, editor of a journal devoted to articles reviewing the state of knowledge in the fields of electron theory and radioactivity. Einstein was writing a review of the theory of relativity which was to be published in Stark's journal in 1907. In that article, Einstein briefly considered the Kaufmann experiment. He, too, said he could see nothing wrong with the experiments, and he admitted that the experimental points were better satisfied by the predictions of the Abraham theory than his own. However, he continued,

> In my opinion other theories [Einstein was also referring to the Bucherer theory of the electron] have a rather small probability because their fundamental assumptions concerning the mass of the moving electrons

are not explainable in terms of theoretical systems which embrace a greater complex of phenomena.

He never referred to the Kaufmann experiment or the Abraham theory again. The Abraham theory of the electron (and the Bucherer theory as well) had been carefully tailored to account for only one thing: the increase in mass of moving electrons. In the Einstein theory, on the other hand, the increase in mass had nothing to do with a specific theory of the electron. In fact it derived only from the measurements of lengths and times and was a kinematic result.

It should be noted that the intuitions which Planck and Einstein demonstrated about the Kaufmann experiment proved to be sound. Within six months of Planck's analysis at the 1906 Stuttgart meetings, there were significant changes in the accepted values of some of the parameters used by Kaufmann in analyzing his data. These changes alone would have been enough to overturn his conclusions. More seriously, work in England and Germany by Hans Geiger and Johannes Stark revealed that the kinds of data which Kaufmann was collecting were unreliable unless the motion of the electrons was measured in an extremely high vacuum. Whereas such vacuums were attainable with the existing technology, Kaufmann, not realizing the implications for his results, had preferred to work at higher pressures apparently because it simplified the building and maintenance of the apparatus. In any event, within a year, Kaufmann's contributions in this region of physics ceased although he remained active in the German physics community for many years.

Let us turn now to the analysis of the Einstein prediction for the change in mass of an object when it is in motion. In one sense, the reasoning behind the analysis is much simpler than in the theories of Abraham or Lorentz. For as with all other parameters that are treated within the theory of relativity, the change in mass says nothing essential about the body itself, but results as an artifact of the way distances and times are measured for the same events by different inertial observers. In another sense, the analysis is very complex and perhaps gives the impression of being forced, arbitrary or artificial. The reason is obvious. The term "mass" evokes the Newtonian concept of a scalar, that is, a single number associated with a body giving either some measure of the net force required to accelerate the body by a specified amount or the weight of the body.

The Einstein analysis of the concept of mass reveals that within the theory of relativity, the meaning of the term *mass* changes. This was also true of concepts like length, time interval and simultaneity. Furthermore, it will be true of almost all other parameters. Although we use the same name for two concepts such as Newtonian mass and relativistic mass, they are not at all the same. The fact that we have preserved the same term reflects a cultural bias that scientific knowledge is cumulative and that successive theories reduce one to the other for the limiting cases. While such reductions are sometimes possible, since logical formalism serves only as a convenient language, the reduction of one theory to the other is trivial and does not signify common meanings between theories.

Even though Lorentz and Einstein had made the same predictions for the transverse and the longitudinal mass, the meaning of the concept of mass in those two theories was decidedly different. Although Lorentz's concept of a deformable electron required a type of non-electromagnetic force to hold the electron together, his electron was the ultimate *source* of mass and played *the* fundamental role in explaining the behavior of matter in all circumstances. For Einstein, mass had no such cosmic significance. After all, it was clear to him that he had produced a theory of measurement, not a theory of matter. In his first paper on relativity in 1905, he calculated the mass of an object while making use of the Lorentz transformations and Newton's second law as applied to electrons in motion. He concluded, as Lorentz had, that the mass of the object was going to be different in the direction parallel and perpendicular to the direction of relative motion between the observer and the electron. For forces perpendicular to the direction of motion, the *transverse mass* (indicated by the subscript t), was given, as Lorentz had given it:

$$m_t = \frac{m_0}{\sqrt{1 - v^2/c^2}}$$

And as Lorentz had concluded, Einstein's derivation for the so-called *longitudinal mass* (indicated by the subscript l) was given by:

$$m_l = \frac{m_0}{\sqrt{(1 - v^2/c^2)^3}}$$

Not only would observers in different frames of reference differ with regard to the magnitude of the mass they determined for an object, the magnitude of the mass would depend on the direction relative to the sense of relative motion between the two frames of reference in which one measured the mass.

From the beginning, Einstein recognized that the expressions for mass he had obtained and the fact that mass did not have the same value in different directions was arbitrary and dependent on how one defined force and acceleration. For the mass to vary, not only with frame of reference, but with direction within a frame, would vastly complicate the derived laws of physics, for example, the conservation of energy and the conservation of momentum.

By 1910 the situation in physics had been considerably simplified by redefining the mass of an object so that it no longer varied as a function of the *direction* of the motion of the object relative to the motion of the frame of reference in which the object was judged to be moving. In other words, the mass was defined in such a way that within a frame of reference, it would again be a number (albeit a number whose value depended on the velocity of the observer relative to the mass being measured). The laws of physics, for example, the conservation of momentum and the conservation of energy maintained the simple form they had acquired in classical physics.

The development of these considerations for mass was by two American scientists, G. N. Lewis and R. C. Tolman. We will consider their contributions in detail in the chapter about the American response to Einstein's innovation. For now, we note that the derivation of the mass of an object which they introduced and whose principles will be developed here was the first example of the application of Einstein's theory of relativity to mechanical situations without reference to electrodynamics.

We begin where we always begin in Einstein's theory: How do we measure a mass? If the mass is at rest with respect to us there are several techniques which we might employ. To measure the inertial mass, we may exert a standard force on the object, observe the object's acceleration and infer the mass of the object from Newton's second law, $\vec{F} = m\vec{a}$: $\vec{F}$ is the force, m is the mass, and $\vec{a}$ the acceleration. Another method is to take a standard mass and allow the mass being measured to interact with the standard mass, perhaps by allowing them to collide. The analysis is simple for two extreme cases. In one case, when the collision is totally elastic, the objects rebound from each other without deformation and without friction loss. In the other case, the collision is totally inelastic, and the two objects stick together after they collide. In both cases, in the absence of external forces, Newton's second law implies that momentum is conserved at all times, so that the total momentum of the system before and after the collision is the same. Knowing the mass and velocity of the standard object before and after the collision and knowing the velocity of the test object before and after the collision, we can solve the equation for the conservation of momentum for the mass of the test object (Appendix 3).

One might object that in this measurement the body being tested is not actually at rest with respect to the observer who is making the measurement. Whereas that is correct, I can always reduce the speed of the object to a value less than any specified limit.

We must consider the measurement of mass within the special theory of relativity from two different perspectives, one in which the mass is at rest with respect to us, and one in which the mass is moving at a high rate of speed relative to us. The simplest situation to consider is the collision of one mass with another under standard conditions and from the behavior of the masses before and after the collision to infer the mass of the object. We invoke two inertial frames of reference Σ and Σ' moving with a speed v relative to each other along the X-X' axis. Each contains an object with the same mass at rest. Before the experiment, the objects were compared directly. Assume the objects to be very hard steel balls which, upon collision, are not deformed and incur virtually no frictional loss. The systems are in communication with each other, and O and O' have arranged it so that each tosses his ball at a very small velocity so the two balls will collide and rebound (Figure 7).

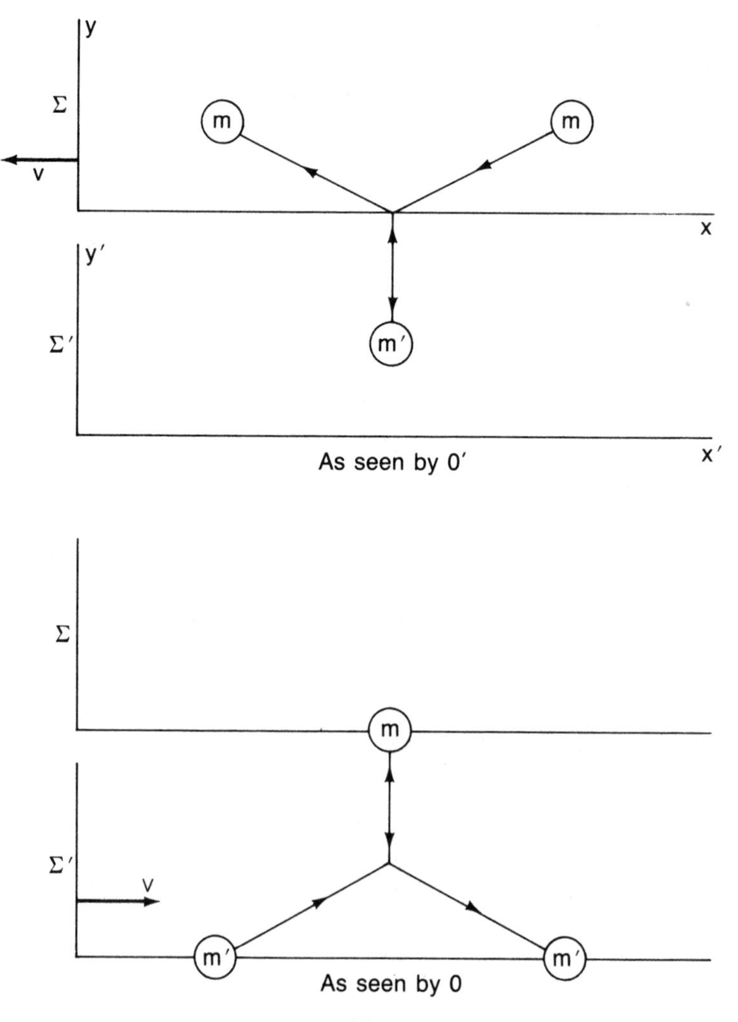

Figure 3.7

In Figure 7, I indicate the situation from the perspective of each observer. Note that from the point of view of O, the ball he throws moves perpendicular to the line of direction of the relative motion of the two frames of reference. Hence, if there is a variation in the mass of that ball as observed by O', it will be termed the transverse mass. And the same is true for the ball thrown by O', as seen and measured by O.

Furthermore, if the same collision experiment at extremely low velocities takes place in O or O' or in another inertial frame of reference, one concludes that the total momentum of the system composed of the colliding balls is conserved before, during, and after the collision. This is basic to Newtonian mechanics, and the result is not affected by the theory of relativity as long as the masses are observed in one frame of reference. This is shown in Figure 8. For simplicity we suppose the objects have equal rest or proper masses. In one frame of reference, we arrange the synchronized clocks and measure the initial and final velocities of two objects. In this instance, not only are the masses the same, but the initial speeds of the objects are also the same, that is, their velocities are equal in magnitude and opposite in direction.

Since Newton's third law specifies that during the collision the objects will exert equal and opposite forces on each other, and there are no external forces, after the collision the velocities of the two objects are unchanged in magnitude and opposite to their initial direction. The problem is solvable for any arbitrary values of mass and initial speeds as long as there are only two objects.

In this experiment (Figure 8) we note that all the observers are in the same frame of reference and that none makes a proper time measurement of the speed of the objects. That measurement would have to be made *from* the objects themselves. In this frame of reference, the momentum is always conserved, and it is appropriate to consider the conservation of momentum to be a law of physics. That being the case, the principle of relativity assures us that the law will have the

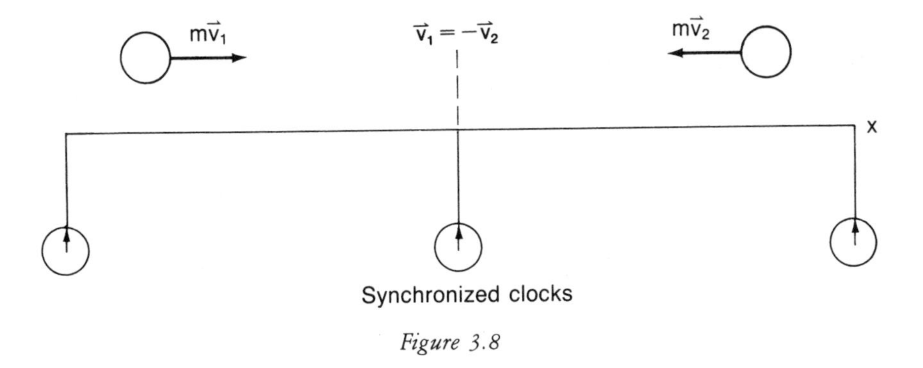

Figure 3.8

same form in all inertial frames of reference. When we consider the collision of the balls thrown at each other by O and O' from their respective frames of reference in uniform motion with each other (Figure 7), we begin with the law of the conservation of momentum. O and O' may not agree on the magnitude of the momentum, but they will agree that whatever its magnitude, for the universe consisting of the two balls, it will remain the same before, during, and after the collision as long as no external disturbances intervene. For the ball thrown by O in Σ, only O can make a proper time measurement of the events: "Ball leaves O, ball returns to O." Observers in Σ' will make nonproper measurements of those events. On the other hand, only O' in Σ' will make a proper measurement of the events: "Ball leaves O', ball returns to O'." Whereas the situation is perfectly symmetrical, it is not completely relative. All observers will agree that they do not make the same kind of measurements for any specified events.

Let us analyze Figure 7 with some care. O throws his ball in a direction exactly perpendicular to the direction of motion between the two frames of reference. The ball travels a distance y, strikes the ball thrown by O' and returns to O. The total distance traveled by the ball will be $2y$ and O will have made a proper time measurement, $2t$. Although from the point of view of O', O did not throw the ball perpendicularly, O' agrees with O about the distance the ball traveled. He or she disagrees about the time required for both the outward journey and return journey of O's ball because O' makes a nonproper measurement of both time intervals. O and O' agreed before the experiment to endow both balls with a certain velocity. From their perspectives, they have done that correctly. But we can see that because of the disagreement in the measured time interval, O' will conclude that the speed of the ball thrown by O is less than that reported by O. O' will also conclude that the *change* in the velocity of O's ball during the collision is less than the change he or she measures for his or her ball. Since O' accepts as the starting point of the analysis the conservation of momentum, and since magnitude of the momentum of the ball he or she threw was the same on the return trip as it was on the outward trip, O' concludes that the mass of the ball thrown by O must be greater than O reported. He or she infers from the momentum he or she measured for his or her ball that the magnitude momentum in the direction perpendicular to the motion of the two frames of reference of O's ball is the same as the magnitude of the momentum of his ball.

This has taken a lot of words because it is very complicated. Since the two observers agree on the distance traversed and disagree on the time in the ratio of proper to nonproper time measurements, O' judges the change in velocity of O's ball to be less than his own by a factor of $\sqrt{1 - v^2/c^2}$. O''s estimate of the mass of the ball thrown by O is greater than the estimate of that mass made by O by the same factor.

$$m' = \frac{m}{\sqrt{1 - v^2/c^2}}$$

In general, the relationship between the proper or rest mass and the nonproper or moving mass will be as follows:

$$\text{nonproper mass} = (\text{proper mass})/\sqrt{1 - v^2/c^2}$$

This is the prediction of Einstein and Lorentz for the transverse mass. The rest mass, which we designate as the proper mass is less than the nonproper mass by the factor $\sqrt{1 - v^2/c^2}$.

The disagreement arises because of our acceptance of the principle of relativity, and confidence in the law of the conservation of momentum and also because we make a proper measurement of the time in flight of the ball thrown by us in Σ, and a nonproper measurement of the time in flight of the ball thrown by our counterpart in Σ', O'. Ultimately, the disagreement about the value of the mass of the object can be traced to our disagreement about judgments of simultaneity.

Should you undertake the same analysis from the point of view of O you would find that, whereas he or she would measure the mass of the sphere thrown in his frame of reference to be m_0, he or she would measure the mass of the ball thrown by O' to be

$$\frac{m_0}{\sqrt{1 - v^2/c^2}}$$

In other words, as with the other parameters we have discussed, the situation between inertial frames of reference is symmetrical. However, the symmetricalness is *not* with regard to the same measurements. That is, only one frame of reference makes a proper measurement of the mass of an object. All other frames of reference make a nonproper measurement of *that* mass. *All* observers in *all* frames of reference will agree on who makes the proper measurement of a given mass and *all* observers in *all* frames of reference, while disagreeing about the magnitude of the mass will agree that the rest mass is less than any other determination of that mass and will agree also on what magnitude that rest mass is. In that sense, rest mass is an invariant identical in behavior to the Newtonian concept of mass. It is an unacceptable simplification to use the theory of relativity to support a view that "all things are relative" even when the phrase "all things" refers to objects in physics.

Does the mass of objects moving with respect to the observer who is measuring actually increase? Within the theory of relativity this question has no meaning. We might coin a phrase: "Actually is as actually measures." When we make a measurement of mass moving with respect to us, we either use the measurements to calculate the force necessary to effect a change in the momentum of the object or to calculate the effects of interaction with other objects in the absence of external forces. From inferences about changes in the velocity we infer changes in the mass. That is precisely the manner in which we determine the changes in the mass of objects in our thought experiment.

Whereas this analysis may appear to be more complicated than those that preceded it, it is of the same character as the analysis of simultaneity, time and length. Using the Einstein theory, even though we make the same quantitative prediction as we would with the Lorentz theory, we have assumed nothing about the nature of matter or mass and we made no electromagnetic assumptions. We assume nothing except the principle of relativity, and the invariance of the speed of light.

The analysis of Lewis and Tolman showed that if one redefined the mass in a given way, one could maintain the law of the conservation of momentum and retain a single number for the mass regardless of the direction in which the mass moved. If we again call the rest mass of the object m_0, the mass of an object is defined as

$$m = \frac{m_0}{1 - u^2/c^2}$$

The variable u does not represent the relative speed of two measuring frames of reference, but rather the relative speed between the object whose mass is being measured and the frame of reference in which the measurement is taking place.

The question is, does mass behave in the way Einstein's or other theories predict? In this respect, the theory of relativity bears a resemblance to attempts to confirm the predictions and expectations of earlier theories.

For example, a key experimental question that would have differentiated geocentric from heliocentric theories about the place of the earth within the system of sun and planets was of stellar parallax. Stellar parallax is to be expected if the earth is moving around the sun and the stars are different distances from the earth. For, as the earth changes position during the year, the line of sight to the closer stars will change relative to the farther stars. The closer the star is to the earth, the greater the effect (See Figure 5 in Chapter 1). Shortly after the Copernican system was seriously proposed, anti-Copernicans called for the evidence of stellar parallax between the planets and the stars. It had never been observed. The Copernicans gave a strange answer. The stellar objects, they said, must be farther away than we thought. There was no empirical evidence that would give anyone, except those already committed for other reasons, confidence in the heliocentric theory. Heliocentric astronomy gained acceptance in the absence of evidence. The discovery of stellar parallax did not occur until 1839, almost three hundred years after Copernicus published his theory. The measurement was made by the German astronomer Friedrich W. Bessel. At the time it was apparently unnoticed that it confirmed the Copernican theory. In histories of astronomy, much is made of the absence of the evidence of stellar parallax in the seventeenth century; but, when these histories describe the discovery, there is no mention of confirming the Copernican theory. The reason is clear. The discovery occurred long after the theory had been accepted. And the acceptance of the theory had had little to do with that kind of evidence.

The same thing happened to the theory of relativity. There was no empirical evidence. The only direct measurements which gave hope of confirming the theory were the kinds of measurements that Kaufmann had used to determine the nonproper transverse mass of electrons moving at high speeds.

Kaufmann's measurements were discredited in 1907. Similar attempts by Adolf C. Bestelmeyer in 1908, although now believed to have been properly done, were also unacceptable methodologically. Alfred H. Bucherer undertook to make the same kinds of measurements in 1908 and they were not accepted either. It was not until 1916 that the measurements of the relationship between the mass of moving electron and its speed were made and gained the confidence of physicists. They were made by Charles – Eugène Guye and Charles Lavanchy. They were not heralded as confirming the special theory of relativity. In fact, although that theory had gained acceptance among many physicists, those who did not accept the prediction of the theory of relativity accepted the Lorentz theory. Since these predictions were identical, the experiment caused hardly a ripple. Since then there have been further confirmations of the theory of relativity (or the Lorentz theory). There are two points to be emphasized. By the time the evidence was available, there was no question about the outcome. Also, there is no way that an experiment can distinguish the predictions of the Lorentz theory and the special theory of relativity. They make identical predictions.

The choice between the two theories was not made, ultimately, on the basis of the answers that they provided, but rather, on the questions that they raised. The theory of relativity proved to be the heuristic theory. It is to that aspect of the theory that we now turn.

·4·

Further Consequences of the Heuristic Nature of the Special Theory of Relativity

IN CHAPTER 3 WE EXAMINED the relationship that must exist between temporal and spatial coordinates as measured by different inertial observers. We also examined how such observers would compare their measurements of inertial mass. The Lorentz and the Einstein theories predict the same consequences because they make use of the same transformation equations for relating temporal and spatial coordinates for different inertial observers. Although the predictions of the two theories were the same, the interpretation of those predictions was different in each of the theories.

Lorentz maintained his belief in an absolutely fixed ether although the frame of reference could not be located. Furthermore, length contractions, time dilations, disagreements over judgments of simultaneity and disagreements about the mass of moving bodies were, for Lorentz, dynamical effects due to the nature of matter and how matter interacted with the ether.

In the theory of relativity, all of the effects and predictions stemmed from the disagreements between inertial observers about judgments of simultaneity. The relativity of simultaneity itself resided in the principle of relativity, the need to stipulate isotropy of space for one signal speed and the postulate of the invariance of the speed of light. There were no dynamical hypotheses about the nature of matter or how matter interacted with ether. It is in that sense that the special theory of relativity may properly be called a theory of measurement.

Certainly this appears to have been Einstein's view. In 1907 the young physicist Paul Ehrenfest published a paper in which he noted that Max Abraham had analyzed the motion of the deformable electron like the one proposed by Lorentz and had concluded that, under the influence of internal and external electromagnetic fields, it would be impossible for such an electron to move inertially. It would accelerate, tumble or otherwise move erratically. Ehrenfest raised the question of whether or not the same difficulty would apply to Einstein's theory. If so, Ehrenfest continued, further hypotheses would have to be introduced into Einstein's theory. Einstein's reply followed immediately.

> The principle of relativity or, more precisely, the principle of relativity together with the principle of the constancy of the velocity of light, is not to be interpreted as a "closed system," not really as a system at all, but rather merely as a heuristic principle which, considered by itself, contains only statements about rigid bodies, clocks, and light signals. Anything beyond that that the theory of relativity supplies is in the connections it requires between laws that would otherwise appear to be independent of each other.*

We recognize, in the first part of this statement, Einstein's assertion that the theory of relativity is a theory of measurement ". . . containing only statements about rigid bodies, clocks and light signals." In the second part, when Einstein states that the only other thing the theory of relativity supplies is ". . . connections . . . between laws that would otherwise appear to be independent of each other. . . ." he is insisting that the theory is descriptive even when it deals with seemingly dynamical situations. That is, even when pertaining to such dynamical quantities as force, energy, or momentum, the theory remains a kinematical, descriptive theory.

In this chapter we give an example of how the theory of relativity related two seemingly independent laws: the conservation of energy and the conservation of

* P. Ehrenfest, "Die Translation deformierbarer Elektronen und der Flächensatz, " *Ann d Phys,* 1907, *23*:204-206. A. Einstein, "Bemerkungen zu der Notiz von Hrn. Paul Ehrenfest: Die Translation deformierbarer Elektronen und der Flächensatz," *Ann d Phys,* 1907, *23*:206. The issue is discussed in M. Klein, *Paul Ehrenfest* (Amsterdam, 1970), p. 151. The translation is by Klein. See S. Goldberg, *Early Response. . . . ,* Chapter 2.

mass. The relationship that ensues, $E=mc^2$, perhaps the most famous equation in all of physics, connects energy and mass but without making claims about the nature of matter, radiation, or how they interact.

In addition to the relationship between energy and mass, we will consider two more concepts in this chapter — the so-called clock paradox of relativity and the four dimensional analysis, which best illustrates the important role that invariants can play within the theory of relativity. These three concepts have received a great deal of attention as a result of Einstein's special theory of relativity. They are examples of concepts which, while latently available within the formal framework of Lorentz's analysis, were not the subject of study within that theory, at least not until after the theory of relativity had brought the concepts into some prominence.

I have chosen these three concepts to illustrate the heuristic power of the theory of relativity. By no means do they exhaust that heuristic power. One of the reasons they were chosen for study here is that in each case, perhaps because the results violate our intuitive notion of the way these things should work, these concepts have received a great deal of attention in the popular literature. Unfortunately, much of that publicity has been misleading and distorted, not only of content but of the nature of the entire theory.

MASS AND ENERGY

We have seen the formula I will be examining many times in many places.

$$E = mc^2$$

It is the most famous equation of twentieth century physics. It is also one of the most widely misinterpreted concepts in physical science.

Before Einstein's theory, the relationship between mass and energy had been examined by several physicists who were committed to an electromagnetic view of nature. The physicists include J. J. Thomson from Great Britain, Max Abraham from Germany, the Austrian Fritz Hasenöhrl, whose career was cut short by his death at the front lines during World War I, and the American D. F. Comstock, who was later to be a central figure in the development of the Technicolor process.

All of these analyses were made by examining the dynamics of a trapped sample of radiation in a container with perfectly reflecting walls, a so-called blackbody, while the container was in motion. Figure 1 is a diagram of such a container. H is a small hole which allows some electromagnetic radiation to enter the body. Note that, since the area of the hole is so small in comparison to the area of the container itself, the chances of the radiation finding its way out of the container again are also very small. We have created a device which reflects back virtually none of the radiation which falls on it, thus the name, blackbody. In the nineteenth century this theoretical device was important to the development of thermodynamics and

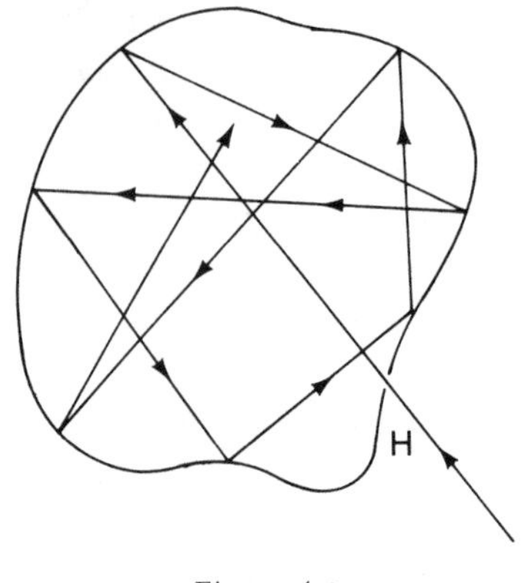

Figure 4.1

statistical mechanics. It had been shown, for example, that the energy distribution of such a body was independent of the nature of the material the walls were made of or of what might be in the cavity. It was dependent only on the temperature of the body.

A good approximation to a blackbody is a blast furnace with a small opening in one side. Measurement of the distribution of radiant energy coming through the opening could be used to estimate very accurately the temperature inside the furnace. Radiant energy is emerging from the furnace and that violates the premise that no radiation emerges, but if the opening is very small, errors are of no practical importance.

The analyses of the relationship between energy and mass by Thomson, Hasenöhrl, Abraham and Comstock rested on the assumption that the radiation moved through an absolutely resting ether. It was also necessary to make certain assumptions about the nature of the interaction between radiation, the ether, and the material the container was made of. All of these analyses were based on the assumption that radiation energy, light for example, possessed momentum and could exert pressure on objects which it struck. It had been Max Abraham who had been the first to propose a quantitative relationship for what he termed *light pressure.* Light pressure is still a respected idea in physics. It is the reason for the orientation of the tail of a comet as it approaches and recedes from the sun. It is often proposed as the power source for certain kinds of satellites which would erect mirrorlike sails to catch sunlight in a manner analogous to the catching of the wind by sails of a sailboat.

Between the years 1880 and 1907 J. J. Thomson, Hasenöhrl, and Comstock had given equivalent analyses of the relationship between the inertia of radiation trapped in moving blackbodies (the radiation has inertia if it exerts a pressure on the walls of the container) and the energy of the system. If the inertia of the trapped radiation is labeled m, and the energy of the radiation is given by E, then according to these analyses

$$E = \frac{3}{4} mc^2$$

where c is the symbol for the speed of light. This equation was generally interpreted as representing the mass content of the energy of the radiation,

$$m = \frac{4E}{3c^2}$$

where "m" was interpreted as representing "the electromagnetic part of the mass of bodies and of radiation." In the light of Kaufmann's experiments designed to reveal the electromagnetic mass of electrons (Chapter 3), it was tempting to conclude that all mass is electromagnetic in origin.

Einstein's analysis of the relationship between energy and mass was not included in his first paper on relativity. It appeared, almost as an afterthought, in the same journal later in the year.* In this note Einstein considered the radiation of energy from a body as seen in two frames of reference. In one frame, in which the body is at rest, equal amounts of energy were radiated in opposite directions. Comparing the conservation of energy in that frame of reference and another inertial frame of reference Einstein was able to show that

$$m_0 - m = E/c^2 \tag{1}$$

where m_0 is the mass of the body before energy radiation, m is the mass of the body after energy radiation and E is the total amount of energy radiated. While it must again be emphasized that Einstein arrived at this result by making no assumptions about the nature of the body, the nature of the radiation or how they interacted with each other or with the measuring instruments, there is no question that he was aware of possible implications.

> *If a body gives off . . . energy . . . in the form of radiation, its mass diminishes. . . .* The fact that the energy withdrawn from the body becomes energy of radiation evidently makes no difference, so that we are led to the more general conclusion that the mass of a body is a measure of its energy content. . . . It is not impossible that with bodies whose energy-content is variable to a high degree (e.g., with radium salts) the theory may successfully be put to the test.†

* A. Einstein, "Does the Inertia of a Body Depend Upon its Energy Content?" in Einstein *et al*, *The Principle of Relativity* (New York, 1923).

† *Ibid.* emphasis in original.

Einstein's comments about this subject are an interesting object lesson for those who are concerned about the relationship between science and technology. It is a common assumption that the inspiration for technological innovation depends on the insights of scientific research and knowledge. On the other hand, there is no question that for most of written history, technological innovation proceeded essentially outside the purview of science and depended only on earlier technology.

Since we keep changing our minds about the fundamental nature of matter and the nature of the universe, it is difficult to maintain the position that scientific knowledge is responsible for technological *progress*. In fact, more often than not, it is technological innovation which makes doing science possible because it provides the means for asking questions which could not have been asked earlier with any expectation of gathering empirical evidence. On the other hand, it would be absurd to maintain that science never provided the inspiration for technological innovation.

In this context it is significant that Einstein proposed that radioactive processes might reveal the connection between mass and energy suggested by the theory of relativity. At the time there was no such field as nuclear physics and the possibility of nuclear fission would not see the light for another thirty years. It is conceivable that the atomic bomb could have been developed regardless of beliefs at the time about the fundamental nature of matter and regardless of hypotheses about the building blocks of matter and the processes to which they are subject. On the other hand, the original suggestion that matter and energy were interconvertible emerged from a program of scientific research. When the details of the problems confronting the builders of the atomic bomb are examined, we find that it was technological problems which had to be solved before a nuclear device could be made. The solution of those problems spawned new areas for the study of science.

Regardless of what one believes about these issues, given the general nature of Einstein's proposal and that his only role in the development of the atomic bomb was to participate in alerting the American government to the possibilities of building such devices, referring to Einstein as the father of the atomic bomb makes no more sense than if we were to refer to Isaac Newton as the father of the intercontinental ballistic missile.

Whereas Einstein arrived at the result $E = mc^2$ in 1905, it was his scientific patron, Max Planck, in 1907, who first derived the result directly from the postulates of relativity without reference to a particular physical system, and in the most general terms. That derivation was made in a debate that Planck and his students engaged in with Fritz Hasenöhrl about the proper assumptions to use in such considerations. We will consider the details of that dispute in Chapter 6 when we examine the German response to the theory of relativity. For now, let us proceed to a modern abstract treatment that leads to the result $E = mc^2$ and allows us to see the relationship between the two postulates of relativity and that equation.

Consider again the collision of two masses as observed in two different inertial frames of reference, and recall that the proper mass, which has a value m_0, is related to the measured value of that same mass, m, in any other frame of reference by

$$m = \frac{m_0}{\sqrt{1 - v^2/c^2}} \tag{2}$$

where v is the relative speed between the two frames of reference and c is the speed of light.

Recall also that earlier, we traced this disagreement to disagreements in the measurement of time which themselves result from disagreements in judgments of simultaneity. At the same time, the principle of relativity requires that each observer maintain his commitment to the law of the conservation of momentum. Without further physical reasoning, the result in (1) can be directly derived from (2) by algebraic manipulation.

The expression

$$\frac{1}{\sqrt{1 - v^2/c^2}}$$

can be written as $1/(1 - v^2/c^2)^{1/2}$ and that, in turn, may be approximated by $1 + \frac{1}{2}v^2/c^2$.*

We use this approximation in equation (2) above, and it becomes

$$m = m_0\,(1 + \tfrac{1}{2}v^2/c^2) = m_0 + (\tfrac{1}{2}m_0v^2)/c^2 \tag{2a}$$

The second term in the last equation has been organized in this way to emphasize the appearance of the expression $\frac{1}{2}\, m_0v^2$ which is termed the kinetic energy of the mass. In Newtonian mechanics applying a force on an object of mass m_0 over a certain distance leads to an increase in the kinetic energy. In all collision processes, if the collisions are perfectly elastic the kinetic energy is conserved. If the collision is not perfectly elastic and the kinetic energy is not conserved, we look for the appearance of energy in other forms—for example, in the deformation of the objects involved in the collision and in the generation of a certain amount of heat energy. It was these kinds of considerations which led to the classical concept of total energy and the emergence of the law of the conservation of energy.

If we call the kinetic energy of the particle K, equation (2*a*) representing the

---

* This approximation has been discussed in a footnote in Chapter 3.

relationship between the nonproper and the proper mass may be rewritten as

$$m = m_0 + K/c^2$$

or

$$m - m_0 = K/c^2 \tag{3}$$

Comparing equation (3) to equation (1) which Einstein and then Planck had derived we see that the difference between the nonproper and the proper mass of an object is proportional to the kinetic energy of the particle. We arrived at this conclusion by algebraic manipulation, but we were manipulating symbols that are endowed with physical meaning. Let us manipulate the equation once more:

$$mc^2 - m_0c^2 = K \tag{4}$$

If we are bold enough to assume that the result of algebraic manipulation in fact has physical meaning, we would say that the term mc^2 represents the total energy of an object, m_0c^2 represents the rest energy of the body and the difference between the total energy and the rest energy is the kinetic energy of motion. Suppose that the object is a billiard ball, which after rolling on a billiard table comes to rest. The energy in excess of the rest mass will then have disappeared. Where might that energy have gone? Into frictional heating of the table and the ball, is the obvious answer. Or consider nuclear fission. The rest mass of the particles after the event are less than the rest mass of the initial uranium atom. Where has the energy gone? Presumably it was the energy responsible for holding the resulting pieces together and appeared at the time of fission in the kinetic energy of the fragments.

Consider a similar case. Two masses having identical rest mass m_0 collide inelastically, that is, rather than rebounding after the collision, they stick together. For example, suppose the masses are lumps of wet clay. Observers in different frames of reference will disagree about the value of the masses prior to the collision and the combined mass after the collision. However, all observers in all frames of reference will agree about the rest (or proper) mass of each object m_0 before the collision and about the rest mass M_0 afterwards (Figure 2). And all observers will agree that the rest mass after the collision is greater than the sum of the rest masses before the collision by an amount represented by the kinetic energy that each of the masses m_0 lost in the collision process. In other words:

$$M_0 = \frac{2m_0}{\sqrt{1 - v^2/c^2}}$$

where v is the speed of each of the particles.

We have used objects described only "like lumps of wet clay." The result is the inverse for the exotic case of nuclear fission. These examples give insight into the

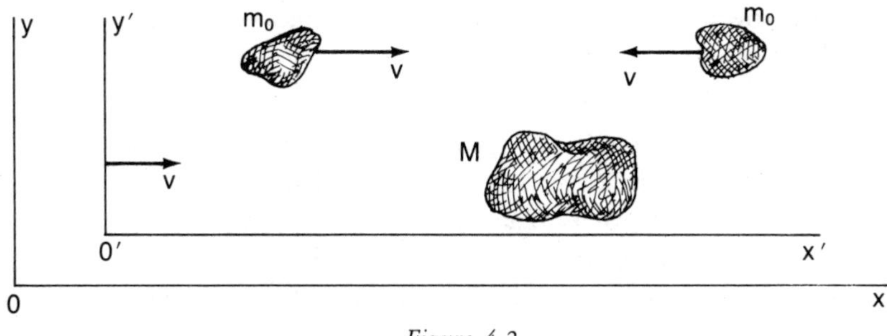

Figure 4.2

relationship of the theory of relativity to such situations. The theory predicts the kinds of circumstances under which energy and mass are exchanged but gives no insight into the nature of the process.

THE CLOCK PARADOX

The clock paradox, sometimes known as the twin paradox, has been known since Einstein's first paper on relativity. It was the subject of intense debate in the literature and provided inspiration for wonderful flights of fancy by writers of science fiction. There are several fascinating aspects of the problem and of the history of the analysis as well. Strictly speaking, the clock paradox is not within the purview of the special theory of relativity because it involves accelerations of at least one frame of reference. Nevertheless, it is possible to treat the problem as a limiting case within the special theory and the results are identical to those of the general theory of relativity (Chapter 5).

Einstein made note of the clock paradox in his first paper on relativity immediately following his discussion of the difference between the proper and non-proper measurement of the time interval between two events. His discussion is quoted in detail:

> . . . . If at the points A and B of frame of reference K there are stationary clocks which, viewed in the stationary system, are synchronous; and if the clock at A is moved with the velocity v along the line AB to B, then on arrival at B the two clocks no longer synchronize, but the clock moved from A to B lags behind the other which has remained at B. . . .
>
> It is at once apparent that this result still holds good if the clock moves from A to B in any polygonal line, and also when the points A and B coincide.
>
> If we assume that the result provided for a polygonal line is also valid for a continuously curved line, we arrive at this result: If one of two synchronous clocks at A is moved in a closed curve with constant velocity

> until it returns to A, the journey lasting t seconds, then by the clock which has remained at rest the travelled clock on its arrival at A will be $\frac{1}{2} v^2/c^2$ seconds slow.*

The notion of a clock can be quite general. A heartbeat, for example, is a kind of clock, as are other metabolic processes. Therefore, the statement by Einstein can be transferred into the following situation: It is "at once apparent" that if we have identical twins and one of them follows any closed polygonal path, eventually returning to the starting point, he or she will have aged less (aging is a clock-like process) than the twin who stayed at home. This idea can be turned into the following problem: Identical twins are selected and one is placed in a rocket and sent from the earth at a speed close to the speed of light. After spending some earth years traveling through space, the twin in the rocket returns, and it is found that he or she has aged considerably less than the twin who remained at home.

There are several remarkable features about Einstein's statement of this problem. First, it was not "at once apparent" to most scientists. Since Einstein wrote his remarks in 1905, there have been many intense exchanges in the literature between physicists and philosophers who have argued the implications of the twin paradox. It becomes clearer that it is not an issue within the special theory of relativity. That theory is restricted to the consideration of measurements of events within *inertial* frames of reference. If a clock or a person follows "any closed polygonal path," he will change direction. Such a problem is properly handled by the general theory of relativity. In order to avoid the problem of accelerations of the frame of reference, we will assume that acceleration which a frame of reference undergoes occurs in an inconsequential time when compared to the total time. The magnitude of the acceleration may be very, very, great, but the assumption is justified because the result agrees with the result when the problem is rigorously analyzed, using the general theory of relativity.

Why is the problem called the twin or clock *paradox*? What is paradoxical? When critics of Einstein's statement of the paradox began speaking out, they argued in the following manner: The analysis from the point of view of one twin could, with perfect justice within the theory of relativity, have been done from the other perspective. For example, the twin who was placed in the rocketship observes that the earth suddenly departs from him or her. After a long period, the earth returns. Therefore, from his or her perspective, it was the earthbound twin who "followed any closed polygonal path whatsoever" and it is that twin's clocks

* A. Einstein, "On the Electrodynamics of Moving Bodies," in A. Einstein *et al*, *The Principle of Relativity* (New York, 1923). A more recent translation by A. I. Miller is the same for these paragraphs. A. I. Miller, *Albert Einstein's Special Theory of Relativity: Emergence (1905) and Early Interpretation (1905-1911)* (Reading, Mass, 1981).

and aging processes which should have shown less elapsed time. Since it is not possible that both twins have aged less, they must remain the same age.

The dispute about these issues became very strong during the 1950s. During that period, the chief protagonists were the physicist W. H. McCrea and the physicist philosopher Herbert Dingle.* Dingle argued that according to the theory of relativity, all things are relative. Therefore, each twin will observe a time dilation of the other twin's clock and they will age the same amount. McCrea argued that, even though strictly speaking, it was a problem to be solved within the general theory of relativity, one could obtain the same result with special relativity. The twin who left the earth and returned would have aged less.

Dingle responded that such a result was at odds with the theory of relativity since all things are relative. McCrea answered that all things are not relative. One of the twins accelerates and one does not. That is, something happens to one twin that doesn't happen to the other. The proof, McCrea continued, is to suppose that on return, the rocket crashes. Clearly all things would not be relative. Something would happen to the rocket twin that would not happen to the earth twin. McCrea then proposed that the experiment be done and that Professor Dingle travel in the rocket ship.

Dingle countered that it was improper to think that one could seriously propose this as a way of extending one's lifetime, therefore, it could not possibly happen. McCrea again responded by showing that, in fact, one would not be gaining by traveling at a high rate of speed for long periods of time and returning to the earth. For example, he continued, suppose that we give two high speed computers a task that required ten years of computation. We send one off in a rocket ship at a speed close to the speed of light and bring it back after ten years' time on earth. The earth-based computer would be finishing the calculation as the rocket returns, whereas the computer on the rocket in whose environment perhaps six years have passed, would have four years of computations to complete.

This can be translated into human terms: Suppose that, wanting to save the genius of young Einstein, we send him on a rocket at a very high rate of speed so that in two hundred years of earth's history, only a few minutes pass by clocks and aging processes on the rocket. In that two hundred years, much technological progress would have been made on the earth, cultural and scientific fashion would have changed several times over, and the still young Einstein would return without friends, relatives, or the two hundred years of civilization on earth.

When Dingle objected that it was immoral to postpone the date of death, the argument lapsed.

---

* The debate took place in 1957-1958 in the pages of *Nature, Discovery,* and *Science.* It has been summarized in H. Samuel and H. Dingle, *A Threefold Cord: Philosophy, Science, and Religion, A Discussion Between Viscount Samuel and Herbert Dingle* (New York, 1961).

The result of this dispute was to realize that one could treat the problem within the special theory of relativity in a fairly straightforward manner. Not only that, but the problem illustrates graphically the fallacy of assuming that within the theory of relativity, all things are relative.

Let us consider another set of twins. One gets into a rocket ship and heads for Alpha-Centuri, four light years away, and the rocket quickly accelerates to a speed equal to ⅘ the speed of light. We have established two inertial frames of reference. One is the rocket frame and the other is not only earth, but the earth-Alpha-Centuri frame of reference. One way to analyze the problem is as follows: From the point of view of the earth-Alpha-Centuri observers, a proper measurement of the distance that the rocket will travel is made. As in all other cases, this proper length will be the longest of the determinations made by observers in every inertial frame of reference for that length. But for observers in the rocket, the earth-Alpha-Centuri distance is given by the contraction formula and is not four light years, but two-and-a-half light years away. As everyone will agree that the relative speed of the two frames of reference is ⅘ the speed of light, as far as rocket travelers are concerned, three years will have passed on the outward journey and three more on return. When the rocket returns, the twin who remained on earth will have aged ten years whereas the rocket twin will have aged only six. Both twins will agree that something strange happened at two points in the journey. The first was when the rocket accelerated, and the second was when the rocket turned around near Alpha-Centuri. That "something strange" can only be handled with complete rigor within the general theory of relativity. Within the special theory of relativity, we see that the difference in aging is the result of how we measure.

Another way of looking at it is as follows. One and only one twin makes a proper time measurement of the events: "rocket leaves vicinity of earth, rocket arrives at Alpha-Centuri" and "rocket leaves vicinity of Alpha-Centuri and rocket arrives in vicinity of earth." That observer is the twin in the rocket. All other observers, for example, observers in the earth-Alpha-Centuri frame of reference, make nonproper measurements of the time interval between these two sets of events. Again, we recognize that there are periods during the acceleration of the rocket when the special theory of relativity cannot be used, where something strange happens. Yet the argument showed that the special theory of relativity could handle the formal argument exactly.

FOUR DIMENSIONAL ANALYSIS

One aspect of the theory of relativity that has always captured the popular imagination is the four dimensional character of the theory. It has not only spawned science fiction accounts of "space-time" existence, but projections by futurists of the nature of the space-time world.

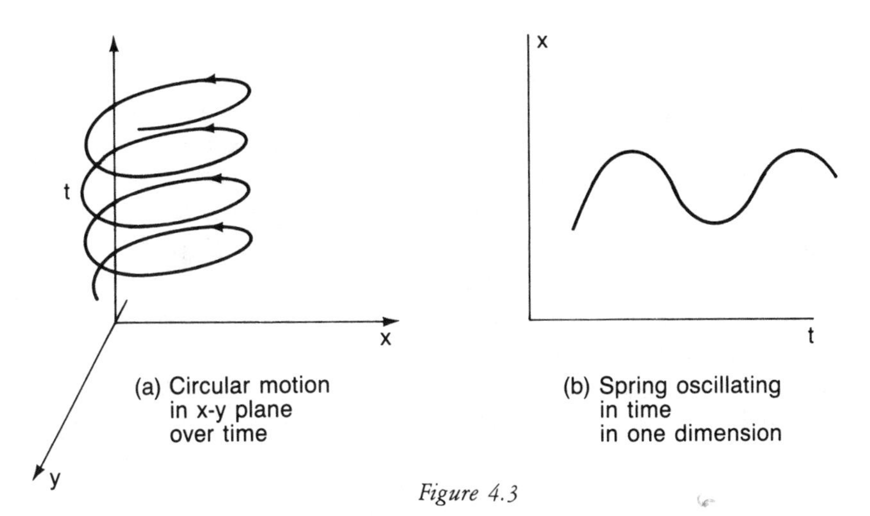

Figure 4.3

It was not Einstein who suggested the four dimensional formulation of the theory of relativity. The idea of four dimensions had long been used for depicting sets of connected events, of which time is a coordinate. Of course, since we are limited to a three-dimensional spatial world, if one of those dimensions is used to depict the passage of time it becomes impossible to depict one spatial dimension. Figure 3 shows several common situations in which time is one coordinate. These depictions are not special and have been with us for a long time.

About the theory of relativity, a more fundamental problem was being proposed. In 1908, Hermann Minkowski, the mathematical physicist who taught mathematics at the Swiss Federal Polytechnic (ETH) when Einstein was a student there, gave a popular lecture on a new relativistic formalism. The main body of the talk was hardly a popularization, as it was a literal translation of the rigorous formalism that had been published earlier. The opening statement of that address was provocative and accounts for the quick spread of the interpretation of the analysis that holds that reality is not lived in three dimensions of space and time, but rather, in a four dimensional space-time continuum. Minkowski's remarks began as follows:

> The views of space and time which I wish to lay before you have sprung from the soil of experimental physics, and therein lies their strength. They are radical. Henceforth space by itself, and time by itself, are doomed to fade away into mere shadows, and only a kind of union of the two will preserve an independent reality.*

* H. Minkowski, "Space and Time," in A. Einstein *et al*, *The Principle of Relativity, op. cit.* p. 75.

It is a remarkable statement. On the surface it appears to herald a revolutionary approach to problems of the description of physical events. But examination of the text of Minkowski's work reveals his view that the development of Einstein's theory was a generalization of Lorentz's and indeed, Minkowski's interest was to understand the dynamics of the interaction of matter and radiation. We have seen that this confusion was a common manifestation of what we earlier termed deep epistemological differences between Einstein and most other physicists.

Even the brief statement I have quoted reveals some of those differences. The suggestion that the views of space and time within the theory of relativity "have sprung from the soil of experimental physics" is not a statement with which Einstein would have been comfortable. And the notion that space alone and time alone would fade into "mere" shadows makes little sense in the perception of ordinary events. Taken as a whole, Minkowski's approach is quite conservative.

The significant aspect of Minkowski's four dimensional analysis is formal. With it one can identify those variables that are invariant, that is, those variables that observers agree have the same value. The formalism also allows presentation of the relationships between variables in a compact, symmetrical, and hence aesthetic manner. We will see, in Part II of this book, that several mathematical physicists first worked on an elaboration of the theory of relativity, not because of Einstein's presentation, but because of Minkowski's. If for no other reason, Minkowski's contribution is important.

Einstein's assessment of Minkowski's contribution was very much in the spirit of what I have said. Einstein pointed out that the notion of a four dimensional space-time continuum had always been a feature of classical physics. The absolute character of the temporal dimension, which had been a feature of classical physics, was removed within the theory of relativity. Einstein said further:

> Minkowski's important contribution to the theory lies in the following: Before Minkowski's investigation, it was necessary to carry out a Lorentz-transformation on a law in order to test its invariance under such transformations; he on the other hand, succeeded in introducing a formalism such that the mathematical form of the law itself guarantees its invariance under Lorentz-transformations.*

Let us take several examples to see how the formalism works. We have seen that within the theory of relativity, the speed of light is postulated to be an invariant. It has the same value for all inertial frames of reference. Suppose we consider a flashbulb located at the origin of a frame of reference. Another frame of reference in motion relative to this frame moves so that the direction of motion is parallel to the X-X' axes. At the instant that the origins O and O' (Figure 4) of the

* A. Einstein, "Autobiographical Notes," *op. cit.* pp. 57-61. The quotation is on p. 59.

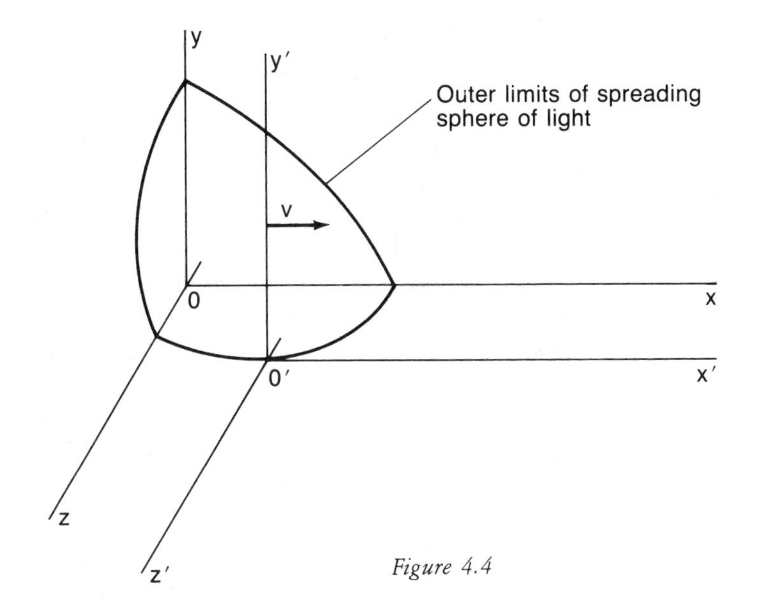

Figure 4.4

two frames of reference coincide, a flashbulb is set off. The question then is how observers in both frames of reference will describe the propagation of light. The situation is described in Appendix 6 where we derived the Lorentz transformations.

Here, we change the emphasis slightly. Although the two frames of reference are in motion relative to each other, they both describe the effects of the explosion of the flashbulb as a spreading *sphere* of light of radius ct in one frame of reference and of radius ct' in the primed frame of reference. The equation of that sphere in each of the frames of reference is as follows:

$$x^2 + y^2 + z^2 = c^2t^2$$

and

$$x'^2 + y'^2 + z'^2 = c^2t'^2$$

Those are both the equations of spheres in which ct and ct' represent the radii of the spheres. We can rearrange each of these equations as follows:

$$\begin{aligned} c^2t^2 - (x^2 + y^2 + z^2) &= 0 \\ c^2t'^2 - (x'^2 + y'^2 + z'^2) &= 0 \end{aligned} \qquad (5)$$

In other words, this particular combination of temporal and spatial coordinates will have the same value, zero, in all inertial frames of reference for the events connected with the propagation of light from the flashbulb. This combination is an invariant and is termed "the fundamental invariant." Classically, all observers agreed on the length of a rigid body. As we will see in Chapter 6, trying to understand the concept of rigidity within the theory of relativity was a key issue which surfaced in the first few years after the theory was introduced. The classical notion of a body whose dimensions are invariant, is not acceptable since the dimension in the direction of motion of the object depends on the frame of reference in which that dimension is being measured. However, although observers will disagree on the dimensions and time intervals, they will agree on the value of the fundamental invariant connecting spatial and temporal coordinates for the events associated with the propagation of light signals. Furthermore, if the right side of equation (5) is not zero, the events are not associated with the propagation of a light signal but with another process. For example, suppose that in one frame of reference the combination is:

$$c^2t^2 - (x^2 + y^2 + z^2) = 5.$$

That combination of variables will have the value 5 in *all* inertial frames of reference. Minkowski's four dimensional formulation of the theory of relativity insures that. The fundamental invariant serves in the formal four dimensional analysis as a type of length with three spatial dimensions and one temporal dimension. As all inertial observers agree on the magnitude of the fundamental invariant, these observers are part of one four dimensional super frame of reference, for the purposes of the analysis.

This suggests defining a new set of universal coordinates which makes use of light signals. This new frame of reference is sometimes referred to as the *light cone.* It is depicted in Figure 5. Although it is sometimes given fanciful interpretations,

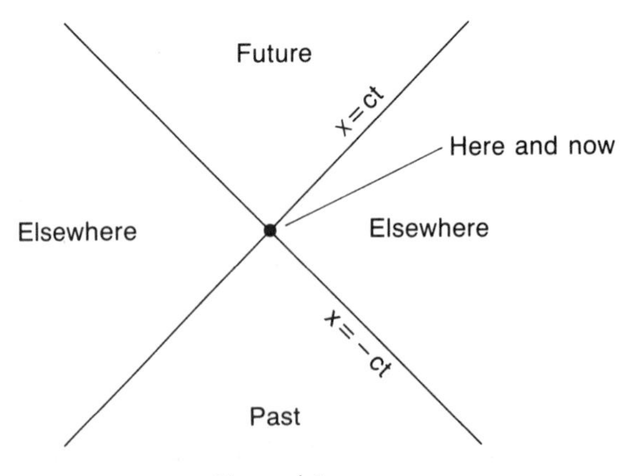

Figure 4.5

it has definite physical implications which give further insight into the meaning of the theory and illustrate the uses of four dimensional considerations.

How is such a diagram to be interpreted? Consider a point in space labeled in the diagram, "here and now." Each such point, and space is filled with an infinite number of these points, may be the potential source of a pulse of light. In Figure 5, a light cone is shown emanating from one "here and now" (We have omitted two spatial dimensions). The cone is defined by the lines labeled $x = ct$ and $x = -ct$. In other words, these would be the spatial points reached by the light pulse in a time t in a frame of reference Σ. All points within the cone are points such that $c^2t^2 - x^2$ is negative. That is, they are points at which a light signal could arrive from the point labeled "here and now" in a time less than t and could be causally related to the point at the "here and now." They are, therefore, within the future of that point because the point at the "here and now" could have a causal effect on all such points. All points outside of the light cone are such that $c^2t^2 - x^2$ is positive. They are points which could not be arrived at by the light from "here and now" in a time less than or equal to t. As light is the fastest signal speed, it is impossible for an event that occurred at the "here and now" to be the cause of events at such points because it is not possible for the information to arrive in a time equal to or less than the time between the events at the "here and now" and these particular points. Because these points cannot be in the causal path of the "here and now," they are labeled "elsewhere." For example, when we synchronize two clocks, one of the clocks is at the "here and now" point and the other is in "elsewhere." All points within the half of the light cone labeled "past" could be reached by a light signal in the time less than t. They could have been the cause of an event at the "here and now" and they are potentially within the past of the "here and now." A diagram may be drawn for each point in space. This description may sound bizarre because of the language, but it is a summary of the information on spatial and temporal measurements which has been discussed.

If the only use of the four dimensional formulation of the theory of relativity were the fundamental invariant, it would not have had such a profound impact. The fundamental invariant in space-time analysis joins the separate dimensional and temporal measurements of objects. There are a host of other invariants that have physical significance. For example, there is a quantity known as the *energy-momentum four vector.* We have seen that observers will agree that the momentum is conserved when moving bodies collide although each frame of reference will differ about the momentum ascribed to each body. We have also seen that observers will agree about the relationship between the mass of an object and the energy content of the object, and that, furthermore, observers agree about an invariant quantity we have called the *rest,* or *proper mass.* Also, observers in all frames of reference will agree on the value of the combination of momentum and energy given by

$$p^2c^2 - E^2$$

and known as the energy momentum four vector, an invariant. It can be treated, as the fundamental invariant is, as a four dimensional vector having three dimensions of momentum and one dimension of energy. The momentum components correspond to the three spatial coordinates of the fundamental invariant and the energy dimension corresponds to the temporal coordinate of the fundamental invariant. Although observers in different inertial frames of reference disagree on the magnitude of the momentum and of the energy, they agree on the magnitude of the energy-momentum four vector.

Similar analyses may be made very simply, for such variables as velocity, force, magnetic and electric field quantities, etc., defining four vectors in each case which contain components corresponding to three spatial and one temporal dimension. In each case, the four-vector is invariant.

While performing this analysis we add nothing new to the meaning of the theory of relativity. It remains a theory of measurement that begins with assumptions about isotropy of space with respect to one signal speed. We have made no new assumptions about the momentum, energy, force, velocity, distance, or time. We have not introduced any dynamical assumptions about the nature of matter or radiation. We have simply recast the theory into a convenient mathematical language that allows us to discover the important property, invariance, in combinations of variables that are otherwise, alone not invariant. The four dimensional formulation also introduces an extremely economic statement of the theory, which to a reader familiar with the language of linear algebra, is symmetrical and more aesthetically pleasing.

·5·

The General Theory of Relativity

THE WORD "SPECIAL" IN "SPECIAL THEORY OF RELATIVITY" means that the theory is restricted to those frames of reference for which the law of inertia holds, that is, inertial frames of reference.* Einstein sometimes referred to them as Galilean frames of reference. At first glance, it does not seem reasonable to expect that the principle of relativity can be extended to non-inertial frames. The reasons are fairly obvious. The principle of relativity that we have been using states that the laws of physics have the same form in all frames of reference to which the principle applies. In other words there should be no way to determine by an experiment which of two frames of reference is in motion relative to the other.

On the other hand, should the frame of reference experience acceleration relative to some other frame of reference, observers or instruments could immediately decide which of the two frames was *really* accelerated. Consider once again the train traveling past the embankment. If the motion of the train is uniform, the question of which is moving has no answer. But if the brakes of the train are suddenly applied, the observer on the train will lurch forward. The observer's counterpart on the embankment will not lurch. Instruments on the train will detect accelerations to which observers will ascribe forces not experienced by instruments on the embankment.

* The following treatment very closely follows Einstein's own treatment in *Relativity: The Special, The General Theory* Chapters 18-32.

It would seem, then, that it is not possible to extend the principle of relativity beyond inertial frames of reference, to any frame of reference. Yet, as Einstein noted in 1917,

> Since the introduction of the special principle of relativity has been justified, every intellect which strives after generalization must feel the temptation to venture the step towards the general principle of relativity.*

For Einstein, the process of "venturing the step" began in 1907 when he tentatively suggested the principle of equivalence. That was the first step in a series that culminated, near the end of 1915, with the completed general theory of relativity.

The principle of equivalence states that it is impossible to distinguish between the effects of acceleration and of gravitational fields. They are equivalent. Therefore, what might be ascribed to a gravitational field could be ascribed to acceleration of the whole frame. A necessary condition for this principle of equivalence is the equality of gravitational and inertial mass. Einstein stipulated gravitational and inertial mass as identical.

In Newtonian mechanics, that gravitational and inertial mass are the same is contingent; it had not come about from a reasoned argument. The equality *was* tested in the nineteenth century. That equality is the starting point of Einstein's general theory of relativity in much the manner that the stipulation of the isotropy of space with respect to the speed of light was the starting point for the special theory of relativity.

In Chapter 2, we saw that the definition of inertial mass was a key element in the elucidation of Newton's second law or axiom of motion. Gravitational mass, on the other hand, determined the force exerted by one object on another when the objects were separated in space.

For Einstein the field concept was far more physical than an abstract notion of action at a distance. Without the concept of field, the effects of a magnet on iron, for example, must remain mysterious. If one assumes that the magnet somehow affects the space, it then becomes comprehensible that iron is attracted to the magnet or that the like pole of another magnet is repelled by the first magnet. It was not necessary that one evoke a fluid or solid such as the ether to fill the space. The concept of "magnetic field" is a precise formulation of "the magnet affects space." In magnetic and electric fields, the acceleration of objects depends on their physical state (what they are made of, how they are charged or magnetized). In a gravitational field the acceleration of objects is independent of their physical state

* *Ibid.* p. 61.

or what they are made of.* The gravitational force is always attractive and depends only on the position of the object relative to any other object.

According to Newton's second law, the acceleration of a body is directly proportional to the net force exerted on the body. Double the force and the acceleration is doubled, and so on. The constant of proportionality is the body's inertial mass. If we represent the force by the symbol $\vec{F}$, the mass by the symbol m_i, and the acceleration by the symbol $\vec{a}$ we write this law as

$$\vec{F} = m_i\vec{a}$$

In the gravitational case, the force on the body is given by the product of what might be termed the gravitational mass and the intensity of the gravitational field.† In symbols this can be written as follows:

$$\vec{F} = m_g\vec{E}$$

where $\vec{E}$ represents the intensity of the gravitational field.

For a force of given magnitude the right side of these two equations may be equated as follows:

$$m_i\vec{a} = m_g\vec{E}$$

or

$$\vec{a} = (m_g/m_i)\ \vec{E}$$

Of this result, Einstein said,

> If now we find [as we do] from experience, the acceleration is to be independent of the nature and condition of the body and always the same for a given gravitational field, then the ratio of the gravitational to the inertial mass must likewise be the same for all bodies.

By suitable choice of units, it is possible to *set* the ratio of the gravitational to inertial mass at unity. That is, we *begin* with a law that the inertial mass is equal to the gravitational mass. This law may be interpreted as stating that the same quality of a body manifests itself as inertia or as weight, depending on the circumstances. That is a necessary condition for the principle of equivalence which states that the effects of gravitation cannot be distinguished from those of acceleration.

With the principle of equivalence in place as a postulate, it is impossible by an experiment to determine the absolute motion of any frame of reference, inertial or

* This ignores the fact that the mass of an object depends on its energy content.

† The source of the gravitational field is other mass, just as the source of electric fields is charge.

otherwise. Non-inertial effects are attributable either to accelerations or to the presence of gravitational fields.

A thought experiment reveals how that works. Consider two observers far removed in "empty" space. One observer is inside an elevator-like box and the second observer may peer in but the observer within cannot see out. At the top of the box is a large hook with a rope attached. The rope pulls up on the box so that the box accelerates upward at the rate of 32 ft/sec^2 (9.8 meters/sec^2). As soon as this motion begins, the floor will come up to meet the observer who finds that to stand erect, he or she will have to exert a force on the floor of the box with his or her legs. Any object, regardless of the material it is made of will, when released, fall toward the floor with an apparent acceleration of 32 feet/sec^2. In short, any effect which the observer outside ascribes to the acceleration of the frame, the observer inside ascribes to being in a gravitational field. Even if the observer inside the box sees the hook and the rope, he or she concludes that the box is suspended by the rope within the same gravitational field.

The extension of the principle of relativity is based on the equivalence of accelerating frames of reference and gravitational fields. This equivalence depends on the equality of gravitational and inertial mass. However, at this point in the development (1911), it was not possible to choose an alternative accelerating frame of reference to account for *all* types of gravitational fields. The uniformly accelerating elevator corresponds to a uniform gravitational field like the one depicted in Figure 1a. The accelerating elevator thought experiment is not enough to account for a gravitational field, such as the earth's, which has radial symmetry (Figure 1b).

There were four more years of struggle for Einstein before he was able to work

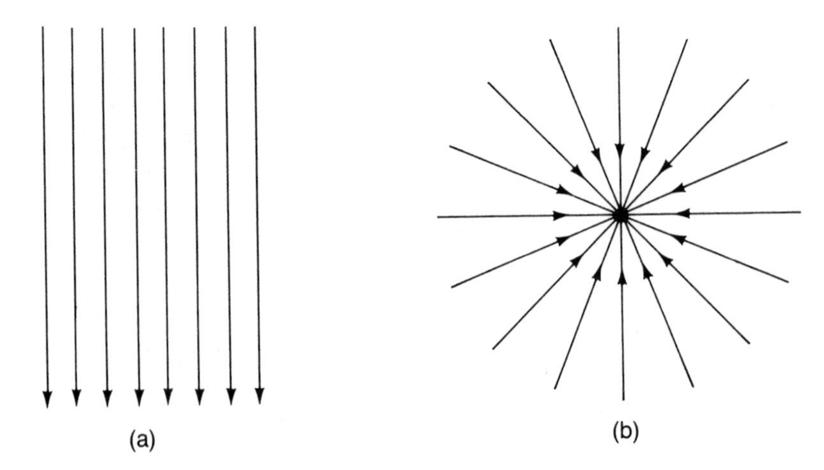

Figure 5.1

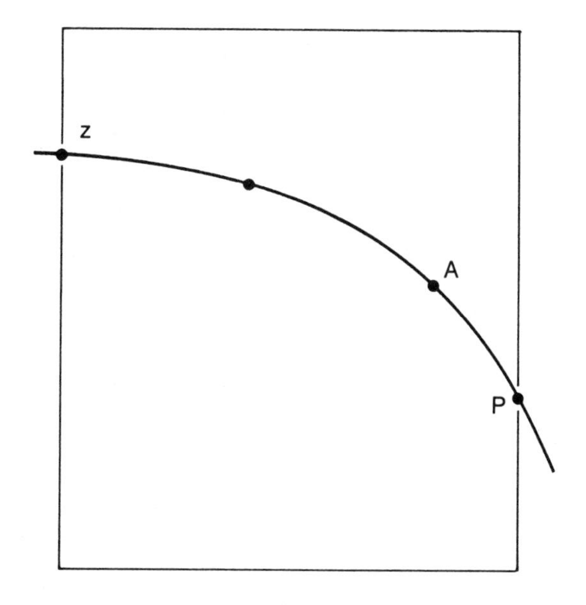

Figure 5.2

out the principles and details of a formalism to allow accelerations and gravitational fields of any kind to be interchanged*. Even at this stage, he developed crucial conclusions from his premises. For example, consider once again the elevator accelerating upward. The observer inside the elevator makes a tiny hole in the wall to admit a beam of light. As depicted in Figure 2, when the light enters at *Z*, and strikes the wall on the other side, the elevator would have moved up so that the light strikes the far wall at *P*.

The elevator is not moving uniformly, but is accelerating. Therefore, it is easy to show, say by temporarily intercepting the light in intermediate planes between *Z* and *P*, that the light would strike points such as *A*. Given that in successive equal time intervals, the distance traveled by the frame of reference *increases* at a constant rate and that, in the perpendicular plane to that motion, the speed of light would be uniform, the path of the light within the box would be *ZAP,* a parabola.†

* The required step was to realize that the principle of equivalence cannot be applied over finite regions of space. It must be applied only to infinitesimal regions.

† Points on the trajectory of the light beam are determined by the resultant of two vectors, one which remains constant and a second, perpendicular to the first, which increases in proportion to the time elapsed. This situation is identical to the situation Galileo described for the case of projectile motion. On the other hand, even if it turned out that the speed of light were not uniform, while the trajectory would no longer be parabolic, neither would it be a straight line.

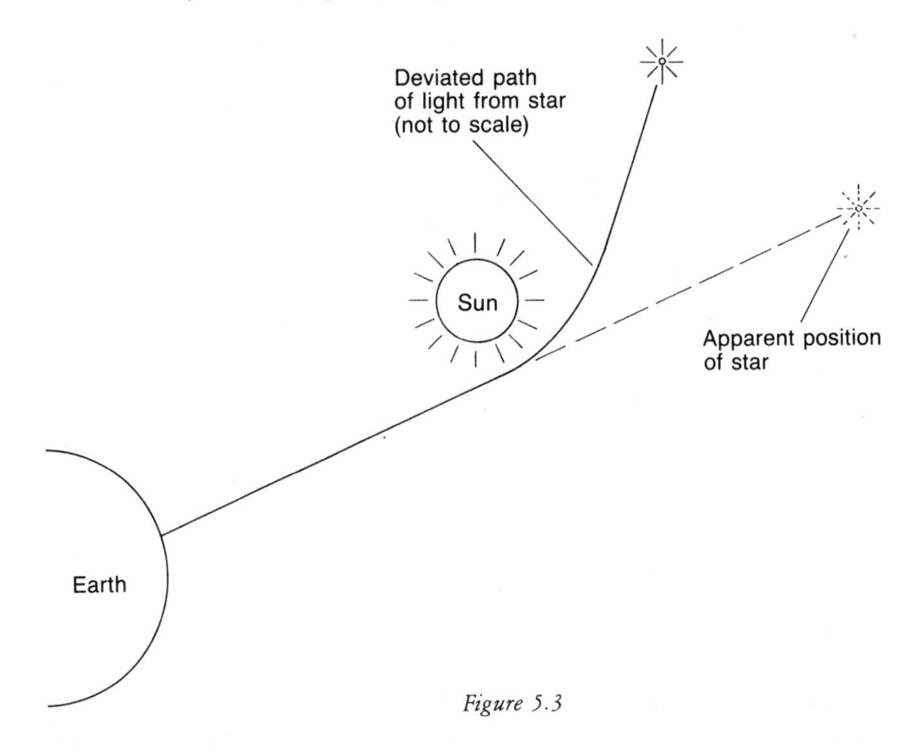

Figure 5.3

From outside the elevator, this effect is ascribed to acceleration of the frame of reference. From within the frame of reference it is ascribed to the effects of a gravitational field. Because in the general theory of relativity gravitational fields are locally equivalent to accelerations, one expects light rays to be bent in the presence of gravitational fields, and the stronger the field, the more the effect.

Once Einstein had perfected the formal structure of the general theory of relativity, he predicted, for example, the effect of the gravitational field of the sun on a beam of light traveling from stars beyond the sun toward the earth. The effect is the apparent displacement of the star when the sun is near the path of the light from the star to the earth (Figure 3).* Of course, it is not normally possible to see the stars in the vicinity of the sun. But during that very brief period of full solar eclipse, such observations are possible. And according to Einstein's calculation the displacement of stars near the sun should be great enough to be measurable. It was the corroboration of this prediction, from measurements taken from photographic plates exposed during the solar eclipse of 1919, that electrified the world and catapulted Einstein into the role of world celebrity.

* Newtonian theory also makes such a prediction, but the value of the apparent displacement of the star would be half that predicted by Einstein.

But before 1915, there were still difficulties to be worked out. We imagine two observers, one in as near to an inertial frame of reference (k) as is possible and the other in a frame of reference shaped like a flat disk (k') which is rotating. From the perspective of k, the angular velocity of the disk means that at any point on the disk, there is a centripetal acceleration toward its center in proportion to how far the point is from the center. The tangential speed of any point on the disk is directly proportional to the distance from the center. The inertial observer concludes that the rate at which clocks run on the disk is slower and slower relative to k as one moves from the center of the disk to the edge. When the observer in k' measures the circumference of the disk with a standard rod, whose length was small in comparison to the overall circumference, he or she concludes that because of the length contraction, the ratio of the circumference of the circle to the diameter is not equal to pi (3.14159. . . .) as demanded by Euclidean geometry. Because the diameter of the disk is always perpendicular to the direction of motion, the length of the diameter is not contracted.

To the observer in k, outside the disk, all of this is ascribable to an acceleration. But accelerations are equivalent to gravitational fields. Therefore, to the observer in k', who experiences a gravitational force the intensity of which increases in proportion to the distance from the center of the disk, the differential rate at which clocks run in his or her universe would be ascribable to the effects of a gravitational field and, as we have already demonstrated, space would be strangely non-Euclidean to a degree determined by the strength of the gravitational field. The principle of equivalence forces us to conclude that gravitational fields not only affect the path (and hence the speed) of light beams, they also affect the rate at which clocks run and the geometry of space.

We are not out of the woods yet. Clocks, no matter how identically built, are not guaranteed to run at the same rate. And we cannot assume Euclidean properties for space. What then of our elaborately built theory of measurement (special relativity) that depends on specifying spatial and temporal coordinates for events?

In order to overcome this new set of paradoxes, Einstein made use of mathematics of a more general character than he had hitherto used. Rather than the analytical geometry we associate with Euclid and Descartes, he adopted the non-Euclidean differential geometry of Gauss's generalized coordinates.* By

* Gauss, who was *the* premier mathematician of the first half of the nineteenth century, was one of the inventors of non-Euclidean geometry. The development of non-Euclidean geometries (i.e., geometries that made use of one or more postulates which differ from those used in Euclidean geometry) is generally credited to the mathematicians Lobachevsy, Bolyai and Reimann working independent of each other. But there is no question that though he did not publish his work at the time, Gauss had anticipated much of this work.

making use of this language, Einstein gave a precise and satisfactory formulation of the relationship between the principle of equivalence and any gravitational field, while maintaining the results of special relativity as a special case. The gravitational field (equivalent to the acceleration of the frame of reference) in that case is reduced to zero. Furthermore, Newton's law of universal gravitation, now equivalent to Newton's second law, appears in a new form. As the strength of the gravitational field increases beyond a certain point, we can predict detection of affects not predicted by the Newtonian theory. For example, not only can we expect that the path of a light beam will be bent in a gravitational field, we should also expect to detect the influence of strong gravitational fields on the distribution, by frequency, of energy from radiant bodies, such as stars. The theory also predicts that the shape of the orbit of planetary bodies should change with time. These effects, while subtle, were measurable in certain cases. Thus, when Einstein completed the general theory of relativity, he could predict that the perihelion of the planet Mercury (the point of closest approach to the sun) is rotating at a rate that had long since been measured and had defied explanation within Newtonian gravitational theory.

Although the mathematical language Einstein found suitable for the general theory of relativity is more abstract than the language he had previously used and, although the general theory modified Newton's dynamical laws, the general theory of relativity can still be considered a theory of measurement. It would be a misnomer to consider the general theory of relativity a kinematical theory (as opposed to a dynamical one), because the theory specifies more than the motion to be expected when certain conditions are satisfied. It also specifies what the conditions are at any given time, and hence, which motion will occur. These specifications make it a dynamical theory.

Recall that in the special theory of relativity, the key question for Einstein was "What does it mean that two events not at the same place are simultaneous?" To answer that question, Einstein realized the need to stipulate that, in one frame of reference, space be isotropic for one signal speed. It made physical sense to choose the speed of light as that signal speed. But since the principle of special relativity required that all inertial frames of reference be equivalent, the stipulation of isotropy had to apply to all such frames.

In the general theory of relativity, the key question became: "How does one define a straight line?" The general principle of relativity, the principle of equivalence which itself hinged on the stipulation that gravitational and inertial mass were the same, shows the way. The best that one can do is to take the path of a beam of light in empty space. As long as there are no masses near the path of the beam, that path will conform to a straight line in the classic, Euclidean sense. If there are masses in the vicinity of the trajectory of the light beam, that trajectory, while no longer straight in the Euclidean sense, is still a straight line in the Gaussian space used as the language of general relativity. The effects of masses (or accelerations) is to warp classical space. Consider, for example, the motion of the planets around the sun. According to Newton's law of universal gravitation, the motion of the planets results from forces exerted between the planets and the sun. According to the general theory of relativity, the effect of the masses is to distort classical space. The planets *are* moving inertially, but they are moving inertially in Gaussian space and that results in what we perceive to be virtually Euclidean elliptical orbits.

The general theory of relativity does not account for the origins of the universe, nor account for the source of mass. It does not explain the nature of the universe. What has been generalized are the concepts of length and time. What has been explained, or rather, explained away, is not only Newton's theory of measurement but the great edifice of Newtonian dynamical theory: The three axioms and the law of universal gravitation.

·II·

THE EARLY RESPONSE TO THE SPECIAL THEORY OF RELATIVITY, 1905–1911

In Part I of this book we examined the substantive content of the special theory of relativity in the context of the historical milieu out of which it was created. We also examined the relationship between Einstein's epistemological and ontological convictions and the structure of the theory.

In studying the structure of the theory of relativity we have from time to time written about how certain scientists thought and felt about the theory. There was nothing exotic in the mathematical formalism which Einstein used to develop special relativity. Most physicists have had no difficulty following that aspect of Einstein's treatment of the problem. Furthermore, the ideas embodied in the theory of relativity are not difficult or obscure. Einstein proposed a new theory of measurement predicated on the notion of a finite ultimate signal speed and he realized that judgments of simultaneity play a central role in all spatial and temporal measurements.

In Chapter 3 we explored the relationship between Einstein's philosophical convictions and the structure of his first papers in special relativity. His insistence on beginning with abstract postulates rather than relying on the results of experiments for the foundation of his theory is unique. Very few physicists had shared his convictions about this. All too often, Einstein's theory of measurement was confused with Lorentz's second order theory of the electron, which made a direct appeal to experiments to justify the nature of the postulates.

In Part II the response to the theory of relativity is examined in a more systematic manner. Although it would be relatively simple to catalogue the response of individual physicists to the theory and its attributes, that would miss the point. I have undertaken to show that the social structure of science is an important factor in the response of individual scientists. This view runs counter to the general notion that science is objective and progressive, and that the successions of theories are closer and closer approximations to the true picture of the working of nature.

At the beginning of the twentieth century, the British writer J. T. Merz reviewed the development of intellectual thought in Europe during the nineteenth century. According to him, one of the changes that took place in the course of the century was the gradual universalization of scientific knowledge.

> A hundred years—even fifty years—ago, it would have been impossible to speak of European Thought in the manner in which I do now. For the seventeenth and eighteenth centuries mark the period in which there grew up first, the separate literature and then the separate thought, of the different countries of Western Europe. Thus it was that in the last century and at the beginning of this, people could make journeys of exploration in the regions of thought from one country to another, bringing home with them new and fresh ideas. . . .
>
> In the course of our century Science at least has become international: isolated and secluded centres of thought have become more and more rare. Intercourse, periodicals, and learned societies with their meetings and reports, proclaim to the whole world the minutest discoveries and the most recent developments. National peculiarities still exist, but are mainly to be sought in those remoter and more hidden recesses of thought, where the finer shades, the untranslatable idioms, of language suggest, rather than clearly express a struggling but undefined idea. . . . We can speak now of European thought, when at one time we should have had to distinguish between French, German, and English thought.*

---

* J. T. Merz, *A History of European Thought in the Nineteenth Century* (4 Vol., Chicago, 1904–1912; repr. New York, 1965), Vol. 1, pp. 16–20.

According to Merz, the eradication of such national differences was a function of improvements in communication and transportation. But if, in fact, the premise which underlay our analysis of the content of the theory of relativity is correct, that the relationship between evidence and belief in science is precisely like the relationship between evidence and belief in other areas of human endeavor, then national differences in science should be no more or no less pronounced than national differences in other spheres. Furthermore, if the social aspects of science, its organization, and its relationship to the social institutions, are important to how a particular scientific theory is constructed, then differences in understanding should be related to differences in the social organization of science.

Therefore, rather than cataloguing the responses of individual scientists, those responses have been organized by country. It is not thereby suggested that physicists in a given country acted together in coming to terms with Einstein's theory according to a preordained party line. We do suggest that one can see in the response of individual scientists manifestations of "national differences." In other words, one can understand the response in terms of traditional national systems and concerns. At the very least one can identify in the nature of the response of individuals a compatibility with other national traits.

Of course such traits do not have mystical origins. They are closely related to the nature of the social customs and social institutions in a country. The drawing of national boundaries cannot in and of itself be the important variable that results in different responses to scientific innovation. But national boundaries do bespeak of differences in social institutions and traditions. The most obvious social institutions to examine *outside* the scientific community would be the educational systems. We will see that there is a close relationship between the nature of the response to the theory of relativity in different countries and the *structure* of the educational systems in those countries. But we will also see that there are close relationships between the views of individual scientists who can be identified with the culture of a country and the traditional concerns of that culture with regard to the status of scientific knowledge.

The situation we have been describing corresponds to the situation in the realm of music. Presumably, musical notation provides a universal language with which musicians and composers the world over can converse and communicate. Given a musical score, anyone conversant with the language will agree about the order of the notes, the tempo, and the key in which the score is written. How does one account then for the difference, say, of Arturo Toscanini conducting a Beethoven symphony and Leonard Bernstein or Georg Solti conducting the same symphony? Attempting to understand the response of an individual conductor to a musical score is a very complex problem and probably not solvable, even in the abstract.

The different interpretations of the score not only represent different emphases, but in some cases different ascriptions of the meaning of the score and what the composer was trying to say to his listeners.

The same situation exists with regard to the responses of individual scientists to a particular scientific theory like the theory of relativity. But although the situation is complex, the threads of a national response are as obvious to one who becomes sensitized to them as is the role that glaciers play in the geology of the earth once one becomes sensitized to evidence like erratic boulders, terminal moraines, etc.

The picture which has been constructed here may be summarized as follows: Knowledge in science is no more and no less subject to social processes than knowledge in other areas of human endeavor. There are many factors that help to shape a scientist's point of view. In the first instance, there will be differences which result from genetic inheritance, upbringing, and experience unique to the individual. There will also be local characteristics identified with local customs or local social structures and beliefs. And there will be national characteristics defined by national customs and social institutions—a kind of national intelligence. In each example, the reader will have to judge for herself or himself how substantial each factor is.

We now examine the response to the theory of relativity in four countries: Germany, France, England, and the United States. Those countries are the source of more than ninety-nine percent of the literature on the subject during the first six years after the publication of Einstein's theory—between 1905 and 1911.

This time represents the period of most intense consideration of the special theory of relativity. 1911 is significant because there is evidence that, at least for the leaders in science, the theory was thought to have been established by then. For example, the first Solvay Conference was held in the fall of 1911 in Brussels. These conferences are held periodically under the sponsorship of the Solvay International Institutes. The Institutes (in chemistry, physics, and sociology) were endowed by the Belgian industrial chemist Ernest Solvay. One of the charges to these Institutes was to organize international conferences periodically to examine in broad context the most pressing fields of science. The topic of the first conference (chaired by H. A. Lorentz, and attended by, among others, Einstein, Henri Poincaré, Marie Curie, Ernest Rutherford, and J. J. Thomson) was not relativity, but rather the nature of light.* In the same year, a month before the Solvay Conference, Arnold Sommerfeld, the great physicist and teacher at the University of Munich, who was earlier very skeptical of Einstein's theory, went to the annual meeting of the

* P. Langevin and M. de Broglie, La Theorie de la reunion tenue a Bruxelles, du 30 october au 3 novembre 1911 (Paris, 1912). Cf. A. I. Miller, *Albert Einstein's Special Theory of Relativity: Emergence (1905), and Early Interpretation (1905–1911)*, (Reading Mass, 1980), pp. 253 ff.

First Solvay Congress, Hotel Metropole, Brussels, October-November, 1911. Seated left to right: Nernst, Brillouin, Solvay, Lorentz, Warburg, Perrin, Wien, Madam Curie, Poincaré. Standing: Goldschmidt, Planck, Rubens, Sommerfeld, Lindemann, de Broglie, Knudsen, Hasenöhrl, Hostelet, Herzen, Jeans, Rutherford, Kamerlingh-Onnes, Einstein, Langevin. Courtesy of the International Institute of Physics and Chemistry.

German Society of Scientists and Physicians and reported that, whereas he had originally intended to speak about the special theory of relativity, he had changed his mind because special relativity was established and the important problems were of quanta and the nature of light.*

It will become clear that for most physicists, it was not Einstein's theory that was accepted by 1911, but his formalism, which could be interpreted in a variety of ways—for example as the second order Lorentz theory. This does not mean that the battle over relativity was ended. Rather, during the next twenty years the terms of the fight changed from physics to other spheres: for example, anti-Semitism. By the time the Weimar Republic fell before the pressure of The Third Reich, relativity and Einstein had become synonymous in Germany with "Jewish and non-Aryan physics" and was not acceptable. Without the anti-Semitic overtones, relativity was attacked in other cultures also as a retreat from standards of absolute truth, or it was interpreted as representing a loss of moral will.† All of these issues did not become apparent in the *popular* literature until after 1919, when Einstein's spectacular prediction that starlight would be bent by the sun (a prediction made on the basis of the general theory of relativity in 1915) was confirmed.

Given that Einstein's theory was not initially acceptable to most physicians, the question considered in Part III will be how the theory was assimilated at all.

* A Sommerfeld, "Das Plancksche Wirkungsquantum und seine allgemeine Bedeutung fur die Molekularphysik," *Physikalische Zeitschrift,* 1911, *12*: 1057–1069. Cf. A. I. Miller, *op. cit.* In the text of the next few chapters there will be little citation of original works. As pointed out in the preface, Part II of this book draws heavily from work I completed in 1969, *The Early Response to Einstein's Theory of Relativity, 1905–1911: A Case Study in National Differences* wherein one may find many of the specific references. The reader may well be interested in comparing the treatment there and here with the work of Miller cited above where many of the same *dramatis personae* and works are discussed. There are however, significant differences of understanding and interpretation which will be pointed out along the way. The reader is also directed to the bibliographic essay at the end of this work.

† Philipp Frank, *Relativity, A Richer Truth* (London, 1951).

·6·

When a Hundred Flowers Bloom: The German Response

THE GERMAN RESPONSE TO THE THEORY of relativity was by far the most extensive, complex, and varied of all the countries being considered. One has to distinguish carefully between the German response and the German-speaking response. German-language physics journals enjoyed the greatest prestige in the world and publication in them assured an important and wide audience for physicists from other countries.

More important than the prestige of the German physics journals is that German-speaking universities stretched from the eastern cantons of Switzerland to universities in eastern Europe, including Warsaw, Prague, and St. Petersburg in Czarist Russia, in addition to the growing physics community in South America. Therefore to discuss the German-speaking response, I have taken into account the country of origin of the individual physicist and the country in which he or she was trained. The issue is further complicated because the national state, Germany, at the turn of the century, was only thirty years old. It had been welded together by Bismarck at the end of the 1860s. Yet, aside from differences in dialect, there were a common language and university traditions stretching back to the Middle Ages. We will see that the common tradition is manifested in communications between physicists about the theory of relativity.

Variety is characteristic of the response to the theory of relativity that one finds in the German physics community and it is absent in the three other countries. It is this variety, alone, that played a large role in the eventual acceptance of the theory. The elaboration of the theory emerged from contentions about interpretation, accuracy, and meaning, and revealed to many of the participants the heuristic nature of the theory.

In Chapter 3 we showed that almost immediately after publication of Einstein's theory, Walter Kaufmann undertook his experiments about the relationship between the mass and speed of beta particles (electrons) and declared that his experiment had proved that the theory of relativity and the predictions of Einstein and Lorentz were incorrect and unsupported by the evidence of nature. We have also noted that it was Max Planck who first actively challenged the conclusions of Kaufmann and Kaufmann's assertion that the experimental result could be used to support the Abraham theory of the rigid electron. It is interesting to note that Abraham had been Planck's student in Berlin between the years 1895 and 1897 when he received his doctorate. He remained Planck's assistant until 1900. Kaufmann was also at the University of Berlin during part of this time, but he left to continue his studies in Munich. Kaufmann and Abraham became colleagues again when both men became *Privatdozenten* (literally "private instructors") at the University of Göttingen.

A *Privatdozent* was a teacher licensed by the university whose only source of income was from student fees. In the German university system it was the first rung up the professional academic ladder. There is no equivalent position in the current American system. It might correspond to a post-doctoral fellow or an assistant professor, although the *Privatdozent* was without official standing within the university. He did not receive funds for research. The *Privatdozenten* have been called the proletariat of the German university system. They were waiting for "the call," as it is known, to a permanent professorial position—a position with great status. It was the source of considerable income within the German university system. But there were many more *Privatdozenten* than there were permanent positions available. At the same time, *Privatdozenten* were essential to instruct the large and growing student population in late nineteenth and early twentieth century physics courses in German universities.*

The problematic situation with regard to the *Privatdozenten* is well illustrated by Abraham's position. He became a *Privatdozent* at Göttingen in 1900. For nine years he remained in this position while he was repeatedly passed over for university positions. The reasons are varied. For one thing, he was a Jew, and there is no question that the proportion of Jewish *Privatdozenten* was much higher than the proportion of Jewish university professors. More important, he had a reputation for having a quick and nasty tongue, and for having no compunction about

* See Alexander Busch, "The Vicissitudes of the *Privatdozent:* Breakdown and Adaptation in the Recruitment of the German University Teacher," *Minerva,* 1963, *1*:319–341.

publicly heaping ridicule on those whom he thought were stupid about problems of physics. It was Abraham who had heatedly challenged Planck, his former teacher, about the interpretation of Kaufmann's data at the September, 1906 meeting of the German Society of Scientists and Physicians (Chapter 3).

Although Abraham was the author of the most respected and widely used textbook on electricity and magnetism, he was still a *Privatdoozent* in 1909 when he decided, like many who were in his position, to emigrate. He left Germany for a position at the University of Illinois. He was so disappointed in the quality of that American institution that, within six months, he returned to Germany. Finally he received an invitation to join the faculty of the University of Milan in Italy. Several years after he left for Milan, one of his former colleagues in Germany asked him how he was getting on. His reply is supposed to have been that he was getting on well with his Italian colleagues because he was not facile in the Italian language.

Abraham returned to Germany with the outbreak of World War I and, during the war, engaged in theoretical work on radio transmission for the *Telefunkengesellschaft*. It was not until 1922 that he received an invitation to a permanent professorial post at the Technical University at Aachen. Tragically, on his way to Aachen, he was stricken with a terminal brain tumor and never was able to fully bask in the glory that was his due in the German university professorial community.

How it was decided that Abraham would not receive a call to a chair before 1922 and why, in 1922, he was invited to Aachen is a clue to understanding the German university system during that period. In the relatively small area of the German state, at the turn of the twentieth century there were seventeen universities and sixteen technical universities. This did not include universities in Austria, Switzerland, Czechoslovakia, and other localities. Each institution had at least one institute of physics dedicated to research, often more than one. Each institute was headed by one professor and in many instances there were several professors who held paid chairs. The chairs were supported by the state ministry of education, and it was the minister, in consultation with university officials, who decided who would receive an appointment when a new position was open. When one examines the records associated with such decisions, one is impressed with the degree to which there was consultation, not only between the minister and officials of the university and the physics institute, but also with the well known members of the wider physics community.

Not only was there consultation, but there was intense rivalry. From the time that Bismarck welded the separate states into a unified country, physicists who were not in Berlin were convinced to the point of paranoia that the Prussian state, the minister of education of the Prussian state, and individual physicists at the

University of Berlin acted solely in their own self-interests in the distribution of funds and appointments, to the detriment of physics institutes at other universities. This battle was to reach its climax just after World War I.

Given this involuted network of communications, once a person such as Abraham got a reputation for being difficult or uncooperative, it is easy to understand how he would be blackballed. It is illuminating to read correspondence between physicists in Germany who were discussing these issues. The moment one position opened, a game of professorial musical chairs commenced. Each person occupying a chair calculated whether it would be in his best interests to remain or to move, and each calculated how it might serve his interests to secure a chair for one of his students.

Add the institutes of chemistry, physical chemistry, organic chemistry, institutes of the other sciences and sub-branches of the sciences, and extend the picture by including the universities in Austria, eastern Europe, and eastern Switzerland, and we have a notion of the vast network of institutions dedicated to scientific research in Germany. This network, going back to the Middle Ages, began to develop in the way we have described during the seventeenth and eighteenth centuries. By the middle of the nineteenth century the German university community was recognized as the leading and most powerful establishment doing research, not only in the sciences, but in history, philology, theology, and almost all other academic fields.

There is a concept guiding German culture during this period which was absent from most other cultures. The concept is *Wissenschaft*.* It is very difficult to describe what is meant by the term because there is no exact English translation. It embodies more than research and scholarship. "Science," as we use the term, is usually rendered as *"Naturwissenschaft." Wissenschaft* was a way of life that can best be described as the scholarly life. However, I do not intend to imply a dry ascetic existence totally removed from the cares and concerns of the day. It did not matter where you did your research. It could be done in industry, in the laboratories of an institute at a university, or in your basement. What was important was how the research was conducted.

It was not out of character to demand the attributes of *Wissenschaft* in one's work and to blackmail the local ministry of education for a higher salary and a new building to house the physics institute. The structure and rivalry between German universities during this period are expected today in the United States in professional sports. There was also cultural prestige analogous to the prestige of today's sports heroes.

Is it any wonder that when the theory of relativity was introduced into this

* See J. T. Merz, *A History of European Thought* . . ., Vol. 1, pp. 168–174.

cauldron, it was attacked and defended, issues were argued back and forth, a variety of elaborations and alternatives were explored, and in the end, the heuristic nature of the theory was revealed? There were more than thirty centers of research in physics, vying for the leading researchers and for supremacy in the realm of ideas. It was in Germany, as in no other country, that the merits of the theory of relativity were debated in many dimensions, and it was in Germany that this debate and elaboration led to acceptance of the theory based more on understanding than on rote application of the equations.

THE CONTRIBUTIONS OF MAX PLANCK

We have examined the relationship between the predictions of the mass of high speed electrons and the actual experimental determinations. In Chapter 3, it was pointed out that the experimental results were, at best, inconclusive and not generally trusted and not an important factor in the acceptance of the special theory of relativity. Max Planck played a central role in that debate.

Planck's early attraction to the theory and his subsequent support for it can be traced to his philosophical and ethical convictions about the ultimate laws of reality. He called this ultimate reality "the physical world picture." It was his view that successive syntheses represented closer and closer approximations to the physical world picture. Planck's support for the theory, then, grew out of his very conservative convictions about the nature of physical reality.

When Einstein submitted his paper to the *Annalen der Physik* in the early summer of 1905, Max Planck was an advisor to the editor, Paul Drude. Planck had an almost day to day knowledge of submissions to the journal. When Planck's former student, Max von Laue, returned to Berlin in the fall of 1905 to serve as Planck's assistant, the first colloquium that he heard at the Physical Institute was Planck's summary of Einstein's paper. von Laue later said that he was so impressed that he took the first opportunity to travel from Berlin to Berne, Switzerland to meet Einstein. When he was in Berlin as Planck's assistant, between the years 1905 – 1908, he made several crucial contributions to the theory of relativity that we will discuss later. (von Laue subsequently left Berlin for the University of Munich where he won the Nobel Prize for his theoretical work on x-ray diffraction.)

Another student of Planck's, Kurt von Mosengeil, was also working on problems of relativity. He died suddenly in the latter part of 1906, and Planck subsequently revised his work and saw to its publication. It was that work which Planck used to upset the analysis of blackbody radiation by Hasenöhrl and others who were committed to the Lorentzian analysis of the electrodynamics of moving bodies, discussed in Chapter 4.

All of this points up the great influence that one individual like Planck can have

Max Planck. Courtesy of the American Institute of Physics Niels Bohr Library, W. F. Meggers Collection.

in an active scholarly community like the German one. He elaborated on the theory; set his students to work on detailed aspects of it, and he defended the theory against those who would have relegated it to the scrap heap during a period when he himself had doubts about its ultimate worth.

Planck was motivated to defend the theory of relativity because of the absolute character with which physical law was endowed under its aegis. Planck found it attractive because the world picture toward which he saw natural philosophy evolving would have laws and variables that are absolute and invariant—the same for all observers. For Planck this represented the supreme objectivity toward which science was moving.

THE RIGIDITY PARADOXES

Between the years 1907 and 1910, the definition of a rigid body within the special theory of relativity was argued in the pages of the German physics literature. It is a good example of how open debate of the theory led to specific elaborations of the theory, which, in turn, revealed the heuristic and enduring qualities of the theory. The issue was first raised in the context of the question of the theory of electrons. Eventually, the analysis of the concept of a rigid body became a key instrument for showing the power of the language of the four dimensional analysis by Minkowski to handle otherwise intractable ideas.

At issue is: Classically, a rigid body is one whose dimensions remain invariant under all circumstances. For example, suppose that we have an absolutely rigid rod of length L. The rod is moving with a speed v as it encounters an equally rigid wall. The rod will rebound from the wall so that its length and shape are not affected by the encounter. However, in the special theory of relativity, that one end of the rod has encountered a wall cannot be communicated to the other end of the rod in less time than that required for light to traverse the length of the rod. Therefore, as one end of the rod strikes the wall and begins to slow down, the other end of the rod, not "knowing" this, continues to move with the speed v for some time. The rod becomes temporarily distorted. Since it is impossible to transmit information faster than the speed of light, the classical concept of a rigid rod has lost its meaning within the context of the special theory of relativity.

During the years immediately after the publication of Einstein's theory, this point was not clear. The beginnings of this debate were an exchange between Paul Ehrenfest and Einstein.* Ehrenfest was then living in St. Petersburg with no

* P. Ehrenfest, "Die translation deformierbarer Elektronen und der Flachensatz," *Annalen der Physik,* 1907, *23*:204–206. A. Einstein, "Bermerkungen zu der Notiz von Herrn Paul Ehrenfest:' Die Translation deformierbarer Elektronen und der Flachensatz,'" *Annalen der Physik,* 1907, *23*:206–208.

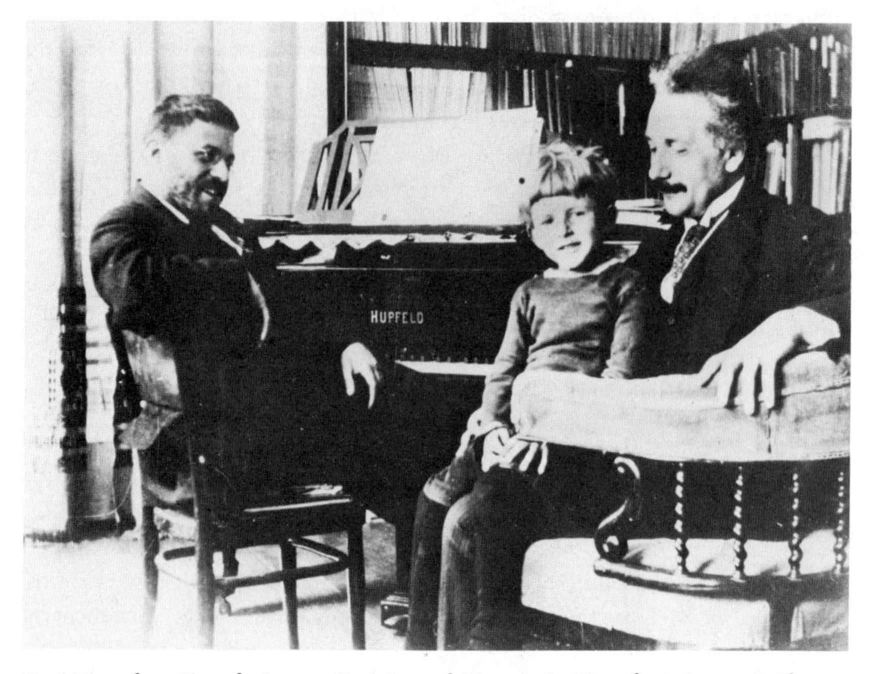

Paul Ehrenfest, Ehrenfest's son, Paul Jr. and Einstein in Ehrenfest's house, Leiden, ca. 1921. Photograph by Willem J. Luyten. Courtesy of the American Institute of Physics Niels Bohr Library.

formal academic affiliation. He had been a brilliant student of the great Ludwig Boltzmann at the University of Vienna and was later to spend most of his career at the University of Leyden in Holland. After 1912, he and Einstein were to become extremely close friends. Before that, as a result of their formal encounters over the question of the theory of electrons and the notion of a rigid body, they quickly gained a healthy respect for each other. Ehrenfest was a consummate teacher and a man whose brilliance was revealed in his ability to probe to the weakest links in the inferential chains of theoretical physics.

This was precisely his role in this instance. In Chapter 3, it was pointed out that Max Abraham had shown, in the development of his rigid electron theory, that a non-spherical rigid electron is dynamically unstable. In 1907, Ehrenfest raised the question of whether or not this analysis applied to the "Lorentz relativistic electrodynamics" in the form published by Einstein. Ehrenfest challenged proponents of the Einstein system to show whether or not the electron could move inertially within the theory. In effect, Ehrenfest was identifying the Einstein theory with the dynamical electron theory of Lorentz and asking proponents to show how deformed electrons would move under specific circumstances.

As we saw in Chapter 3, Einstein's reply was immediate. He pointed out that his was not a system in the sense that Ehrenfest was implying but rather a set of heuristic principles saying nothing about the nature of the electron. His theory, Einstein continued, involved only rigid rods, perfect clocks, and light signals. To answer Ehrenfest's question, Einstein concluded, one would require a separate theory of rigid bodies. There was no such theory as yet.

An attempt to discover such a theory was made several years later. The chief architect of the attempt was Max Born. Born had been a student of Hermann Minkowski and was on the threshold of an extremely distinguished career. We have discussed Minkowski's four dimensional analysis; as pointed out in Chapter 4, Minkowski's view of Einstein's contribution was that he had generalized Lorentz's theory. Minkowski was thinking about the mathematical formalism and certain aspects of Lorentz's theory of matter and radiation and he did not appreciate the kinematical character of Einstein's work. Minkowski died in 1909, we will never know whether or not he would have changed his mind.

At the time of Minkowski's death, Born was his assistant. He undertook to see through publication Minkowski's last works, and in the same terms that Minkowski might have used also undertook an analysis of the concept of the rigid body with the four dimensional formulation of the theory of relativity. This was intended to be the basis for a fundamental extension of the electron theory.

Indeed, an analysis of Born's first efforts to define a rigid body, confirmed by the reports of his colleagues,* support the views that he was more interested in the mathematical aspects of the theory than in the physical concepts and that he did not recognize the difference between Einstein's kinematical approach and Lorentz's dynamical interests.

Born's new definition of a rigid body was that all points on the body had to satisfy the equation for the fundamental invariant:

$$c^2t^2 - x^2 = F$$

F is the fundamental invariant for the system. Born used the idea of a rigid body to investigate the motion of the electron, and arrived at the Lorentz formulation for the mass of the electron. Since he had used only electromagnetic assumptions about the electron itself, he thought that his four dimensional definition of a rigid body had placed the Lorentz theory of electrons on a purely electromagnetic basis.

In the two years since his exchange with Einstein, Ehrenfest also had been

* Philipp Frank, personal communication. December 7, 1964.

thinking about extending the classical notion of a rigid body to the theory of relativity. Within a few weeks of the publication of Born's article, Ehrenfest published a brief paper on the subject. He showed that the use of a definition such as Born's led to an interesting paradox—a paradox that became known as the *Ehrenfest paradox.*

Take a cylinder at rest (Figure 1a). It has a radius R and a circumference $2\pi R$. Suppose that the cylinder is "rigid" in the sense that Born suggested. Once the cylinder is rotating (Figure 1b), from the rest frame of reference its radius R' equals R since the direction of motion of the cylinder is perpendicular to the length of the radius. But if one considers any segment of the circumference, one will conclude that $2\pi R'$ is less than $2\pi R$. These results are contradictory and Ehrenfest concluded that it is impossible to set any relativistically rigid cylinder into rotation—a paradox.

A close friend of Ehrenfest's from their student days, the young, brilliant mathematician Gustav Herglotz, gave a slightly different interpretation of the problem. He pointed out that Born had allowed for only three degrees of freedom for his rigid body. That is, the body could move along the x, y, and z axes, but there were no degrees of freedom for rotation. (In classical physics a simple point mass has six degrees of freedom, three degrees of translation along the three axes, and three degrees of rotation around the three axes.) The effect, Herglotz concluded, was as Ehrenfest had suggested. Within Born's analysis, it is impossible for a body at rest to be placed in uniform rotation. Born, while acknowledging these shortcomings, maintained that such an analysis of rigid bodies was a necessary basis for a dynamical theory of electrons. Even so, by 1910 he had attempted and published yet another definition of a rigid body within the theory of relativity that allowed for six degrees of freedom. He persisted in his belief that

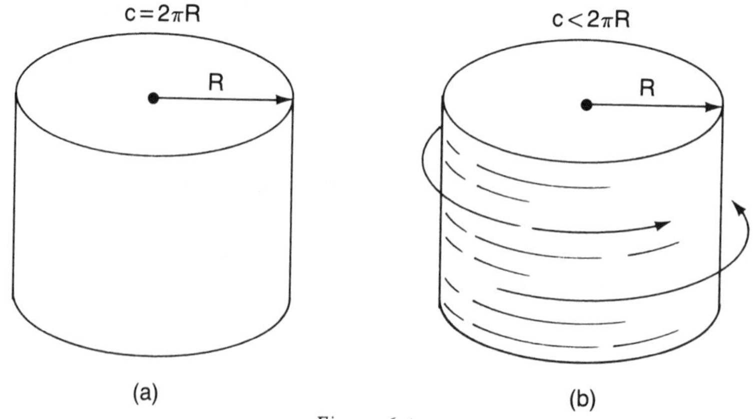

Figure 6.1

such a definition was essential for a theory of electrons. His new definition was also subjected to intense examination by colleagues and found insufficient. Shortly thereafter, Born himself realized that the problem he was struggling with was not dynamical but kinematical and had nothing to do with the theory of electrons.

Independent of Born's struggle with these problems, there were major contributions by Abraham, Planck, and von Laue. Almost coincident with Born's first reply to Ehrenfest, Planck suggested that the problem was not related to the theory of relativity, but rather to the theory of elasticity. Max Abraham used the dispute to illustrate the shortcomings of the theory of relativity.

The most significant contribution to the discussion was in two exchanges—the first between a relatively unknown mathematical physicist, W. von Ignatowsky, and Ehrenfest, and the second between a certain V. Varičak and Einstein.

In an article published in 1910, Ignatowsky defended the classical notion of a rigid body in relativistic mechanics. He began by invoking the principle of relativity and, in place of Einstein's postulate of the invariance of the speed of light, another stating that the maximum signal speed in any frame of reference is given by the ratio of the square of the speed of light to the speed of the frame of reference. Ignatowsky maintained that the true form of a body could only be determined when the body is at rest. In a rather sharp response, Ehrenfest challenged this idea and showed that the *measured* length of a body says nothing about the body, but depends on the frame of reference in which it is measured. The Lorentz contraction, Ehrenfest had realized, is an artifact of how we measure. A significant aspect of this exchange was that Ignatowsky's article was published in the *Annalen der Physik* that was now co-edited by Wilhelm Wien and Planck. (They had taken over in 1906, after the tragic death of Paul Drude.) Curiously, neither seems to have recognized that the two postulates of Ignatowsky are contradictory as the second provides a criterion for determining absolute motion in contradiction to the first. In Ignatowsky's theory the absolutely resting frame of reference is the one in which the maximum signal speed is c^2. Although Ehrenfest was properly critical of the body of Ignatowsky's work, he too failed to see this contradiction. It was not until a year later that Arnold Sommerfeld and Born criticized Ignatowsky when he again presented this postulate. They pointed out that he could not be talking about a signal speed. Signal speed means the speed with which information is transmitted. As we will see later, there are speeds that can exceed the speed of light as long as they do not entail the transfer of energy or information.

That Ignatowsky's work received a serious hearing in a journal such as the *Annalen der Physik* suggests that at the time of a radical innovation like the theory of relativity, there may be lapses in standards, even by an editor as perceptive as Max Planck.

The 1910 exchange between Varičak and Einstein is significant for several reasons. First, Varičak, proceeding on a point of view recently published by two Americans, Lewis and Tolman, tried to distinguish between a real length contraction of the Lorentz theory and a psychological contraction of the Einstein theory. Varičak tried to rationalize the paradox as psychological rather than physical. Einstein wrote to Ehrenfest and invited him to respond. The published response finally came from Einstein, who noted that Ehrenfest had shown very nicely with his paradox that the measured length of an object was not a subjective psychological perception, but was real because length is an artifact of how we measure.*

The notion of a rigid body and the status of the contraction was finally summarized in 1911 by Max von Laue, who showed definitively that the concept of "rigid body" as it had been understood in classical physics, is capable of instantly transmitting information throughout its dimensions. It had no place within the theory of relativity. Points on an object must move through space to satisfy Einstein's postulates. von Laue made use of Minkowski's four dimensional formulation of the theory of relativity because of convenience in expressing Einstein's ideas, but he was clear and incisive about the meaning of the theory. In the same year, von Laue published the first monograph on the theory which became the standard reference on the subject in Germany for ten years.

The significance of the issue is how the ideas of the theory of relativity were manifested in a clear resolution of the rigidity paradox. The transition of Max Born's views is paradigmatic here. His initial concerns were with the mathematical formalism and the dynamical problems of his mentors Minkowski and Lorentz. After the debate, his understanding of the theory of relativity was transformed. In a popular account first published in 1920, Born demonstrated clearly that the measured length of an object was an artifact of how the measurement was made.† Without the arena provided by the German physics community, it is conceivable that the process might never have occurred.

WAVE AND GROUP VELOCITY

Another interesting elaboration of the theory of relativity was made between the physicists Wilhelm Wien and J. J. Laub. Wien was Professor and Director of the Institute at the University of Würtzburg. He had occupied the chair since 1900 when he succeeded Wilhelm Röntgen, who had accepted a more lucrative position at the University of Munich. Wien had been a student of Hermann von

* The rigidity paradoxes have been discussed in a somewhat different context by Martin Klein in *Paul Ehrenfest* (Amsterdam/London, 1970), pp. 150–156.

† Max Born, *Einstein's Theory of Relativity* (Revised edition, New York, 1965).

Helmholtz at Berlin and later accepted a position with him when Helmholtz became the first President of the Royal Physical-Technical Institute in Berlin. It was that institution on which the United States Bureau of Standards was modeled. The PTR (*Physikalische Technische Reichsanstalt*) as the Institute was called, was started with an enormous gift from the industrialist Ernst Werner von Siemens, who was related to Helmholtz by marriage. There is little doubt that von Siemens saw the PTR as a means of providing an appropriate position for Helmholz, but there is also considerable evidence that von Siemens was chiefly moved by his commitment to *Wissenschaft.** Significantly, while a major responsibility of the Institute was to set industrial standards and insure uniformity of industrial products, it was also devoted to the kinds of research we have described as *Wissenschaft.* While he was at the PTR Wien made major contributions to our understanding of the distribution of radiant energy. For this work he won the Nobel Prize in 1911.

In 1905 Wien was one of the recognized leaders in the physics community in Germany. He was in an enormously powerful position. He had many leadership qualities and was one of the guiding lights of the German physics community until his death in 1928. During the first World War, for example, he was instrumental in shaping the official posture of the German physics community. He assumed the responsibility for communicating with other countries about German policy, and after the war, although in good standing with the disillusioned and embittered upperclass, conservative intelligentsia, he led the way to reestablish communications with "enemy" physicists.

In 1905, J. J. Laub studied physics with Wien. Although he already had a doctorate, this was not unusual in Germany. From Laub's account, on the day that Einstein's article was published Wien appeared in his rooms and asked him to prepare a seminar on the subject. Laub was a logical choice since most of the people working for Wien were experimental physicists and Laub was a theoretician.

Laub was captivated by Einstein's work. His understanding, if one can judge by his publications of 1907 and 1908 about relativistic optics, was clear and deep, and it penetrated to the core of Einstein's argument. Years later when Laub reminisced about the seminar in the fall of 1905, he recalled how puzzled he had been because almost no one seemed to understand what he was talking about. For him, the problems of synchronizing clocks and the subsequent relativity of simultaneity were simple and straightforward. To others they were arcane and

* David Cahan, "Werner Siemens and the origins of the Physikalische-Technische Reichsanstalt, 1872–1887," *Historical Studies in Physical Science* 1982, 12:253–282.

mysterious. We have already seen, with Lorentz, how difficult it was to confront deepseated preconceptions of these questions.*

In particular, Wien found it impossible to accept the conclusion of the theory that the speed of light is the ultimate speed. He pointed out that there are transparent media such as liquid carbon disulfide, in which the speed of light is greater than its speed in empty space. For example, if we construct two hollow glass prisms and fill one with carbon disulfide and the other with water and shine a beam of white light on the water prism, we obtain a spectrum from red to blue. That is, the red wave lengths are diffracted much less than the blue ones. The breadth of the resulting spectrum is a measure of the dispersive power of the water—that is, the power of water to spread the colors. Most transparent substances have different dispersive powers from water, but the order of the spectrum they create when white light passes through them also runs from red to blue.

In liquid carbon disulfide, the order of the colors is reversed. Blue light is refracted less than red; the phenomenon is referred to as *anomalous dispersion.* It is accounted for by assuming that, as light enters the carbon disulfide, it speeds up depending on the wavelength. (The physical correlate of our perception of color is the wavelength of the light). It was because of this that Wien rejected the theory. For Wien, that light moved faster in carbon disulfide than in a vacuum contradicted Einstein's second postulate.

In 1907, at the ninety-seventh annual meeting of German Scientists and Physicians, Arnold Sommerfeld presented a very short paper on this subject. He pointed out that the speed of light of a given wavelength is different from the speed of the packet of waves which is the sum of the simple waves added up in space at a point. The simple waves are symmetrical, smoothly shaped *sine* waves stretched to an infinite distance in any direction. They are abstractions. We do not perceive a single wavelength of light arriving from a single point. There is always a spread of wavelengths and other waves from different points, and it is the summation of the *group* for which the speed is measured.

There is another way of thinking about the problem. A smooth unmarked *sine* wave has no distinguishing features. Unless there is a mark placed on it, it is impossible to distinguish one maximum from another. One way to mark the wave is to add other waves to it which are out of phase, or of different wavelengths to create a hump in the wave *train.* The speed of this hump caused by a *group* of waves is measured. The speed of a simple *sine* wave, an abstraction, can be calculated but not measured. Energy is not transmitted from place to place at the

* For more about Laub, see Lewis Pyenson, "Einstein's Early Collaborators," *Historical Studies in Physical Science,* 1976, 7:83–124, esp. pp. 92–99.

so-called *wave velocity,* but rather is transmitted at the speed of the hump of waves, or *group velocity.* We can calculate the velocity of a pure sinusoidal wave from a point and compare it to the *group* velocity. If the velocity of the group of light waves is measured c, the individual wave velocity is calculated to be c^2/v. It was this abstraction that Ignatowsky spoke about. Both Sommerfeld and Born were unsuccessful in explaining this point to Ignatowsky. It is not an easy concept to understand and, in the first decade of the twentieth century when the distinction between wave velocity and group velocity was gaining importance, it was obviously difficult.

An example of the difference between wave and group velocity may be found between the rate at which light travels *from* a searchlight to a point in space as opposed to the rate at which the circle of light sweeps *across* the space (Figure 2). On leaving the source, the motion of the light is independent of the motion of the source. The information that the source is rotating cannot be transmitted faster than the speed of light.

When Sommerfeld presented his paper in 1907, he made a specific reference to

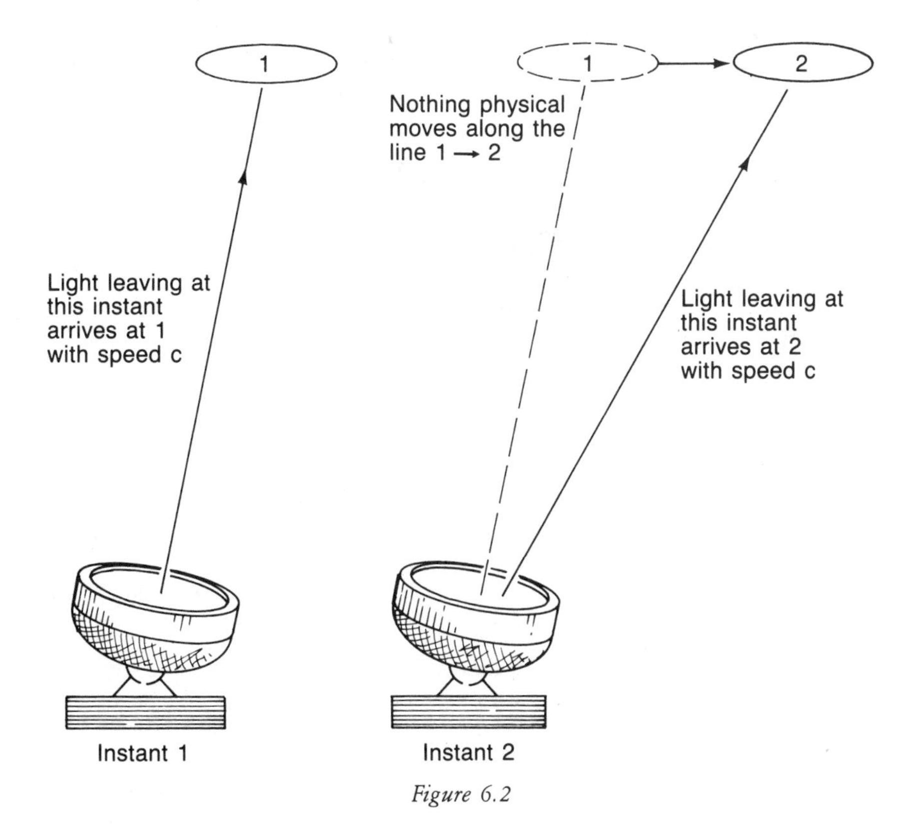

Figure 6.2

a recent article that Einstein had published about the inertia of energy. Einstein very clearly showed that the ultimate speed refers to the ultimate *signal* speed, or the speed at which energy is transmitted, *not* the wave speed, an abstraction. Sommerfeld also pointed out that Wien had not yet understood the distinction. And Wien, sitting in the audience, corroborated that, saying that not only did he not understand the distinction, it did not appear to be intuitively possible.

Sommerfeld and Wien were good friends. They corresponded and they frequently spent vacations together in the forests of southern Germany. When Wien moved to Munich in 1920, they became faculty colleagues. There is some evidence that between the time of their exchange about group versus wave velocity in 1907, and 1909, much effort was spent to educate Wien about these questions.

During this period, at Wien's suggestion, Laub visited and studied with Einstein and he was Einstein's first scientific collaborator. They published a number of jointly written papers. Laub once wrote to Sommerfeld to ask for help in educating Wien about the distinction between group and wave velocity, and he suggested that Einstein might also be of some help.

Just what transpired along these lines is not known. However, in 1909 Wien saw into publication a second edition of a short monograph entitled *On Electrons,* first published in 1906. The text of the first edition had followed Lorentz's point of view. In the second edition, Wien did not change the text but he added a set of footnotes. In them, he wrote a sparkling, delightful and straightforward account of the basis for the postulates of the special theory of relativity from Einstein's point of view and how they led to the transformation equations. He then used the transformation equations to show how they entailed length contraction, time dilation, and the relationship between the energy content of a body and its mass. Remarkably, although he recognized that the experimental evidence did not lend clear support to Einstein's theory, he was confident that the support would be forthcoming:

> The most important point in its favor is the inner logical consistency that makes it possible to lay the foundation for the entire body of physical appearances whereby all the older conceptions undergo a transformation.*

It was a remarkable change of attitude. Others in the German physics community also had exhibited similar, dramatic changes in their attitudes toward the theory. Sommerfeld himself had. Arnold H. Bucherer changed his mind in the midst of a polemical dispute about the correctness of the theory. It was almost as if the theory had suddenly become fashionable; the formalism had not changed. However, in some instances there is little evidence that the new proponents gave

* W. Wien, *Ueber Elektronen* (Leipzig, 2nd ed., 1909), p. 32, fn. The translation is mine.

more than lip service in support of the theory while continuing to use the Lorentz conceptions. Yet, by 1911, the special theory of relativity was well enough established that Arnold Sommerfeld declined to give a talk about it at the first Solvay Conference. The emphasis had shifted from relativity to questions of quanta.

There is an interesting and important sidelight about Wilhelm Wien's conversion to belief in the theory of relativity. Laub apparently played a central role. In 1910, Laub left Würtzburg and Wien's institute to become an assistant to Philip Lenard at the University of Heidelberg. Lenard, of Hungarian birth, had been trained in German universities and had won the Nobel Prize in 1905. He was renowned as an experimentalist, but it was well known that he had little aptitude for abstract mathematics and little patience with theoretical physics. When Laub went to Heidelberg, he was the only theoretician working for Lenard.

It is also well known that Lenard was very temperamental. We do not know precisely, but apparently Lenard asked Laub to read and criticize the proof sheets of a monograph Lenard had written entitled *Ether and Matter*. Lenard did not accept the theory of relativity. Laub undoubtedly tried to instruct Lenard about it, and, for his pains, was discharged. Earlier, Einstein had written Laub urging him to take the position at Heidelberg on the grounds that Lenard was a very clever experimentalist. After Laub was discharged, Einstein wrote to Laub and called Lenard "crazy."

Laub was unable to get another position in Germany. He applied for positions in other countries and finally, having been unsuccessful in eastern Europe and the United States, took a job at the German-speaking university in La Plata, Argentina. Eventually, he left physics to join the Argentinian diplomatic corps.

Lenard's behavior toward Laub had been extreme. After 1923, thoroughly disillusioned with postwar Germany and at odds with the Weimar regime, Lenard became one of the early proponents of National Socialism. But earlier, in 1911, there is little evidence that he carried more than the German distrust and dislike for Jewish physicists such as Laub and Einstein. He associated "theoretical" and "Jewish" physics. Theoretical physics was, of course, bad. It was only later that he tumbled into a kind of madness which found its expression in Nazi anti-Semitism. In 1940, he published a three-volume work under the title *German Physics,* little more than a virulent, bitter, angry and anti-Semitic tract.

To a large extent the German physics community was an old boys' club. Being "different" was ill-advised if one wanted to become a member. Einstein's acceptance into the club was based on recognition of his genius by those in power; men like Planck, Wien, and Sommerfeld. They could protect him, up to a point, from men like Lenard. Laub, without protectors, was helpless.

In the six years following the publication of the theory of relativity, although

Einstein receiving the Planck medal from Planck, July, 1929. Courtesy of the American Institute of Physics Niels Bohr Library.

Abraham, Lenard, and Hasenöhrl remained obdurate, the worth of the theory was established in the eyes of most physicists and use of the formalism became commonplace. The situation was summarized by von Laue in the introduction to his monograph on the special theory of relativity, published in 1911:

> In the 5½ years since Einstein's establishment of the Relativity Theory, this theory has found an ever growing amount of attention. To be sure, this attention is not always agreeable. Many researchers, among them the carriers of very well-known names, maintain that the theory's empirical establishment is not sufficient. . . . Much greater however is the number who cannot satisfy themselves with the intellectual content, to whom, in particular, the relativity of time with its many seemingly paradoxical consequences appears unacceptable.*

I have suggested that this situation could only have been possible in a social setting such as that of the German scientific community, within and outside the structure of the academic community. It has been well characterized by J. T. Merz:

> The migration of students as well as of eminent professors from one university to another is one of the most important features of German academic life. . . . No one university has been allowed to retain for any length of time the supremacy in any single branch. The light has quickly been diffused all over the country when once kindled at one point. . . .
> . . . . no nation in modern times has so many *schools of thought* and learning as Germany. . . . the university system, in one word, not only teaches knowledge, but above all it teaches research. This is its pride and the foundation of its fame.†

With the kind of intellectual competition and alertness that one would expect to find in that environment, it is not surprising that Einstein's theory received a variety of reactions from men at different centers. It is also not surprising to find those who, like Planck, were attracted to the theory, and elaborated the ideas of the theory in an effort to counter or confirm the criticisms and comments of colleagues at other institutions. As we have seen, men like Minkowski, Planck, Born, Ehrenfest, and Sommerfeld were attracted to the theory for different reasons. Often these reasons had little to do with Einstein's view of his work. But the theory was discussed and taken seriously and this would ensure acceptance or rejection on the merits of the elaboration.

---

* M. von Laue, *Das Relativitätsprinzip* (Braunschweig, 1911), p. v.

† J. T. Merz, *A History of European Thought . . .*, Vol. I, p. 62, fn 1.

·7·

As If It Never Happened: The French Response

IN CONTRAST TO THE SITUATION we have described in the German literature, the French response to the special theory of relativity in the years 1905 to 1911 was almost complete silence. Until Einstein visited France in 1910, there was hardly a mention of him in the French literature and no mention in the context of the physics and mathematics problems we now associate with the special theory of relativity. The contrast with Germany is startling, and at first glance, very puzzling.

The leading French theorist in electromagnetic theory and the person who had devoted himself more than anyone else to the problems associated with the Lorentz electron theory and ether theory in general was Henri Poincaré. In 1905, he was at the height of his powers, clearly the most outstanding mathematician in France, if not the world, widely recognized for his work in theoretical physics, known to the general public for his many publications on the philosophy of science.

We have already mentioned Poincaré as a collaborator of Lorentz's. In recent years, several histories of the development of electromagnetic theory at the turn of the century have ascribed the theory of relativity not to Einstein, but to Lorentz and Poincaré. The most remarkable of these, and certainly the most influential, was that of the British physicist-historian E. T. Whittaker.

The second volume of Whittaker's *History of the Theories of Aether and Electricity,* published in 1953, contains a chapter entitled "The Relativity Theory of Poincaré and Lorentz." In that chapter, Whittaker refers to Einstein's contribution to the theory in this way:

> In the autumn of . . . [1905] in the same volume of the *Annalen der Physik* as his paper on the Brownian motion, Einstein published a paper which set forth the relativity theory of Poincaré and Lorentz with some amplifications, and which attracted much attention. He asserted as a fundamental principle *the constancy of the velocity of light* i.e., that the velocity of light *in vacuo* is the same in all systems of reference which are moving relatively to each other: an assertion which at the time was widely accepted, but has been severely criticized by later writers.*

There have been several other authors who, over the years, have agreed with Whittaker's assessment of the relationship between the work of Poincaré and Lorentz and the contribution of Einstein. Clearly, Whittaker's assessment is not acceptable here. In the next section, when we discuss the British response to the special theory of relativity, we will consider the reasons why Whittaker made this assessment. We will also consider why he thought that earlier there was wide acceptance of Einstein's second postulate, and only later was it subject to severe criticism, when, in fact, the reverse seems to have been the case. For now, we turn to an examination of Poincaré's contribution and influence on the French physics community.

HENRI POINCARÉ

There can be little question that by 1900, Poincaré had in his hands all the elements necessary to build the special theory of relativity. By the special theory of relativity we mean a theory which, like Einstein's, begins with two postulates, the invariance of the speed of light and the principle of relativity. Poincaré made his first suggestions about the need for a principle of relativity as early as 1895, and by 1898 had analyzed the role that the signal speed plays in synchronizing clocks. Yet he never saw that one could combine the two elements as Einstein was to do later. Given his interests and outlook, he probably would not have been able to make such a synthesis.

In 1895, Poincaré published a series of papers on the state of available electromagnetic theories of bodies in motion. He emphasized that they give an account of first order ether drift experiments and be compatible with Newton's laws of motion—in particular, Newton's third law, which demands that the action of one object upon another be equal and opposite to the reaction of the second body upon the first. According to Poincaré, no theory at that time was satisfactory. He was more inclined toward the Lorentz theory than to others because it was the least defective.

* E. T. Whittaker, *A History of the Theories of Aether and Electricity* (2 Vols; New York, 1953) Vol. 2, p. 40.

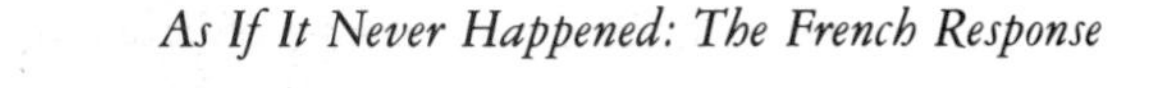

Henri Poincaré. Courtesy of the American Institute of Physics Niels Bohr Library.

Lorentz had been developing his theory step by step with experiments. But according to Poincaré, the theory did not accord with Newton's laws. He illustrated the problem with the following example: Suppose electromagnetic radiation traverses the absolutely fixed ether of the Lorentz theory. When it encounters a conductor (a piece of copper, for example) a force will be exerted on the conductor. All other things being equal, this force causes the copper to accelerate. Since there is no other ponderable matter in the vicinity and the ether is fixed, there is no reaction to this force.

Poincaré urged Lorentz to modify his theory to change the result, while recognizing that:

> It is impossible to measure the absolute movement of ponderable matter, or better the relative movement of ponderable matter, with respect to the ether. All that one can provide evidence for is the movement of ponderable matter with respect to ponderable matter.*

This statement, Poincaré thought, summed up a number of experimental facts and we recognize it as a statement of the principle of relativity. For Poincaré it *was* a summary statement of experiment. In 1895 he was disturbed because the Lorentz theory predicted that this principle of relativity is only true for first order experiments, and yet there was evidence, the Michelson-Morley experiment for example, that the principle applies to second order experiments as well.

As early as 1895, Poincaré had mapped out a program for the completion of the electromagnetic theory of bodies in motion which was to sustain him in mathematical physics for the remainder of his career. He advocated modifying and adjusting the Lorentz theory of electrons by making the principle of relativity an integral part of the theory, including not only first order, but second order experiments, and by uniting Lorentz theory with classical Newtonian mechanics. Each of these strands merits examination.

THE PRINCIPLE OF RELATIVITY

In 1899, in a series of lectures that Poincaré gave at the Sorbonne on electricity and magnetism, he returned to the question of the principle of relativity. He was emphatic in his belief that experiment showed that all higher order ether drift experiments gave a null result. Referring to the contraction hypothesis that Lorentz had recently proposed, Poincaré said that the strange property seemed to be a "fudge" factor from nature to prevent the detection of the motion of the earth through the ether. And a year later, in 1900, Poincaré said that the principle of relativity was not imposed as a postulate but rather it was verified by experiment.

In 1904, Poincaré gave a widely heralded address at the St. Louis exhibition. When he spoke of the principle of relativity, he said that it demanded that:

> . . . the laws of physical phenomena should be the same whether for an observer fixed or for an observer carried along in uniform movement of

* *Oeuvres d'Henri Poincaré* (11 Vols.; Paris, 1934-54) Vol. 9, p. 412.

> translation; so that we have not and could not have any means of discerning whether or not we are carried along in such a motion.*

Again, he cited experiment as the source of his confidence in the principle.

It was Poincaré's recognition of the principle of relativity which led some scholars to conclude that he had anticipated Einstein's theory. But we saw in Part I that a principle of relativity was important in physics since at least the time of Newton, and that during the nineteenth century, there were attempts to extend the principle from mechanics to all experience. Had the program behind the mechanical world view been successful, the extension would have been automatic, since the principle of relativity for mechanics would have been applicable to all phenomena.

More important, when one compares Poncaré's attitude toward the principle of relativity with Einstein's attitude, the contrast is startling. In Einstein's view, whereas the principle was *suggested* by experiment, it was *a priori.* It was a postulate in a theory of measurement. No experiment that was analyzed within the theory could confirm or deny the principle.

According to Poincaré, the principle of relativity was simply a summary of experimental facts. Indeed, after Kaufmann published the results in 1906 of his comparison, by experiment, of the predictions of Abraham, Lorentz and Einstein for the mass of the moving electron, Poincaré immediately suspected that the principle of relativity might not have the "rigorous value" previously ascribed to it. Later he said that a single experiment like Kaufmann's might upset all of electricity, magnetism, and optics.

This is very puzzling. Poincaré not only published actively in mathematical physics, theoretical astronomy, and pure mathematics, he also wrote popularizations of science in which he strongly advocated an epistemological stance now given the label "conventionalism." In that context he spoke of the principle of relativity as a convention that was not amenable to the test of experience. It was agreed to because it was the most convenient and simplest way of conceiving of reality without direct experience.

Whereas Poincaré used the conventionalistic stance when *talking* about the nature of physics, he used naive realism, when *working* in theoretical physics. Poincaré the physicist required that the principle of relativity be supported, indeed proved, by induction from experiment. Whereas the principle of relativity was a postulate for Einstein's theory, for Poincaré it was only another summary of experimental results within a larger and more ambitious theory. It was a theory of

* H. Poincare, "The Principles of Mathematics," *The Monist,* 1905, *15*: 1-24, p. 5.

electrons which would account for the interactions of matter with matter, radiation with radiation, and matter with radiation. The contrast between Poincaré and Einstein about the status of empirical information is best shown by discussing how each analyzed the synchronization of clocks.

POINCARÉ AND SECOND ORDER THEORY

In Chapter 3 we analyzed the synchronization of clocks and indicated the logical connection between that problem and Einstein's statement of the second postulate of relativity: the velocity of light is an invariant. Einstein may have thought about this problem as early as 1895, but he published nothing about it until his first paper on the special theory of relativity in 1905.

Poincaré, on the other hand, had begun analyzing publicly the problem of synchronizing clocks as early as 1898. He repeated the analysis in 1900, 1905 and 1912. Poincaré's treatment differs from Einstein's in only one crucial way. Poincaré assumes the existence of an ether frame of reference.

He stated that it was impossible to measure the speed of light without the postulate that it has the same value in all directions. He said further that, whereas it could not be verified by experiment, an experiment could contradict that postulate. Poincaré suggested that the validity of the constancy of the velocity of light in a frame of reference can be inferred by attempting to show that its adoption leads to predictions that do not conform to experiment.

Poincaré then invited his audience to picture two observers, *A* and *B*. Both have clocks and they wish to synchronize them with light signals. *B* sends a signal to *A* and *A* sets his or her clock to the time when he or she receives the signal. Poincaré noted that the process would build a systematic error into the setting of the clocks since it takes a time, given by the ratio of the distance the light traverses to the speed of light, for the light to traverse the space between the observers. The error is easy to correct. We assume that the speed of light is the same in both directions. Poincaré implied that this assumption is satisfactory only if the observers are at rest. That is, that the observers be in the frame of reference of the absolute ether. When the signal from *A* is received by *B*, *B* starts his or her clock, set ahead by a factor d/c. D is the distance between the observers and c is the speed of light in the ether frame of reference.

Poincaré asked what would happen if the observers, while at rest with respect to each other, shared a common motion, relative to the ether. He said that the problem becomes complicated because "the speed of light is the same in all directions" is only correct if the observers are at rest. Of course, he meant at rest with respect to the fixed ether. The idea can have no other meaning. But, if the principle of relativity is true, the observers have no way of knowing that they are in a moving frame of reference. They would not synchronize their clocks properly. Yet, they would not know that the speed of light was not the same in all directions relative to them because of the compensatory effects of the Lorentz-Fitzgerald contraction.

The bases of Poincaré's analysis were the principle of relativity and the Lorentz-Fitzgerald contraction. The conclusion of the analysis which began with those statements was the invariance of the speed of light. That analysis can be contrasted with Einstein's, which begins with the principle of relativity and the invariance of the speed of light. The question of absolute motion relative to a resting ether frame of reference is not relevant. The contraction is one of the kinematic conclusions of the postulates.

It is clear that long before Einstein published his first paper on special relativity, Poincaré had found the framework which was to prove heuristic to Einstein. Poincaré had considered those elements which Einstein was to use as the building blocks of his theory. Poincaré was not as bold as Einstein. To Poincaré, both the principle of relativity and the invariance of the speed of light in *vacuo* were statements induced from experience. The fundamentals of his theory were ether and electrons.

THE THEORY OF ELECTRONS

In one sense, Poincaré had a more ambitious vision than Einstein about these problems. Whereas Einstein was constructing a very tight, limited theory of measurement, Poincaré was an active supporter of Lorentz's attempt to account for all the interactions of matter and radiation using the electron theory. He saw defects in the early attempts of Lorentz. In particular, he did not like what he perceived to be the *ad hoc* nature of Lorentz's early explanations of the second order ether drift experiments by invoking the contraction hypothesis. Poincaré publicly urged him to change the manner in which the hypothesis had been introduced. Meanwhile, between 1900 and 1904, Poincaré attempted to rationalize the theory of electrons to satisfy his own attitudes. He tried to rescue the principle of action and reaction by assuming that electromagnetic energy acts as a kind of fluid (He referred to it once as a "fictitious fluid.") which carried part of the momentum of the system.

When Lorentz published his second order theory in 1904, Poincaré was well pleased. Lorentz had pointed to Poincaré's criticism of the *ad hoc* nature of the contraction hypothesis as one of Lorentz's motivations. Poincaré apparently thought that by assuming the transformation equations and deriving the contraction, Lorentz had generalized his theory.

Poincaré also expressed more and more confidence in the view that all of the electron, and hence, all mass, was electromagnetic in origin. By 1905, Poincaré viewed mass as a kind of electromagnetic induction of the Lorentz electron, and by 1908 stated boldly that "Beyond electrons and ether there is nothing."*

POINCARÉ'S VISION OF A GOOD THEORY

I have devoted considerable time to describing the views of Poincaré because he was the leading theoretical physicist in France who was working with the same problems as Einstein. He had anticipated many of these problems, although he did not put the puzzle together as Einstein later did. There is little question that Poincaré knew the literature in this field. In fact, one of his responsibilities to the French Academy of Sciences was to report periodically on advances in electromagnetic theory. Yet, until his death in 1912, Poincaré never mentioned Einstein's name in connection with the problems associated with the special theory of relativity, and in fact did not refer to the special theory of relativity at all. Rather, he spoke of "Lorentz's theory," "The New Mechanics" or "Non-Newtonian Mechanics." It was a strange situation, made even stranger because no other French physicist mentioned Einstein or the theory of relativity before Einstein's visit to France in 1910, either.

It is always difficult to explain why something did not happen. We can, however, sketch in the elements that we know were important to Poincaré in the development of good theories and perhaps indicate why he thought that Einstein's theory did not meet those criteria. Then we will discuss why a single individual in the French physics community could exert such an enormous influence, one not possible within the German physics community.

Poincaré had three requirements for "good" scientific theories: simplicity, flexibility and naturalness. Each of these will be discussed and then related to Poincaré's refusal to acknowledge the existence of the special theory of relativity.

SIMPLICITY AND INDUCTION

Poincaré, like most scientists, maintained that nature should be described by simple laws. It is difficult to define "simple." On the one hand, Poincaré often appeared to suggest that a criterion of simplicity be imposed on physical laws. This was in accord with his widely publicized view that the laws of nature are agreed-upon conventions. For example, he had argued that only Euclidean geometry would ever be used to describe physical space because that geometry was simpler than any other.

* H. Poincaré, *Science and Method* (New York, n.d.) p. 209.

But a contradiction appears between Poincaré's views about conventionalism and his beliefs about the proper criteria in choosing among electromagnetic theories. Poincaré apparently reserved his conventionalism for mathematics and problems closely related to mathematics, but for problems in physics—for instance, the choice of the proper laws of nature—Poincaré believed that nature informs us of the truth.

There is a direct connection, for example, between simplicity and nature, according to Poincaré. He believed that of all the events that may occur, those that are the most *interesting* are more likely to recur most often, because it is those events on which we base the laws of nature. And, he believed that of all the possible events, the ones most likely to recur are also the simplest. Poincaré did not define *fact* or *event*. But from his contexts it is probable that he meant "phenomenon," or "perception."

Poincaré was not an empiricist who relied solely upon induction to form a bridge between data and theory. For at the same time that he argued for the supremacy of experimental evidence and experience, he also stated that theories are the free creation of the human mind.

Apparently, when Poincaré *talked about* physics, he invoked notions of conventionalism. When he *did* physics, he acted like an empiricist. For example, his view that the principle of relativity and the contraction hypothesis were generalizations from experience fit well with his thinking about simplicity and induction. The most likely laws to be induced are the simplest because the phenomena they describe have the highest chance of recurring.

FLEXIBILITY AND GRADUALISM

Gerald Holton has identified "gradualism" as a tendency in Poincaré's thinking about the problems of the electrodynamics of moving systems.* Holton had in mind an attitude which Poincaré displayed many times when he urged against rejection of Lorentz's theory. As early as 1895, although Poincaré recognized serious defects in Lorentz's analysis, he maintained that the theory should be carefully examined, modified little by little, and "perhaps everything will arrange itself." This gradualistic approach was in keeping with another feature which Poincaré maintained that good theories should exhibit: They should be flexible.

For example, when a theory like Lorentz's is first proposed, there are likely to be serious difficulties and objections. The best theories maintain their validity, according to Poincaré, despite such objections, by having the flexibility to incorporate serious modifications without destroying their essential character. "The objections thus serve, rather than to harm the theory, to permit it to develop

* G. Holton, *Thematic Origins of Scientific Thought* (Cambridge, 1973) p. 188.

all the latent truth which it contains."* There is little question that Poincaré would have considered Newton's theory *good,* exhibiting a high degree of flexibility. In 1889 Poincaré received his first prize from the French government for showing how to modify Newtonian theory in order to solve the so-called "three body" problems. If it were thus modified, Newtonian mechanics alone could probably account for all astronomical phenomena pertaining to the motion of celestial objects. His colleagues characterized his work as the culmination of three centuries of research. Poincaré's student and friend, Paul Langevin, later recalled that Poincaré was disquieted in 1904 at the prospect that Newtonian physics, for which he had earned his prize, might be superseded by a new mechanics.

As late as 1909, Poincaré had not definitively rejected Newton's theory. He asked: "Are the general principles of dynamics which have served since Newton's day as the foundation of physical science and appear immutable on the point of being abandoned or, at the very least, profoundly modified?"†

It was premature to decide definitively. Even if Lorentz's theory triumphed, Newton's mechanics would continue to be applicable to mechanical phenomena at slow speeds compared to the speed of light.

NATURALNESS AND THE INCREASE OF HYPOTHESES

Whereas theories had to be sufficiently flexible to be modified in the light of new experimental results, there was a counterbalance, in Poincaré's view, to this attribute; the theory had to exhibit "naturalness." Behind Poincaré's objection to Lorentz's *ad hoc* attempts to account for the second order ether drift experiments was Poincaré's conception that the theory should exhibit naturalness. In 1900, when he commented about Lorentz's proposal for a length contraction, Poincaré objected to the introduction of a new hypothesis to handle every successive change brought about by more sensitive experimental results. The need for the contraction hypothesis had been forthcoming because Lorentz's first order theory was not adequate to explain second order experiments like that of Michelson-Morley. Poincaré maintained that:

> A well made theory ought to permit the demonstration of the principle of relativity with a single thrust, in all its rigor. The theory of Lorentz does not allow for all that yet. . . . One can only hope for the perfectly

* *Oeuvres d'Henri Poincaré,* Vol 9, pp. 464-488.

† H. Poincaré, *La mechanique nouvelle* [an address delivered at the French Association for the Advancement of Science, 1909], (Paris, 1924).

> satisfactory rendering of this relationship without too profound a modification.*

This appeal, in 1900, was repeated in 1904 just before the publication of the final version of Lorentz's second order theory. Lorentz had paid close attention to the criticisms of Poincaré. In the Introduction to his 1904 work Lorentz remarked,

> Poincaré had objected to the existing theory of electrical optical phenomena in moving bodies that, in order to explain Michelson's negative result, the introduction of a new hypothesis has been required and that the same necessity may occur each time new fact will be brought to light. Surely this course of inventing special hypotheses for each new experimental result is somewhat artificial. . . . I believe it is now possible to treat the subject with a better result.†

Lorentz introduced, not only the Lorentz contraction as a hypothesis, but as Holton has pointed out, approximately ten more hypotheses to explain specific phenomena.‡ Remarkably, Poincaré was satisfied. That satisfaction was apparently because Lorentz had: (1) extended the Lorentz contraction hypothesis to all bodies moving with respect to the ether, and (2) generalized the transformation equations to situations other than those in electrodynamics. That Lorentz had made these extensions by *fiat* did not disturb Poincaré.

The key to Poincaré's satisfaction appears to have been that, regardless of his methods, Lorentz had generalized his theory. Poincaré must have thought that the theory was natural because it applied to all phenomena. Even after 1905, Poincaré invariably treated the Lorentz contraction as one of the fundamental postulates of the theory.

POINCARÉ'S SILENCE ABOUT EINSTEIN'S THEORY

There are two reasons why I devoted so much attention to Poincaré's work in the theory of electrodynamics and its relationship to his general attitudes toward theories. There has been an important minority view that the theory of relativity was anticipated by Poincaré in association with Lorentz. I do not subscribe to that view. Therefore, to illuminate the differences between Poincaré and Einstein about this area of physics, and the general relationship these two men saw between ideas and evidence, we examined Poincaré's physics and his philosophical positions about the status and meaning of theories. This comparison also serves as a nice test

* H. Poincaré, *Electricité et Optique* (Paris, 1901) p. 536.

† H. A. Lorentz, "Electrical Phenomena in a System Moving with any Velocity less than that of Light," in Lorentz *et al*, *The Principle of Relativity* (New York, 1923) p. 13.

‡ G. Holton, *Thematic Origins of Scientific Thought* pp. 165-183.

of the views that having the same mathematical formalism does not necessarily mean that two theories are identical, have the same power, or the same usefulness. That two theories have the same mathematical formalism merely means that different sets of postulates may be used to arrive at the same conclusions.

There is another reason to write about Poincaré's contributions. He did not mention Einstein's theory of relativity from the time of its introduction in 1905 until his death in 1912; few French physicists spoke of it either.

One can only ponder the reasons for Poincaré's silence. He died prematurely from post-operative complications in 1912, and we are deprived of his reactions to the theory in later years when it was assimilated throughout the world.

Everything that we know about Poincaré's character belies a motivation as petty as jealousy. Again and again, men who knew him as a colleague, teacher, or as a distant, but great man, have testified to his intellectual integrity, generosity to others, and his humble lack of concern with status.

It may have been that Poincaré thought that Einstein's theory was trivial— perhaps trivial because Einstein had only provided a theory of measurement to account for observations that were made in inertial frames of reference. Poincaré and Lorentz had been working on a comprehensive theory of electrons which was going to subsume all of physics.

There is something more concrete and important to be said about Poincaré's silence. We have examined three features which Poincaré held to be important for good theories: they should be simple, flexible, and natural. It is likely that Einstein's theory failed on these three counts in Poincaré's eyes. It lacked simplicity because it did not begin by generalizing from simple "facts" as the Lorentz theory did. Einstein began by questioning notions of simultaneity, noting certain symmetries in nature. He rarely cites experimental evidence, or the Michelson-Morley experiment. It is also likely that Poincaré was not attracted to the logical tightness of Einstein's theory. There are only two postulates in the special theory of relativity, and one cannot alter either significantly without greatly altering the conclusions. In that sense, Einstein's theory was not flexible. Finally, given Poincaré's views of natural theories and *ad hoc* hypotheses, Einstein's second postulate of the invariance of the speed of light must have seemed artificial.

On all three counts then, simplicity, flexibility and naturalness, Einstein's theory of relativity would have fallen short in Poincaré's eyes.

OTHER FRENCH RESPONSE

There was virtually no response to the theory of relativity *per se* in France during the years between 1905 and 1911. Later in his life, Einstein commented that, had he not developed the ideas of the theory, they would have been developed by his close friend Paul Langevin, who as a young man had been Poincaré's student and subsequently his colleague. There is little evidence to support Einstein's contention.

As late as 1904 Langevin had doubts about the principle of relativity, which he expressed by suggesting that perhaps experiments would discover the absolute motion of the earth. At the same time he developed a theory of the electron on the assumption that, although the electron was deformable, it maintains a constant volume. Such an electron, Poincaré had immediately pointed out, was not compatible with the principle of relativity.

In 1906, Langevin's position was similar to that of Poincaré. Although he had not embraced the principle of relativity with as much confidence as Poincaré, he was more confident of the electromagnetic nature of the mass of the electron and of the *material* existence of the ether.

After Einstein's visit to France in 1910, there was a noticeable shift in Langevin's view. He emphasized that the principle of relativity was a conclusion found by experiment, and he recognized that Einstein was the first to obtain Lorentz's transformation equations by the idea of the invariance of the speed of light. It is clear from Langevin's writings that he held the invariance of the speed of light to be a principle obtained from experiments like the Michelson-Morley experiment. As late as 1920, although it played no active role in his analysis, he retained the concept of ether as the preferred, if unknowable, frame of reference within which electrodynamic processes proceeded.

The physicist E. M. Lémeray also became aware of the theory of relativity after Einstein's visit to France. And although he understood the distinction in the analysis made by Einstein and Lorentz about the sources of the transformation equations, he maintained that the ether and absolute space are essential. Later Lémeray became one of the few outspoken proponents of special relativity in France.

The only other important work in French physics on those questions between 1905 and 1911 is that of G. Sagnac, who devised ingenious techniques to detect the motion of the earth through the ether. The conclusion of his experiments was that the principle of relativity is a valid experimental generalization. It was the validity of the principle, according to Sagnac, that prevented the detection of the earth's motion through the ether.

It is difficult to believe that the silence in France was not influenced by Poincaré's attitude. The leading theoretician in France, who was probably the most eminent member of the Academy of Sciences at that time and the leading expert on electrodynamics, had opted for the Lorentz theory of the electron and never changed his position. On the other hand, to the extent that Poincaré's attitudes toward Einstein's theory were in part cultural, the same factors might have shaped the attitudes of other French physicists, as well.

AFTERMATH

In trying to understand French attitudes, it is very instructive to examine later French reconstructions of the development of physics after 1911. In 1906, Lucien Poincaré, a cousin of Henri Poincaré, had published a book about the evolution of modern physics, a considerable portion of which was devoted to discussing the relationship between ether and matter. It is not remarkable that in 1906, L. Poincaré insisted on the need for the luminiferous ether, that Kaufmann's experiments had confirmed the predictions of Max Abraham, and that the electron was a determined volume—a region of the ether possessing special properties, or even that in the entire book there is no mention of Einstein or the special theory of relativity. It is remarkable, however, that when the book was republished in a second edition in 1920, none of these features of the first edition had been altered. At the time of the publication of the second edition, its author was director of the French ministry of public instruction. The book also received an *imprimateur* from the French Academy of Sciences.*

Not every French author was neutral. A. and R. Sartory published a survey of scientific opinion in France about the theory of relativity, probably between 1920 and 1930.† Although they claimed that opinion was divided, they pointed to only one supporter of the special theory of relativity, Paul Langevin. In opposition to the theory, the authors presented a variety of statements from eminent men of French science. Whereas some saw beauty in the structure of the theory, according to the Sartorys the theory was rejected by these leaders of French science because it did not have a firm basis. These were the kindest comments on Einstein's theory in the book.

One finds considerable hostility toward Einstein in this later French literature also. For example, in 1921, Charles Nordmann quoted texts of Poincaré to show that he had indeed anticipated Einstein with the relative measurements of space and time, and added:

> All of this Poincaré and others have indeed sustained long before Einstein, and it is a harm to the truth that it be attributed to him.‡

There was some support for Einstein's theory in late French literature. Earlier we cited E. M. Lémeray as one of the few men in France who had openly discussed Einstein's theory before 1911. In 1916 he published a small supportive textbook on the theory of relativity. It was followed in 1922 by a second book, *L'ether actual,* which was also supportive of Einstein's theory and which denied the

* L. Poincaré, *La Physique Moderne Son Evolution* (Paris, 1st ed., 1906; 2nd ed., 1920).

† A. Satory and R. Satory, *Vers le monde d'Einstein* (Strasbourg, n.d.).

‡ C. Nordmann, *Einstein et l'universe* (Paris, 1921).

existence of the ether. In France, this must have been considered remarkable. Even more remarkable is that the 1922 work contains a very hostile preface written by Leon Francois Alfred Le Cornu, one of the most honored and eminent mechanical engineers in France.

Le Cornu objected to Lémeray's conclusion that the ether does not exist. According to him the only important question was to determine the properties that the ether possesses to account for all of the facts. Lémeray, Le Cornu maintained, retreated from the forefront of physics by denying the existence of materials like the ether on the grounds that such a material presented seemingly insoluble problems concerning its nature and motion.

There was a more serious and significant issue raised in the argument between Le Cornu and Lémeray which perhaps reveals the core of the French uneasiness with Einstein's theory and the reasons the theory of relativity remained outside the mainstream of French physical science. Earlier, Henri Poincaré had urged caution in accepting Lorentz's theory as a replacement for Newtonian physics because it might create serious difficulties for science instruction in French schools. Teachers might no longer teach elementary mechanics to their students; they might replace it with a new mechanics in which mass and time no longer had the time-honored meanings.

It was to this issue that Le Cornu and Lémeray returned in 1922. In the text of his book, Lémeray lamented that textbooks in France too closely resemble ancient treatises and that new ideas are not found in them. Lémeray thought that the authors made little effort to comprehend the new ideas. He compared the lethargy in curriculum reform with regard to the special theory of relativity to the difficulties at the beginning of the nineteenth century with the introduction of Lavoisier's theory of combustion and the rejection of the concept of phlogiston. Lémeray urged a thorough revision of the curriculum to return joy and understanding to the students of physics in France.

In his preface Le Cornu rebutted Lémeray by urging that Newtonian mechanics continue to form the basis of scientific education in France. Le Cornu wrote that he knew from conversations with him that Lémeray intended the introduction of the theory of relativity to be restricted to those pursuing advanced degrees. Le Cornu urged restraint until it were ascertained if the ideas of the theory of relativity would receive an unqualified and universal endorsement, "beyond the shadow of a doubt," characteristic of Newtonian mechanics.

The fears of Poincaré and Le Cornu were unwarranted. As late as 1955, in the preface to a textbook on relativistic physics written by Henri Arzeliès, one finds the following remarks:

> . . . In France more or less, there exist few general expositions of relativity theory and most of these are addressed to a restricted audience

> (elementary level or, on the contrary, mathematically very sophisticated). Outside of rare and very commendable exceptions, the teaching of fundamentals illustrates the ignorance of our *Faculties* with regard to the theory of relativity. The certificate of rational mechanics, for example, does not contain, in general, a single lesson on the mechanics of large velocities (and indeed still less on quantum mechanics). All is passed as if at the start of this century, an ill-natured genie had petrified French mechanics (that which one teaches) in a statue of mathematical salt. . . . *

Even today the curricula of French universities is codified at the national level. This type of organization of the educational system may explain the slowness with which the French responded to the special theory of relativity. It suggests that the diffusion of an innovative scheme, be it in science or other disciplines, can be severely retarded when control of the curriculum is exercised nationally, instructors must follow a predetermined syllabus and the examinations are devised at the national level. It is tempting to identify the "wicked genie" referred to by Arzeliès as Henri Poincaré. But more accurately, the genie which petrified French education into a "statue of mathematical salt" was more likely the conical oligarchical educational system which allowed one person to so influence the entire system.

* H. Arzeliès, *La cinematique relativiste* (Paris, 1955), p. vii.

·8·

Defending the Ether: The British Response

THE BRITISH RESPONSE TO THE THEORY of relativity during the years 1905-1911 had a different character from the responses we have analyzed in Germany and France. For the first two years of the period, there was virtually no recognition of the theory in the literature. Following that, there was some controversy, not as much as in the German literature. However, there was not the wide diversity of opinion, the elaboration, or the debate so characteristic of German response. The acceptance of the theory hinged upon making it compatible with the concept of the ether. As paradoxical as that might be, there was almost unanimous agreement within the British physics community about such a program. Almost all British physicists who worked in electrodynamics behaved in similar ways about these issues.

THE TRADITION OF BRITISH ETHER THEORY

The concept of an ether, a space filling medium responsible for the propagation of various kinds of forces, was not a concept of interest only to the British. It had been the French philosopher René Descartes who had constructed the universe of light and matter out of a type of ether and, in the eighteenth century, natural philosophers on the Continent, in Great Britain and in the United States had made use of several different kinds of ether for the transmission of several types of phenomena whose propagation through space could not otherwise be understood. There was an electric ether, a magnetic ether, and a gravitational ether. It was not until the nineteenth century, as the wave theory of light gained acceptance, that a luminiferous ether became important. Later, with the identification of light as an electromagnetic disturbance, three of the four phenomenas—light, electricity and magnetisms—could be recognized as propagated within one medium, the electromagnetic ether. This suggested a research program in which scientists tried to unite gravitational effects with electromagnetic effects. This was to be part of Lorentz's program; the possibility of a unification has been basic to much of natural philosophy. It motivates many of today's theoretical high energy physicists.

Whereas there was widespread acceptance of the *concept* of the ether in the nineteenth century, the detailed investigation of the properties of the ethereal medium became a uniquely British enterprise. Work on the ether in Britain during the nineteenth century was concentrated in "The Cambridge School," a group of mathematical physicists who studied or worked at Cambridge University and included George Green, James MacCullagh, William Thomson (Lord Kelvin), and Charles Stokes, among others. "Ether Mechanics" was the investigation of these unearthly properties and how they related to ordinary matter and radiation. The investigation was an all-consuming occupation to some British physicists. In the words of Arthur Eddington, one of the leading British theorists of the twentieth century:

> The nineteenth century is littered with debris of abortive aethers—— elastic solids, jellies, froths, vortex networks. . . . *

It is, then, not surprising to learn that between 1905 and 1907 no notice was taken of Einstein's theory. As late as 1923 the physicist and later philosopher Norman R. Campbell wrote:

> . . . It remains an indubitable fact that, in spite of the attempts to enlighten him, an average [British] physicist—the man in the laboratory as I have ventured to call him—is still ignorant of Einstein's work and not very much interested in it.†

In 1907, Lord Kelvin, considered by many to have been the greatest of Britain's nineteenth century physicists, made his last appearance at the British Association for the Advancement of Science. In a paper on the motion of the ether, produced by the collision of atoms, he assumed an ether that is an elastic, compressible, non-gravitational solid. The physicist Oliver Lodge also read a paper on the motion of the ether. According to Lodge the density of the ether must be on the

* A. S. Eddington, "Larmor, Sir Joseph," *Obituary Notices of Fellows of the Royal Society,* 1942, 4: (#11).

† N. R. Campbell, *Relativity* (Cambridge, 1923), p. v.

order of magnitude of thousands of tons per cubic millimeter and every part of the ether must be "squirming with the velocity of light."

Such considerations of the character of the ether at the end of the nineteenth century are qualitatively different from similar investigations on the Continent. In Lorentz's ether the substantial reality of it was not an issue. More and more frequently in the Lorentz analysis, the ether became a symbol for absolute space. Lorentz did not work very hard to find the existence of the ether and did not worry about the detailed properties of the medium which propagates electromagnetic energy.

In British physics, at the end of the nineteenth century and during the first decades of the twentieth century, a considerable effort was devoted to discovering the detailed properties of the ether. When thinking about these properties, it was common to compare the ether to ordinary things like glass, motor oil, glue, or gelatin.

OLIVER LODGE

As his work is an extreme example of British preoccupation with the ether, and he was widely respected in British physics, let us examine Oliver Lodge's work on the ether. From 1881 to 1900, Lodge was Professor of Physics at the University of Liverpool. In 1900, he became the first Principal of the new university in Birmingham. He was knighted in 1902. He was very active in the British Association for the Advancement of Science and was the president in 1913. His major research interests were in the propagation of electromagnetic waves and it is commonly accepted that, had Heinrich Hertz not discovered the phenomenon of the transmission of those waves through space, Lodge would have done so. In any event Lodge made many fundamental contributions to wireless telegraphy.

It does not follow that someone who has devoted himself to studying the transmission of electromagnetic energy through space would have a singleminded focus on the ether and its properties but neither is it surprising. With passing time, Lodge became a mystic and an outspoken defender of beliefs in ESP and other parapsychological phenomena. His belief in the ether also grew more and more mystical. In 1925 in a book entitled *Ether and Reality* he wrote as follows:

> Touch seems to be purely material sensation, the result of direct contact with matter: it is indeed what we call 'contact.' But when we come to analyze touch, we learn that atoms are never in contact. They approach each other within an infinitesimal distance; but there is always a cushion, what may be called a repulsive force between them—a cushion of ether. Hence even our apparently most material sense is dependent on this omnipresent medium on which alone we can directly act, and through which all our information comes. It is the primary instrument of the

> Mind, the vehicle of Soul, the habitation of Spirit. Truly it may be called the garment of God.*

Seven years later in a book intended to sum up his philosophic outlook he wrote:

> The Ether of Space has been my life study and I have constantly urged its claims to attention. . . . I always meant someday to write a scientific treatise about the Ether of Space; but when in my old age I came to write this book, I found that the Ether pervaded all my ideas, both of this world and the next. I could no longer keep my treatise within the proposed scientific confines; it escaped in every direction and now I find has grown into a comprehensive statement of my philosophy.†

These are extreme statements. They represent the views, not only of a person committed to the concept of the ether, but also of a mystic who utilized the ether concept as a central feature of his mystical beliefs. Whereas almost no other British physicist might have shared Lodge's mystical beliefs about the ether, almost all of them shared his conviction of the centrality of the concept of the ether for understanding the propagation of electromagnetic theory. Lodge's writings, then, are useful to examine the centrality of the ether in the thinking of British physicists.

In 1889 Lodge published a textbook entitled *Modern Views of Electricity,* which was extremely popular and widely read. He published three editions, the second appearing in 1892 and the third in 1907. In this book Lodge made explicit the mechanical model for electricity and the ether which was only implicit in the writings of other British physicists. It is remarkable how constant Lodge's views remained through the three editions. In his introduction Lodge wrote that the "doctrine expounded in this book is the etherial theory of electricity." The analogy he used for understanding the relationship between electricity and ether was that, as heat is a manifestation of the energy of material, (the energy of the random motions of the atoms making up matter), so electricity is a manifestation of the ether, magnetism is another manifestation and light is yet a third. The eye may be called the etherial sense organ as the ear is an "aerial" sense organ.

The underlying philosophy and metaphysics for these views was basically an appeal to common sense. Lodge maintained that there were two possibilities: space either is or is not empty. Those who argued that there is empty space did so because of the need to provide room for the motion of the fundamental entities that make up the material world.

---

* Oliver Lodge, *Ether and Reality* (New York, 1925) pp. 178-179.

† Oliver Lodge, *My Philosophy: Reporting My Views of the Many Functions of the Ether of Space* (London, 1933), preface.

On the other hand, those who argued that space is not empty based their considerations on the enormous distances between objects in interstellar space and the need to transmit energy from one place to another. According to Oliver Lodge:

> If anyone tries to picture clearly to himself the action of one body on another without any medium of communication whatever, he must fail. A medium is instinctively looked for in most cases; and if not in all, as in falling weight or magnetic attraction, it is only because custom had made us stupidly callous to the real nature of these forces.*

Lodge summarized the entity that satisfies the need for a medium. "One continuous substance filling all space, which can vibrate as light, which can be sheared into positive and negative electricity, which in whirls constitutes matter; and which transmits by continuity and not by impact every action and reaction of which matter is capable."†

These ideas were originally expressed in a paper delivered in 1882 about the nature of the ether and its functions. They were incorporated seven years later into the first edition of *Modern Views of Electricity* and were repeated, essentially unchanged, in the 1892 and 1907 editions of that book and in almost all other documents Lodge published about the ether, its function and nature. Lodge's belief in the ether was in the tradition of his British predecessors and contemporaries including Kelvin, MacCullagh, Green, Stokes, Oliver Heaviside, G. F. Fitzgerald, and Joseph Larmor, all of whom had theorized about the nature of the ether. Most of those men were theoretical physicists, but experimentalists like J. J. Thomson also shared Lodge's enthusiasm for the ether. Lodge's initial convictions about the ether and its nature stemmed from a metaphysical inability to conceive of two bodies affecting each other without a mechanical connection.

The preface to Lodge's 1907 edition of *Modern Views of Electricity* indicated the direction toward which British physicists would move to incorporate the recent developments in the electrodynamics of moving bodies. Lodge wrote that it was noteworthy that he had been required to make very few corrections in the text which had been written twenty years earlier. This indicated to him that newer discoveries were supplementary rather than fundamental. Lodge was not explicit about these newer discoveries but it is likely that he meant discoveries about the electron theory of matter. It was under that banner that the special theory of relativity was introduced into British physics, by another, younger, British physicist, Ebenezer Cunningham, a student of Larmor.

* Oliver Lodge, "The Ether and Its Functions," Delivered at the London Institution, December 28, 1882. Cf. Oliver Lodge, *Modern Views of Electricity* (1st ed., London, 1889) pp. 328-332. The same statements may be found in the 2nd (1892) and 3rd (1907) editions of the book.

† *Ibid.* p. 338.

THE MECHANICAL ETHER

Larmor was one of the outstanding British theoretical physicists of the last quarter of the nineteenth century. He had rejected the view of the ether that Lodge and others held in favor of a more abstract mathematical model. Lorentz was one of the initiators of that point of view. In a prize winning essay published in 1900 entitled *Aether and Matter* Larmor maintained that the only thing one need know about the ether can be formulated in the equations describing the propagation of electromagnetic energy through it. Larmor thought that to attempt further explanation of the nature of the ether by complicated mechanical structures was superfluous. He maintained this view until his death in 1940. Clearly his thinking was compatible with the Lorentzian analysis. In fact, the work of Lorentz and Larmor so paralleled each other that in Britain, often, one referred not only to the Lorentz theory or the Lorentz equations, but to the Larmor-Lorentz theory or the Larmor-Lorentz equations.

In refusing to speculate and develop a substantial view of the ether based on a mechanical model, Joseph Larmor was in the minority in British physics. For example, J. J. Thomson's attitudes toward the ether were similar to those of Lodge. Thomson expressed these views most clearly in the Adamson lecture at the University of Manchester in 1907. In order to insure the strict conservation of energy and momentum, he hypothesized an "invisible universe" intimately connected with the real universe, possessing the properties of inertial mass and storing momentum and energy when they appeared to be lost during certain electromagnetic processes. This invisible universe, Thomson continued, may be called *the ether*. In his Presidential Address to the British Association for the Advancement of Science in Winnipeg in 1909, he elaborated further on this model by asserting that the ether is the seat of all electrical and magnetic forces, that it is not "the fantastic creation of the speculative philosopher; it is as essential to us as the air we breathe." He went on to suggest that this invisible universe be regarded as a bank in which energy may be deposited and withdrawn.

Throughout the period in which the theory of relativity was being elaborated upon in Germany, British physicists dealing with the same set of phenomena remained preoccupied with the nature of the ether and concentrated on understanding the null result of second order ether drift experiments by hypothesizing definite interactions between matter and ether. There is ample evidence that British physicists who were aware of Einstein's theory had a great deal of difficulty understanding what he was talking about. For example, in 1909, at the urging of a colleague, Einstein had sent the British astronomer F. C. Searle a paper on relativity. In writing to Einstein to thank him for the paper Searle said:

> I have not been able so far to gain any really clear idea as to the principles involved or as to their meaning and those to whom I have spoken in England about the subject seem to have the same feeling.
>
> I think it would help us if you were to write a short account of the subject which could be translated and published in some English journal. . . . If I could give any assistance in the matter I should be glad to do so, but I say that I am not an expert in German.*

But the difficulty was not merely a matter of translation from the German. Of the three physicists whose ideas we have presented, Lodge called for a return to Newton, J. J. Thomson thought that Newton had not been abandoned and the biographers of Larmor referred to him as a child of the nineteenth century "with the names of nineteenth century heroes ever on his lips."†

Nor was it only the physicists who were in positions of power and responsibility who thought along those lines; it was a matter of culture and training. Almost all British physicists at the turn of the twentieth century were committed to the kind of mechanical model building we have described. J. T. Merz, writing at the end of the nineteenth century, characterized the model building as part of "the scientific spirit" in Britain:

> Continental thinkers, whose lives are devoted to the realization of some great idea, complain of the want of method, of the erratic absence of discipline which is peculiar to English genuis. . . . The Englishman of science would reply that it is unsafe to trust exclusively to the guidance of the pure idea. . . . ‡

Merz expressed the belief that the kinds of differences were much more pronounced at the beginning of the nineteenth century than at the end. He was convinced that improvements in transportation and progress in the ease of long distance communication had eradicated largely the "national peculiarities" in science.

But at about the same time that Merz was completing his comparative intellectual history from which we have been quoting, the French physicist-philosopher Pierre Duhem wrote that there is one element in the English treatment of physics which French physicists find astounding; namely, the use of physical models.§ Duhem discussed the fact that on the Continent, physicists were content

* F. C. Searle to A. Einstein, May 20, 1909. Einstein Archives, Princeton University.

† Eddington, *op. cit.*

‡ J. T. Merz, *A History of European Thought*. . . . Vol. 1, pp. 251-252.

§ Pierre Duhem, *The Aim and Structure of Physical Theory* (New York, 1962) pp. 69-72.

to imagine the abstraction of electric point charges and to calculate, abstractly, the forces between the charges. Furthermore,

> . . . the French or German physicist conceives in the space separating two conductors, abstract lines of force having no thickness or real existence; the English physicist materializes these lines and thickens them to the dimensions of a tube which he will fill with vulcanized rubber. In place of a family of lines of ideal forces conceivable only by reason, he will have a bundle of elastic strings, visible and tangible, firmly glued at both ends to the surfaces of the two conductors. . . .

Duhem then turned to Oliver Lodge's *Modern Views of Electricity.* His contempt for Lodge's use of mechanical models was extreme:

> Here is a book intended to expound the modern theories of electricity. . . . In it there are nothing but strings which move around pulleys, which roll around drums, which go through pearl beads, which carry weights; and tubes which pump water while others swell and contract; toothed wheels which are geared to one another and engage hooks. We thought we were entering the tranquil and neatly ordered abode of reason, but we find ourselves in a factory.

There is no doubt that British physicists relished the use of mechanical models in a process of explanation. Perhaps the most outspoken defender of that style was Lord Kelvin. He once told an American audience, "I never satisfy myself until I can make a mechanical model of a thing. If I can make a mechanical model, I understand it. As long as I cannot make a mechanical model all the way through, I cannot understand. . . ."* And because he could not build a mechanical model Kelvin rejected the electromagnetic theory of light and its propagation, even in the face of the success of Hertz and others in transmitting electromagnetic signals from place to place without wires.

It was not only model building which Duhem found distasteful, it was the total disregard that British physicists displayed for consistency in the building of such models. Such a penchant for model building has persisted throughout the twentieth century. Thus, as late as 1951, the British physicist-historian, E. T. Whittaker, wrote that James MacCullagh had devised an ether which satisfied the requirement that all vibration within it would be perpendicular to the direction of propagation without a component of vibration in the direction of propagation.† Whittaker called such a medium one which displayed "rotational elasticity."

* W. Thomson [Lord Kelvin], *Lectures on Molecular Dynamics and the Wave Theory of Light* (Baltimore, 1884), p. 270.

† E. T. Whittaker, *A History of Theories of the Aether and Electricity.* Vol 1, pp. 144-145.

According to Whittaker, one of the problems the medium had was that when MacCullagh devised the equations for it in 1839, there was no evidence of such behavior on the part of any known substance. According to Whittaker, "this difficulty was removed in 1889" when Kelvin designed a mechanical model displaying the desired property. It was not an ordinary mechanical model:

> Suppose . . . that a structure is formed of spheres, each sphere being the center of the tetrahedron formed by its four nearest neighbors. Let each sphere be joined to caps at their ends so as to slide freely on the sphere. Such a structure would, for small deformation, behave like an incompressible perfect fluid. Now attach to each bar a pair of gyroscopically mounted flywheels, rotating with equal and opposite angular velocities, and having their axes in the line of the bar; a bar thus equipped will require a couple to hold it at rest in any position inclined to its original position, and the structure as a whole will possess the kind of quasi-elasticity which was first imagined by MacCullagh.*

The model is for the luminiferous ether, a substance which not only must have the strength and rigidity to transmit luminous disturbances of wavelengths of the order of magnitude of one ten-thousands of a centimenter at speeds of three hundred thousand meters per second, but also must offer no resistance to moving matter and whose motion relative to an observer must be undetectable.

A favorite of mechanical models in nineteenth century British physics was the vortex atom. Both Lord Kelvin and the great German physicist Helmholtz had produced brilliant mathematical studies of the behavior of vortices in fluid media. Such vortices persisted over time and moved as a unit—almost as an entity. The mathematical description of the movement was sophisticated and yet simple. Kelvin thought that perhaps matter was the manifestation of the vortex motion of the ether. Eventually, however, Kelvin abandoned the model because it could not explain all of the properties of matter. J. J. Thomson, on the other hand, never lost his enthusiasm for the model. He won the 1882 Adams prize for an essay on the action of two vortex rings on each other. In his 1907 book entitled *The Corpuscular Theory of Matter,* he wrote that he held the vortex theory to be "fundamental." And as late as 1936 he claimed that he did not know of any phenomenon which could not be explained by making use of the vortex model. He added, "I regard the vortex-atom explanation as the gaol [sic] at which to aim."† Yet, as the Lorentz model for the electrical nature of matter and the role of the electron became known in England, Thomson had no difficulty tailoring his model and his

* *Ibid.,* p. 145.

† J. J. Thomson, *Recollections and Reflections* (London, 1936) p. 183.

belief in the fundamental aspects of the vortex-atom to meet the situation. The vortex atom might be fundamental, but the origin of all mass was, in Thomson's views, electrical, and resided in the total electrical nature of the electron.

THE INTRODUCTION OF RELATIVITY THEORY

We have already seen, in Part I, that the British interpretation was that the Lorentz theory was successful even when it explained the null result of first and second order ether drift experiments by using postulates like the Lorentz-Fitzgerald contraction. To the British such a theory was not *ad hoc*. Rather it represented an activity that British physicists were familiar with: the tailoring of a model to fit the data.

It is not surprising, then, to learn that it was in this manner of model tailoring that the formalism, if not the spirit of the theory of relativity, was introduced into the British literature. In 1907, Ebenezer Cunningham took part in a public, acrimonious, polemical debate with the German physicist A. H. Bucherer about the nature and import of the theory of relativity. What emerged from the dispute was that according to Cunningham it had been proven, presumably by experiment, that for a given set of phenomena the electromagnetic equations of Maxwell were equally applicable from the point of view of observers at rest with respect to the ether, or from the point of view of observers moving at a uniform speed with respect to the ether. Since the speed of light in both cases has the same value, Cunningham said, we have to explain that, regardless of the state of motion of the inertial observer relative to the ether, the speed of light has the same value. Cunningham then showed that if the speed of light has the same value for all inertial observers, the Lorentz transformations could be derived as shown in Appendix 6. That the speed of light is invariant is a consequence, Cunningham said, of the principle of relativity.

One of the fascinating features of Cunningham's analysis was that unlike Poincaré in France, he was not willing to ignore Einstein's theory. In fact, by 1909, he referred to the transformation equations as the "Lorentz-Einstein" equations. On the other hand, he was not willing to relinquish the concept of the ether as an active component. Rather he redefined the ether so as to make it compatible with the principle of relativity and then he made the principle of relativity a supplementary principle to the theory of electrons.

It was simple and very much in the British tradition of mechanical model building. Following his teacher Larmor (and Lorentz), Cunningham took the position that the ether had no substantial reality other than the equations which defined it. Those were the Lorentz-Einstein transformation equations, and therefore it follows that the luminiferous ether was *always at rest regardless of the frame of reference in which the measurement was made.* Strictly speaking, of course, this meant that one part of the ether is moving relative to other parts and it was on those grounds that Larmor rejected the work of his student Cunningham. On the other hand, the model that Cunningham proposed had made the ether undetectable in principle and the principle of relativity was no longer a statement about experience but rather a statement about the ether.

Not only did Cunningham reduce the theory of relativity to the principle of relativity, a statement of the undetectability of the ether, he made it auxiliary to the electron theory of matter. One had to distinguish, Cunningham asserted, between "modes of measurement" and "the nature of space and time." The *principle* of relativity demarcated practical limitations of the (then) current theory of electrons.

> It is because the experimental evidence extends into regions where existing electrical theory is insufficient, that the principle of relativity becomes of importance as a supplementary and independent hypothesis.*

Those regions which Cunningham referred to were the ether drift experiments which were designed to detect the earth's motion through the ether. Cunningham's principle of relativity did nothing more than state that the detection was impossible. Cunningham's analysis which has been outlined here began in 1907 and reached its most sophisticated level in 1914 when his monograph, *The Principle of Relativity* was published by Cambridge University Press. It was the first British book length account of the special theory of relativity.

In addition to Cunningham, there were only a few other young British physicists who were interested enough to write about the subject. One was Harry Bateman, who emigrated to the United States in 1910. He essentially agreed with Cunningham that the principle of relativity was supplementary to the electron theory. In his view the mathematical theory on which the principle of relativity was based had been introduced by Lorentz in 1892 and had been "developed considerably" by Einstein, Planck, and Minkowski. Bateman disagreed with Cunningham about one major point. The principle of relativity allowed, he asserted, the satisfactory account of the null result of ether drift experiments "on the supposition of a stationary ether."

Another physicist who wrote about the theory of relativity was Norman R. Campbell, the author of several very popular books on the philosophy of physics. In his philosophy, Campbell was a strict empiricist who not only emphasized the importance of measurement but believed that all of the apparent paradoxes which relativity appeared to generate would disappear "if attention is kept rigidly fixed

* E. Cunningham, *The Principle of Relativity* (Cambridge, 1914) p. 41.

upon quantities which are actually observed." It is clear that such a philosophical position was an exception in British physics. Campbell's was the only voice raised during this period that questioned and rejected the validity of the concept of the ether and indicated forthright support for the theory of relativity.

Campbell's views were expressed very vigorously in a series of three papers beginning in 1910. One of them, "The Aether," was immediately translated and republished in a German journal. The other two, "The Common Sense of Relativity," and "Relativity and the Conservation of Mass," appeared a year later, in 1911.

Campbell rejected the concept of the ether as unsatisfactory. Taking note of the fact that those who were committed to the concept of the ether had consistently redefined its properties to account for the lack of detection, he wrote that it was tantamount to the following situation:

> If a man tells me that his watch weighs 100 grams, his statement is for me a significant proposition, because the ordinary definition of 'weight' can be applied to a watch; but if he tells me that the colour of his watch weighs 100 grams, and refuses to tell me how a colour is to be weighed, I can only conclude that he is uttering meaningless nonsense.
>
> It is probable that the future historians of physics will be astounded that the vast majority of physicists should accept a system of such bewildering complexity and precarious validity rather than abandon ideas which seem to have their sole origin in the use of the word 'aether'. . . .

Campbell then explained that one of the reasons he was writing the paper was to expedite the ether's "relegation to the dust-heap where 'phlogiston' and 'caloric' are now mouldering."*

Whereas Campbell expressed his support for the theory of relativity, he was so deeply committed to the notion of measurement conferring meaning that he rejected attempts to analyze physical situations by employing thought experiments. Recall our analysis of proper and nonproper measurements of mass in Chapter 3. That analysis has been popular ever since it was introduced in 1909 by the Americans, G. N. Lewis and R. C. Tolman. Campbell objected strenuously. Such experiments, he said, were impossible to perform. Therefore, Lewis and Tolman

> . . . can make any assumption that they like, happy in the confidence that they will never have to make any more, unless they want to do so, because their theory cannot possibly come within the range of experiment.†

* N. R. Campbell, "The Aether," *Philosophical Magazine,* 1910, *19*:181-191, pp. 185-186, 188.
† N. R. Campbell, "Relativity and the Conservation of Mass," *Philosophical Magazine,* 1911, *21*: 626-630, p. 628.

Campbell apparently did not realize that the entire edifice that Einstein had created was founded on thought experiments. It is an excellent example of the difficulty to which the empiricist position can lead. In one sense, he was doing the same thing that he had criticized Lewis and Tolman for: accepting those analyses which suited his needs and rejecting others.

On the other hand, Campbell had no objection to the use of models and he considered them essential to theory. He thought that in attempting to understand the nature of light and its propagation his collegues had erred, not in building models, but in insisting that the models be mechanical. He had no doubt that a consistent model for the propagation of light could be constructed to conform to the principle of relativity.

I have described here the response to the special theory of relativity in Great Britain during the years between 1905 and 1911. Only Campbell raised his voice against the concept of the ether and yet he was unable to comprehend the kind of physical intuition that Einstein had exhibited and that others, chiefly Einstein's German-speaking colleagues, had learned to appreciate.

Ebenezer Cunningham, despite his interest in "The Principle of Relativity," said that he "was surprised to be asked to speak to the British Association in 1910" about the subject; he had thought there was little interest. In 1911, the British Association featured a panel discussion about "The Principle of Relativity." The participants included an American, G. N. Lewis, a famous Dutch physicist, Pieter Zeeman, and a very young British physicist, W. F. G. Swann, who, within two years, emigrated to America. According to the anonymously written account of the meeting the discussants' attention focused first, on the scope of the principle and second, on its mathematical aspects. The audience response to the discussion was reported:

> Dr. C. V. Burton, after expressing his satisfaction that no one had confessed a disbelief in the aether urged the importance of the search for residual phenomena not falling within the electromagnetic scheme. Conceivably gravitation is such a phenomenon. There is the further question as to whether neighboring electrically neutral masses exert forces upon one another in virtue of their motion through the aether.*

British physicists, as a group, appeared to have shared Oliver Lodge's inability to think of a theory which did not account for apparent action at a distance in terms of direct mechanical pushes and pulls.

One physicist wrote in 1910:

> Sir Oliver Lodge seems to strike the correct note when he . . . calls upon us to explain everything in the world by pushing and pulling, or

* "Mathematics and Physics at the British Association, 1911," *Nature,* 1911, 87:498-502, p. 500.

> indeed, by pushing and not pulling. . . . Carry the pushing view to its logical conclusion. The now familiar perfect fluid is alone left from which to extract our bricks and straw wherewith to build the world as it appears.*

Whereas many publicly shared such sentiments, other candidly confessed to being unable to understand the foundations of the theorys of relativity.

We have quoted F. C. Searle's letter to Einstein in 1909 saying that neither he nor anyone he had talked with could understand his theory. Joseph Larmor is reported to have remarked at about the same time, "This relativity business is all right for you young fellows, but it's not for these old brains. I was raised and trained on absolute time and I just can't believe this relative time business."† According to the mathematical physicist A. A. Robb, the idea that Einstein proposed "to preserve symmetry" was that events might be at the same time simultaneous to one observer and not to another. Robb equated this with a serious logical fallacy:

> This remarkable suggestion [relativity of simultaneity] was at once seized upon with it apparently not being noticed that it struck at the very foundations of logic. That a thing cannot 'be and not be at the same time' has long been accepted as one of the first principles of reasoning, but there it appeared for the first time in science to be definitely laid aside. . . . [To some people] this view of Einstein's appeared too difficult to grasp or analyze, and to this group the writer must confess to belong.‡

We are not speaking of scientists who were incapable of understanding sophisticated abstract mathematical descriptions of physical systems. We are speaking of highly trained, competent experimental physicists, theoretical physicists, mathematical physicists and mathematicians.

The foundations and principles on which the theory of relativity is built are simple, straightforward and not at all arcane. Although the results are in some instances counter-intuitive, they flow directly from those principles. Any interested person can follow the reasoning and a slight knowledge of algebra allows for an analytical account.

Yet here is a subculture, highly trained and conversant with the general area,

* A. McAuly, "Spontaneous Generation of Electrons in an Elastic Solid Aether," *Philosophical Magazine,* 1910, *19*:129-152, p. 135.

† Interview with D. F. Comstock, Cambridge, Massachusetts, December 10, 1964. Comstock placed the date of the remark at about 1912. Comstock's contributions will be discussed in Chapter 9.

‡ A. A. Robb, *A Theory of Space and Time* (Cambridge, 1914) p. 2.

who are, almost to a man, unable to follow that reasoning, and virtually proud of it. As late as 1923 N. R. Campbell commented:

> [British] physicists of great ability, who would be ashamed to admit that any other branch of physics is beyond their powers will confess cheerfully to complete inability to understand relativity. . . .*

Campbell's claim that the reason for the failure to understand was because of the lack of emphasis on the experimental aspects of the theory of relativity, reflects Campbell's biases about useful theories more than about anything else.

CONCLUSIONS

There are two aspects of the British response which require our attention. The first is the seemingly uniform response of British physicists. It was not unlike the response in France. However, whereas there had been silence in France, in Great Britain there was an effort to make the theory compatible with traditional ether theory.

The second aspect of the British response is the slowness with which the theory became part of the literature and began to appear in textbooks, to become second nature, as it were, to practicing physicists.

Developments such as those associated with the theory of relativity appear first in research papers and are read and actively worked on by only a few individuals at first. Later they become generally known and later still, they become part of an advanced curriculum in graduate seminars and courses. Eventually, the innovation filters down to undergraduate courses. It is finally recognized that the theory is not only amenable to undergraduates but to interested laymen as well. The time required for the various steps in this process for an innovation varies widely.

In Germany, with relativity, diffusion happened much faster than it did in France, England or the United States. In fact, the assimilation of the special theory of relativity in the French and British cultures appears to be still in progress whereas diffusion and assimilation appears to have occurred faster in the United States. As might be expected, as with the nature of the response to the theory, the rate at which the theory diffuses through the culture is culturally dependent. Having noted the phenomenon, we will return to it in more detail in Part III.

How is it that British physicists were so uniformly willing, on the one hand, to concentrate on the ether and ignore the theory of relativity, and on the other hand, when that posture was no longer acceptable, to reinterpret the theory in terms of the ether and reduce it to the principle of relativity and a subsidiary to the electron

---

* Campbell, *Relativity* p. v.

theory of matter? For three hundred years British physicists had speculated and experimented about the nature and structure of the ether. In the eighteenth and early nineteenth century, it had been the name of Newton which had inspired this activity. In the middle of the nineteenth century, a new hero, Lord Kelvin, had come to the front ranks to lead the search. But even Kelvin had, by his own admission, failed. At the very time that Kelvin was declaring his failure to understand the nature of the substance beneath electromagnetic phenomena, J. J. Thomson, Ernest Rutherford, and others were making assaults on the constitution and configuration of the atom; and it was to this activity that British physicists turned. If it were impossible to determine the structure of the ether, it was possible to determine the structure of the atom, held by some in Great Britain to be itself the product of ether.

The various techniques in Britain for avoiding the special theory of relativity were to reinterpret the theory to conform to an ether theory, to reject the theory as inadequate on experimental grounds, to ignore the theory as incomprehensible or unimportant. One or all of these tactics were used by Lodge, Cunningham, J. J. Thomson, Larmor, and almost all other British physicists of that era except N. R. Campbell.

As with France, when one looks for reasonable cultural and social explanation of such uniformity, the structure of the educational system immediately suggests itself. It is not surprising to find that theoretical physicists in Britain in the nineteenth and early twentieth century who were interested in electrodynamics were trained to do ether mechanics. It is what they *had* to learn and it was what they knew best.

During that period, most theoretical physicists in Britain were trained at Cambridge University. The Mathematical Tripos examination at Cambridge was the most competitive field for mathematicians and mathematical physicists in Britain until 1910. The name of the examination originated from seventeenth century custom in which a three-legged stool was used in the ritual of granting the bachelor's degree. By the middle of the nineteenth century, the Mathematical Tripos examination was a two-part nine-day trial to determine honors for those taking degrees in mathematics and mathematical physics.

Whereas a wide range of topics was covered on the Mathematical Tripos, there were a significant number of very detailed questions in the mathematical physics section on ether, ether mechanics, and the mathematics of the related subjects of "jellies, froths, and vortices." The details may be obscure to us, but consider the import of questions such as the following from the 1901 examination:

> Obtain the energy function of an isotropic elastic medium and assuming that waves of dialation are propagated through the medium with an indefinitely great velocity and that the difference between different media

> is one of density only, find the intensities of the reflected and refracted waves when plane waves are incident on a plane interface separating two media.

or

> Waves of light are incident on a face of a uniaxial crystal cut perpendicularly to its axis. Find on MacCullagh's theory the intensities of reflected and refracted waves (1) when they are polarized in the plane of incidence, (2) when they are polarized perpendicularly to it.

Presumably, in order to answer the second question one would have committed MacCullagh's theory to memory. This theory, about which the best students in physics in Britain were being examined in 1901, was created in 1839 and required an elastic solid ether which exhibited the property of "rotational elasticity." Is it surprising to find that Whittaker, himself a Tripos veteran, was still enchanted with MacCullagh's ether in 1951?

But MacCullagh's was not the only ether theory required of the student taking the 1901 Tripos:

> Give an account of Helmholtz's and Lord Kelvin's theory of vortex motion.

Or, read the following questions from other years during the period 1900-1909:

> Investigate Helmholtz's expression for the velocity of a fluid in vortex motion in terms of a vector potential.
>
> Assuming that the spin at any point within a circular vortex ring of finite cross-section is proportional to the distance of the point from the axis of the ring, prove that the vector potential at any point P is directed at right angles to the axial plane through P, and is equal in magnitude to the value at P of the gravitational potential of the ring, supposed to have at any point Q a density-proportional to the projection on the axial plane through P of the perpendicular from Q to the axis of the ring.
>
> Reduce expressions for the stream-function in the case of Hill's spherical vortex, and express the velocity of translation in terms of the strength of the vortex.
>
> Discuss generally evidence for the conclusion that the velocity of the earth's motion does not sensibly affect observable optical phenomena. Show however that if two independent sources of light could interfere, a simple experiment would at once reveal the earth's motion in space.

These are samples of the kinds of knowledge advanced students of physics were expected to acquire in order to qualify for advanced university positions. Obviously a student would have to be extremely dextrous with mathematical manipulations and to memorize not only the range of ether theories, but the range of all physical theories thought to be relevant. J. J. Thomson's description of the taking of the Tripos is worth quoting in some detail.

> The examination for the Mathematical Tripos when I sat for it in January 1880 was an arduous, anxious and a very uncomfortable experience. It was held in the depth of winter in the Senate House, a room in which there were no heating appliances of any kind. The ink certainly did not freeze in the wells as it is reported to have done. The examination was divided into two periods. The first lasted four days. In the first three days we were examined on the elementary parts of geometry, conic sections, algebra, and plane trigonometry, statistics, and dynamics, hydrostatics, and optics, Newton and astronomy. Five papers were set on these subjects, each paper containing about twelve questions, each question consisting of a piece of book-work and a rider which was a question not supposed to be taken from a book but one whose solution was closely connected with the piece of book-work. In addition to these five book-work papers, there was a paper called the problem paper in which there was no book-work. . . . At the end of the fourth day there was an interval of ten days in which the examiners drew up a list of those who, by their performance on the first three days, had acquitted themselves so as to deserve mathematical honors. These and these only, could take the second part of the Tripos, which lasted five days. . . . The questions ranged over all branches of mathematics and there were two problem papers. The questions set on the first two days of this examination were on the parts of mathematics which had been included in the Tripos almost since its foundation, and were subject to the restriction that the book-work papers set in the first two days of the examination should not contain more questions than a well prepared student might be expected to answer in the allotted time: this did not apply to the papers set after the second day. As each paper contained twelve questions and each question consisted of two parts. . . . this was pretty good going but I believe that a few men each year did get a large percentage of the maximum marks obtainable on these papers. . . . Accuracy in manipulation was perhaps the most important condition in this part of the examination and the most difficult to impart. . . . Another quality which played a great part was concentration on the question in hand, and ability to get quickly into stride for another question as soon as one had finished with the old. These qualities, having one's knowledge at one's finger-ends, concentration, accuracy and mobility owed their importance to the examination being competitive, to

> there being an order of merit, to our having to gallop all the way to have a chance of winning.*

Indeed, as Thomson writes, the examination was highly competitive. Those who ranked the highest on the second part of the examination were termed "Wranglers," the name probably meant that a Wrangler is one who debates or wrangles. To be first ("Senior Wrangler") or second Wrangler in any year was to insure one's future. Some of the greatest names in British physics in the nineteenth century were first or second wranglers.

Preparing for the examination was a major undertaking. The entire time allotted, according to J. J. Thomson, was three years and a term. If you were serious, you hired a "coach" or private teacher. Thomson himself "coached" with Routh, a tutor who had seen many successfully through the Tripos:

> Routh's system certainly succeeded in the object for which it was designed, that of training men to take high places in the Tripos; for in the thirty-three years from 1855-1888 in which it was in force he had 27 Senior Wranglers and he taught 24 in 24 consecutive years. . . .
>
> Routh was Senior Wrangler in the year when Clerk Maxwell was second. Perhaps no other man has ever exerted so much influence on the teaching of mathematics; for about half a century the vast majority of professors of mathematics in English, Scotch, Welsh, and Colonial Universities, and also the teachers of mathematics in the larger schools, had been pupils of his and to a very large extent adopted his methods. In the textbooks of the time old pupils of Routh's would be continually meeting with passages which they recognized as echoes of what they had heard in his classroom or seen in his manuscripts.
>
> . . . Routh like Maxwell studied mathematics under Hopkins, the great "coach" at that time, who had taught Stokes and William Thomson [Lord Kelvin] and scored 17 Senior Wranglers before he retired.†

There existed then a natural filter for the processing of mathematicians and mathematical physicists in Britain. It is understandable why so many of the people we have considered acted with such uniformity when confronted with the theory of relativity. They had studied, indeed, they had been conditioned, under Hopkins or one of his pupils. They had been trained, in this region of physics, to master the equations of waves moving through "froths, jellies and vortices."

The situation did not remain static. According to Ebenezer Cunningham, the Tripos he sat for in 1902 hardly dealt with electrodynamics.

---

* J. J. Thomson, *Reflections and Recollections,* pp. 56-58.
† *Ibid.,* p. 35.

> Maxwell's work was too recent and had not reached the textbook stage. Abraham's book was unknown. . . . In the Tripos as I knew it, the physical subjects were Geometrical Optics, Positional Astronomy, Mechanics (Routh) and Electricity and Magnetism. . . . This was a period of transition which culminated in 1910 when the order of merit in the Tripos was abolished and teaching in all subjects took a much more theoretical turn, while the Cavendish Laboratory [at Cambridge University] moved on into atomic theory. . . . Modern physics was in its infancy, I had to find for myself the work of Lorentz, Planck, etc., after I had graduated. Eddington, Jeans, and others were still to come.*

The British system of higher education was in its own way as rigid and narrow as the French. And, as in the case of the French response to Einstein's special theory of relativity, that narrowness and lack of variety were products of typically strong British cultural traditions. In physics they included the appeal to theories of past heroes of British science and a metaphysic which required a model for the consummation of a physical theory. The result is that, not only was Einstein's special theory of relativity not elaborated on during the period between 1905 and 1911, its diffusion and assimilation has taken a very long time.

* Personal communication.

·9·

Defending the Practical: The American Response

IN TURNING TO THE AMERICAN response to Einstein's theory we introduce several new variables to our study. First, there is the great physical gulf between America on one side of the Atlantic, and the three European countries on the other side. Second, in comparison to the European communities, American culture is extremely young and the American physics community was, until the last decades of the nineteenth century, virtually nonexistent. Certainly, American physics had none of the character with which it is associated today.

Despite these factors, the *differences* between the American response to the special theory of relativity and that of physicists of Great Britain, Germany and France are about the same as the differences between any two of the European countries we have examined. It is not the national boundries that are important. It doesn't matter if the boundary is an imaginary line or a two thousand mile stretch of water. It is the different political and social institutions, customs, and expectations that are significant. With the British, French and German response to the special theory of relativity, belief and understanding within a community of scientists are functions of social custom and fashion.

One may not expect the physical distance between the United States and the other countries to be an important factor. On the other hand, the newness of the American culture and physics community were central to the nature of the American response to Einstein's innovation.

THE AMERICAN TRADITION IN SCIENCE

The most pervasive theme in the literature about American attitudes toward science and scientific research during the nineteenth and early twentieth century is the so-called *indifference* theme. Beginning in the second third of the nineteenth century it was argued by American scientists like Joseph Henry that, while there was a great deal of activity devoted to inventions and the creation of practical implements, Americans were indifferent to basic science and basic research. The language was somewhat different from that we are accustomed to today. Rather than basic science or research, one referred to "pure science" or "pure research" by which was apparently meant that researchers were motivated solely by the desire to know and understand the workings of nature rather than by baser motives of increasing material comforts or wealth. Indeed, these distinctions are still important in governmental scientific institutions which are responsible for giving support for scientific research.

In support of claims of American indifference to pure science, the adherents of that view list numerous inventions by Americans which shaped our technological prowess and point to the absence of private or public support for basic science except that which leads to immediate practical applications.

Modern commentators on nineteenth century American science who have pointed out *indifference* have often been guilty of circular reasoning to account for the supposed indifference. "Americans are not interested in basic science because Americans are practical."

The indifference thesis is not uniformly accepted by scholars of American science. Recent studies indicate that nineteenth century American scientists were, in fact, engaged in basic research.* In part this lack of consensus is because the major proponents of the indifference thesis have concentrated on fields of study other than the natural sciences, for example physics and astronomy, rather than botany or geology. It is not my purpose to judge the indifference literature. We can, however, make use of that literature to understand American attitudes toward science during the latter part of the nineteenth century.

The source of much of the talk of indifference was a small group of nineteenth century American scientists who thought that it was very difficult to gather material and moral support for their activities compared to their observations, and in some cases experiences, in Europe. This group included Joseph Henry, the first Secretary of the Smithsonian Institution, Simon Newcomb, an astronomer and perhaps the best known American scientist of his time, who had revolutionized

* N. Reingald, "American Indifference to Basic Research: A Reappraisal," in G. H. Daniels (ed), *Nineteenth Century American Science: A Reappraisal* (Evanston, 1972) pp. 38–62.

observational astronomy at the United States Naval Observatory, and Henry Rowland, a young physicist who was to revolutionize the measurement of electromagnetic spectra by introducing a new standard of accuracy for the manufacture of the required tools. Henry was the elder statesman of the group. As early as 1839 he made public pleas for recognition of the work of pure science and chastized his countrymen for their concern with practical work. These themes were echoed by Newcomb, Rowland and others.

In the early 1870s there was a dramatic shift in popular attitudes about science, based partly on confusion between *science,* that is, investigating the way the physical universe works, and *technology,* that is, manipulating the physical universe. That confusion exists in the American culture to this day.

There were many nineteenth century examples of invention of technological wonders independent of scientific investigation. These included the perfection of the steam engine, the invention of the telephone and the electric light. In these three instances and many others, the creation of the device did not depend on superior insights into the workings of nature, but rather on being skilled and clever, having the patience to withstand repeated failure, and being able to maintain a vision.

SCIENCE AND AMERICAN TECHNOLOGY

The shift in attitude which occurred in the 1870s in American culture was from indifference about the enterprise of science, to calls from all sides for building a culture based on science. As I have said, very often such a vision was based on confusion between technological innovation and scientific investigation. In practical terms this meant little change in the material support received by those who were interested in scientific investigation.

In the winter of 1872–73 Simon Newcomb had organized the visit of British physicist John Tyndall to help to marshall support for pure science in the United States. To that end Tyndall barnstormed the country giving a series of popular lectures on light. Science had become very popular and everywhere that he went, Tyndall spoke to packed houses. The cream of American society had begun to consider the popularization of science as fashionable. Another measure of popularity was the amount of space that journals like *Harpers* and *Scribners* devoted to science, and the successful emergence of journals like *Popular Science Monthly.* The rise in the popularity of science among laymen did not mollify those practitioners who had complained of a lack of interest in pure science in America. They saw the public interest as reflecting confusion between gadgeteering and science.

In his "Farewell Lecture," given in New York in February, 1873, Tyndall reminded his American listeners of the words of Alexis de Tocqueville, who, fifty years earlier, warned that he saw no science in this country. To de Tocqueville there was something distasteful about "collecting the treasures of the intellect without bothering to create them." He was referring to the American genius for turning ideas into gadgets, inventions and implements. The ideas that the inventions came from, de Tocqueville maintained, originated in Europe and it was only the proximity of Europe which prevented the American culture from remaining in a state of barbarism. The moral force of de Tocqueville's remarks about American science were heavily tinged with *Wissenschaft:*

> If Pascal had had nothing in view but some large gain or even if he had been stimulated by love of fame alone, I cannot conceive that he would ever have been able to rally all the powers of his mind, as he did, for the better discovery of the most hidden things of the creator.*

For de Tocqueville, as for his German contemporaries who espoused such scholarly activity, it was clearly a state of mind, a way of life, that made the difference.

The British view, reflected by Tyndall, did not include the moral imperative of German *Wissenschaft*. Tyndall, like many of his British contemporaries, maintained that one did pure research *because* of its vital importance for technological advance. For example, in his "Farewell Address" to America, he cited the works of Michael Faraday and H. C. Oersted for the actual discovery of the telegraph and lamented that those who gained scientific prominence in America, like Joseph Henry, were diverted to administrative jobs.

Tyndall's major theme, that one needs to create a pool of ideas gleaned from pure research in order to advance technology, was enlarged upon by commentators like Newcomb and Rowland. They admitted that there was a curious paradox involved. Pure research was perceived as research with no end product; to acquire knowledge was the sole motivation. Yet it was this research that maintained technological progress. After Tyndall's visit, the appeal by American scientists for support of pure science in order to provide for technological growth was reinforced by similar appeals from university administrators and, curiously enough, some elements of the business community.

In fact, those peculiarly American phenomena, the independent research laboratories which began to appear in large numbers at the turn of the century often explained their mission to a curious public with copious references to both Tyndall and de Tocqueville. But something was added to these explanations. Technological growth, they said, required the fertilization of knowledge acquired

* A. de Tocqueville, *Democracy in America,* trans., Henry Reeves (Philadelphia, 1841), Vol 2, p. 42.

from pure research. But since technological growth meant economic profit, one did pure research for economic profit.

Arthur D. Little, founder of one of the most successful research laboratories in the United States, referred to the industrial researcher as a member of the fifth estate. He saw the fruit of the marriage of research to industry as "the making and the saving of money." He took pleasure in pointing out that a problem undertaken by the Mellon Foundation, which had been solved in eight hours, had saved a certain shoe manufacturer every week the cost of the entire project. Whereas this might not sound like the fruit of pure research, it was a type of common rhetoric near the turn of the century and represented a popular view about the relationship between science and technology that has persisted to the present day.

These attitudes were not restricted to a small segment of the business community and the operators of the laboratories. They were wide-spread and deeply believed—an expression of the American concern and preoccupation with the useful and practical. For example, in 1907, William McMurtrie, an industrial chemist, gave the dedicatory address at the opening of a new chemistry laboratory at Rensselaer Polytechnic Institute. McMurtrie called for the "scientific use of imagination," illustrated by the work of Henry Bessemer, the discoverer of the Bessemer process for the manufacture of steel. He quoted, with obvious approbation, a recent editorial in *The Wall Street Journal.* The editorial was entitled "Science as a Financial Asset" and it is illuminating to consider it in detail:

> Science as a source of strength in promoting private wealth and public welfare is the one thing that draws the line of demarcation between ancient and modern times. . . . The declaration of the German chemist [is] that scientific research is the greatest financial asset of the fatherland. . . . Every commercial transaction in the civilized world is based on the chemist's certificate as to the value of gold, which forms our ultimate measure of values. Faith may move mountains, but modern science relies on dynamite. Without explosives our greatest engineering works must cease and the Panama Canal, no less than modern warfare would become impossible.
>
> The work of science which probably needs most development in the present day however is not so much the application of knowledge already acquired to the increase of wealth. . . . Fundamental research is by far our greatest need. Common clay is full of commodities which if they could be extracted economically would probably solve for centuries the question of metal supply. . . . It was this thought that led an American journal to say that the accidental killing of Professor [Pierre] Curie, the discoverer of radium, in the streets of Paris last April was a greater loss to the world than the Earthquake in San Francisco where over a thousand people lost

> their lives . . . and property losses estimated at five million dollars occurred. If that be true, it is neither numbers nor wealth, but scientific talent that gives power of mastery to nations because of its capacity to unlock the secrets of nature in which lie the sources of material wealth.*

The contradiction between the concept of pure or fundamental research for knowledge, and the motivation of financial wealth permeates many of the writings that one encounters during this period in the American culture. In their enthusiasm about the awarding of the Nobel Prize for 1907 in physics to Albert A. Michelson in part for his invention of the interferometer which we analyzed earlier in this book, the editors of *Popular Science Monthly* suggested that were the terms of Nobel's will strictly heeded, the award would go first to Alexander Graham Bell and second to Thomas Alva Edison. These examples have been introduced to illustrate a tradition of the role of research in the American culture by the turn of this century and the relationship which was perceived between science, technological progress and economic power. It is perhaps a mixture of German *Wissenschaft* and British utilitarianism, with an added profit motive. Although it was somewhat confusing, it was very effective in capturing the public imagination and firing the enthusiasm of industry for "basic," "pure," or "fundamental" research.

SCIENCE AND AMERICAN UNIVERSITIES

One might call the developments I have described the pragmatic tradition for scientific research in America. The business and industrial community were not the only source of this tradition. During the same period, the last quarter of the nineteenth century and the first decades of this century, higher education in America was undergoing a rapid and thorough revolution. The education of proper gentlemen and gentlewomen (at separate institutions, of course) was being replaced with the education of professionals. By 1860 some science was required in the curriculum of most institutions of higher learning in order to provide for the proper balance of the character of the individual. The distinction between practical science and abstract science was quite pronounced and the science taught in the colleges was almost exclusively experimental. Virtually no theoretical science was taught in American colleges and universities. When Charles Elliot became president of Harvard in 1869, science became a more important part of the Harvard curriculum, reflecting the fact that Elliot was trained in and had taught science. But the emphasis was on laboratory science, not theoretical study.

* William McMurtrie, "Address at the Dedication of the Walker Laboratory of Rensselaer Polytechnic Institute," *Science,* 1907, *26*:329–332. Of course, it was *Marie* Curie, assisted by Pierre, who discovered radium.

To give some measure of how small the American scientific community was, Yale, the first American institution to grant the Ph.D. degree, granted only two in the 1860s. In 1871 there were 198 Ph.D. candidates in American institutions in all fields. But things were changing at a fairly rapid rate. There is much evidence that the German model for higher education and research affected the outlook of American scholars, intellectuals, and university officials during the last quarter of the nineteenth century. They became increasingly convinced that the study of science, philosophy and history were aspects of *Wissenschaft* and should be pursued. Philosophy became a distinct field, no longer identified solely with moral philosophy and ethics. Science became a separate, specialized field and excluded all that was metaphysical and non-scientific. That attitude about science fit well with the growing tradition of experimental and practical work.

In 1900 graduate education in the United States was fewer than thirty years old. Johns Hopkins, the first university in the United States which was patterned explicitly after the German model of higher education, was founded in 1873. It was the first institution in America where research was acknowledged to be as important as teaching. The example of Johns Hopkins quickly had an effect on other institutions of higher education. President Elliot of Harvard, for example, stated that it had been the example of Johns Hopkins which forced the Harvard faculty to put strength into the development of graduate education. Within four years of the founding of Johns Hopkins, Harvard, Yale, Princeton, and Columbia had mounted vigorous programs in graduate study. By the turn of the century they were joined by Catholic University, the University of Chicago, M.I.T. and Clark University.

This is not to suggest that by the first decade of the twentieth century American educational and research ideals had been transformed from British influence to German. The change was gradual, but the result was eclectic rather than transforming.

THE AMERICAN PHYSICS COMMUNITY

From the middle of the nineteenth century to the beginning of the twentieth, the growth of the American physics community paralleled the growth of the American community of scientists as a whole. In 1870, there were no more than seventy five people in the country who called themselves physicists. Of these, only one can be identified as a theoretical physicist. He was Josiah Willard Gibbs, who taught at Yale and lived in total obscurity. The importance of his work in theoretical thermodynamics, statistical mechanics and vector analysis was discovered in Europe chiefly by Clerk Maxwell, who was responsible for bringing it to the attention of the scientific world.

As interest in science in the American universities developed, and as the popular identification of the importance of science in technological progress emerged, there was a concomitant increase in physics in the universities. By 1890, there were fifty-four Ph.D.s in physics in the United States. They were almost all experimentalists.

This growth continued unabated throughout the first decade of the twentieth century. In the year 1909 alone, twenty-five Ph.D.s in physics were granted to Americans. As the number of physicists in the country grew, the need for a separate social organization emerged. Not only would such a forum provide a mechanism for internal recognition, it would provide a forum whereby individuals could keep abreast of recent advances and innovations and where individuals in similar areas could identify each other. Thus the American Physical Society came into existence. In 1900 there were fewer than one hundred members. By 1910, the number of American physicists in the American Physical Society had increased to more than five hundred. At the same time, the *Physical Review,* the journal which became the research organ of the American Physical Society, was largely ignored by the world community of physics, and American physicists preferred to publish in British journals such as *Nature* and the *Philosophical Magazine.* With growing frequency, Americans wrote articles which appeared in German publications and the flow of graduate students of science to German research laboratories quickened to almost flood proportions.*

AMERICAN SILENCE, 1905–1907

The American response to the theory of relativity between the years 1905 and 1911 was, given the state of the American university physics community, rather meager. The response, however small, was in accord with the pragmatic tradition of American science.

During the years 1905–1907 there was no notice taken of Einstein's theory. Rather, there was a preoccupation with ether drift experiments, a strong resistance to suggestions that the ether be dismissed, and skeptical renunciations of those Europeans who were perceived as undermining the mechanistic philosophy. It was chiefly in the United States, during this period, that the Michelson-Morley experiment was repeated over and over again, almost as if those doing the experiment knew what the result *had* to be and they were going to continue to do the experiment until it came out "right." For example, according to the Lorentz theory, one expects a fringe shift in the pattern of striations seen by the observer in

* For a detailed account of the history of the American physics community see D. J. Kevles, *The Physicists: The History of a Scientific Community in America* (New York, 1978).

the Michelson apparatus of about one third of a fringe. Michelson and Morley had expressed confidence that their apparatus was capable of detecting a shift of a hundredth of a fringe. Nevertheless, E. B. Brace, Professor of Physics at the University of Nebraska, in a review of such experiments in 1905, concluded that more accuracy was needed before a definitive judgment could be made. In the same year, he himself had redone one of the ether drift experiments of the nineteenth century French physicist Fizeau. Although he claimed to have improved the sensitivity of the apparatus by several orders of magnitude he found no detectable effect of the earth moving through the ether.

Also in 1905, Michelson's collaborator Edward Morley, working with D. C. Miller, reported a new series of trials of the Michelson-Morley experiment. For each trial, the base of the apparatus was constructed from different materials. Morley and Miller said that they could state with confidence that the "Fitzgerald-Lorentz" contraction was the same for wood as for iron and that both materials responded in an identical fashion as a base made of sandstone. Recalling that the expression for the Lorentz contraction contains no term which depends on the nature of the material, it is difficult to understand how Morley and Miller had done such an experiment.

The sense one gets in reading Brace, or Morley and Miller, of the sensible reality of the ether, and of loss in the face of null results, is reinforced when we learn that Morley and Miller thought that perhaps the null result was because the interferometer had been cloistered in the basement of a brick building. Perhaps, the suggestion was, the ether through which the earth, planets, and stars moved freely was incapable of penetrating the brick. They moved the apparatus from the basement laboratory at the Case School of Applied Science (now part of Case-Western Reserve University) in Cleveland, Ohio, to a high hill overlooking the city. Surrounding the apparatus with glass in the fall of 1905, they reported that they only awaited good weather to carry out the experiment. A year later, they reported that the experiment had yielded no measurable ether drift.

Obviously, such experiments would not be done unless the concept of the ether was that of a very substantial material, perhaps a fluid like water, or a solid like glass. When one compares European attempts to account for the ether with the examples presented here, one is impressed with the unwillingness of the Americans to allow the ether to be an abstract entity.

By 1907, such attempts at measuring the drift of the ether had all but vanished from the public literature. But concern with the properties of the ether persisted. There was much comment about recent publications of the British physicist Oliver Lodge, in which he had estimated the density of the ether to be about 10^{11} grams per cubic centimeter. When one considers that the density of mercury is about 13.6 grams per cubic centimeter and that a density of 10^{11} grams per centimeter is a density of a hundred billion grams per cubic centimeter, we realize just how special the ether would have to be. This did not seem to phase most American commentors.

Carl Barus, Professor of Physics at Brown University, saw far reaching implications to Lodge's model. If Lodge were correct, Barus maintained, then by considering the effects of the earth's motion through such a dense material, one would expect the earth to spin in the ether. That one did not detect such a rotation did not signal to Barus that there was something wrong with the expectation or Lodge's concept of the ether; rather "the electronist gets around this [null result] by the principle of relativity."

On the basis of what Lodge had written, the editors of *The Scientific American* saw the following as reflecting scientific opinion about the ether in early 1909:

1. Matter comes from undifferentiated ether;
2. Ether as a whole is stationary with fine grained, internal circulation;
3. There is no etherial viscosity relative to matter.*

These kinds of remarks are not examples of isolated aberrant viewpoints in the United States. Although the amount of relevant written work in America was small compared to the production of German scientists, almost all of it until mid-1907 was about ether mechanics, ether structure and the practical problem of detecting the earth's motion through the ether.

That something else was happening in the world of electromagnetic theory was only dimly perceived by American commentators. The only hint of what was to come was from American physicists who were publishing in American journals while working or studying overseas. One such physicist was D. F. Comstock. Comstock had studied in Berlin and Zurich and had received his Ph.D. in Basel in 1906. In the last half of 1906 and the first half of 1907, he was an International Fellow at the Cavendish laboratory in Cambridge University. He subsequently returned to the United States to accept a teaching position at M.I.T. which he held from 1907 through 1917. (He left M.I.T. to become vice-president and then president of an independent research laboratory, Kalmus, Comstock, and Wescott Inc., in Cambridge, Mass., which, among other things, developed the "Technicolor" process.)

Writing in *Science* near the end of 1907,† Comstock said that he detected a distrustful attitude among scientists who were not physicists, about ether. The ether could not, he argued, be dispensed with. The speed of light was always the same relative to space and this meant, Comstock continued, that a moving

* C. W. Rafferty, "Scientific Opinion," *Scientific American Supplement,* 1909, 68:198.

† D. F. Comstock, "Reasons for Believing in an Aether," *Science,* 1907, *25*:432–433.

observer in a windowless box would be able to determine his motion by measuring the speed of light in different directions. For Comstock to propose such a thought experiment suggests that he knew nothing of the second postulate of relativity or, for that matter, the result of the Lorentz analysis in which the speed of light is shown to be the ultimate speed. Nor did he have a clear concept of the principle of relativity. He later remarked that he knew nothing about the work of Einstein and did not learn of it until 1910. The evidence suggests this may have been true for many American physicists.

In June of 1907, at the end of his stay at the Cavendish, Comstock published a paper entitled "On the Relation of Mass and Energy."* The derivation, although done independently, was identical to that by Hasenöhrl and Abraham in which the conclusion is that $E = \frac{3}{4} mc^2$. As Abraham and Hasenöhrl had, Comstock called this mass the "electromagnetic part of the mass." He said that he could not answer the question of whether or not the inertia of matter (that is, inertial mass) was completely electromagnetic although he thought that Kaufmann's experiments on the variation in the ratio of the charge to mass of high speed electrons appeared to suggest this to be the case for single electrons.

Comstock's public ruminations give us an insight about how isolated American and British physicists were from these issues. Whereas Comstock was unaware of the work of Hasenöhrl, Abraham or Einstein, he did know of the Kaufmann experiments, made between the years 1901 and 1906. On the other hand, Comstock was uninformed of Max Planck's forcible critique of the Kaufmann experiment or his definitive paper about the relationships between radiation, matter and the concept of energy, derived from the theory of relativity.

Many years later Comstock recalled that he had shown his paper on mass and energy to colleagues at the Cavendish. J. J. Thomson had found it interesting as it paralleled some of his own work. According to Comstock, Thomson "was always interested in that sort of thing." Another colleague said that he thought he had heard about a German who had found similar results.†

It is perhaps understandable that there was, in effect, no response to the theory of relativity in the United States before 1908. The activity we have described comes closest. At the same time, the similarity of American and British concerns about the physics of radiant energy and the ether is noteworthy. The similarity is to be expected since the two physics communities shared a common language as well as many journals. The Americans relied heavily on British publications such as *Nature,* the *Philosophical Magazine* and the *Philosophical Transactions of the Royal Society.*

---

* D. F. Comstock, "On the Relation of Mass to Energy," *Philosophical Magazine,* 1908, *15*:1–21.

† D. F. Comstock, Personal Interview, December, 1964.

THE CONTRIBUTIONS OF LEWIS AND TOLMAN

Beginning in 1908, the situation in the United States began to change in a unique fashion. It was in that year that Gilbert N. Lewis, then a young physical chemist at M.I.T., later to become one of the most important and influential chemists in America, published a remarkable paper which led him directly to the theory of relativity. The paper was motivated, according to him, by recent experiments which suggested a "review of Newtonian mechanics" was in order. The experiments he was referring to were those of Kaufmann and the most recent attempts by Arnold H. Bucherer to make the same measurements of the ratio of the charge to the mass of high speed electrons. Lewis entitled this paper, "A Revision of the Fundamental Laws of Matter and Energy."* Within weeks after it appeared in the British physics journal *The Philosophical Magazine* it was translated for publication in the German journal, *Annalen der Naturphilosophie,* whose editor, the physical chemist Wilhelm Ostwald, was a leader of the so-called energeticist school. Energeticists were committed to the proposition that the fundamental reality of the universe was not matter, but energy. The journal often served as a platform for the energeticist point of view. Lewis was not a member of the energeticist school but his work in electrolytic conduction would have certainly brought him into contact with Ostwald, who had earlier made major contributions in that area of physical chemistry.

Throughout his career, Lewis displayed a very active and useful imagination for dreaming up representational schemes. These schemes often made little physical sense but proved to be useful predictive instruments. It was the case in this instance where Lewis thought that he had attained new insights about the relationship between matter and energy.

In introducing his subject, Lewis cited the recent publication of his colleague D. F. Comstock (Comstock's paper on the relationship between mass and energy) and the original paper on special relativity by Einstein. He did so, he said, not because they had given him any insight into the problem, but because they had emboldened him to make public some ideas he had entertained for several years; ideas, he said, that were stimulated by the ruminations of chemists concerned with changes of weight in chemical reactions. Lewis's goal, in fact, was to build a mechanics in which energy, mass and momentum are conserved at every instant in every process. The processes that Lewis was most concerned with were chemical, and he wanted to establish that, regardless of what was occurring in chemical reactions—for example, rapid heating or cooling, the emission or absorption of

* G. N. Lewis, "A Revision of the Fundamental Laws of Matter and Energy," *Philosophical Magazine,* 1908, *16*:707–717.

light, radical shifts in the physical state of the chemical reactants—the mass, energy and momentum of the system would always be conserved.

Lewis began his analysis by assuming the validity of the concept of light pressure. As Maxwell had suggested, Lewis accepted the notion that when a beam of light interacted with a material system, the light exerted a force on the system in proportion to the energy received by the system. If a force was exerted on a system by a beam of light, the system will, according to the laws of mechanics, accelerate, that is, there is a change in the momentum of the material system. But if the material system is acquiring momentum, another system, in this case, the beam of light, is losing momentum if the conservation of momentum is to be preserved. These considerations led Lewis to a mathematical derivation in which the speed of light (c) is equal to the ratio of the energy of the light (E) to its momentum (p). In mathematical terms

$$c = E/p \tag{1}$$

In order to conserve, not only momentum and energy at each instant of time in all physical processes, Lewis adopted a standpoint which he said was contrary "to the prevailing view":

> In such a beam [of light] something possessing mass moves with the velocity of light and therefore has momentum and energy.

Lewis designated the mass of the "something" with the symbol m. Using the standard concept of momentum as the quality of motion, that is, the product of the mass and the velocity, it follows that the momentum of a beam of light is given by

$$p = mc$$

Putting this directly into (1) and rearranging the terms immediately gives

$$E = mc^2$$

Lewis never justified this analysis. He never explained why he treated the speed of light as a constant. He remarked that since mass is not a property but a function of state (that is, the mass of the body depends on such things as the energy of the body) we will have a mass change when energy of any kind is absorbed or admitted. "The mass of a body is a direct measure of its energy." Lewis noted that Einstein had come to the same result, but, Lewis said, only as an approximation. Although he was aware of the discrepancy between the relationship he proposed and the one earlier proposed by Comstock ($E = \frac{3}{4} mc^2$), he was not willing to explain the difference. His only comment was that it would lead too far into electromagnetic theory "from which in the present paper I wish to hold entirely aloof."

The theory Lewis had proposed also allowed him to derive an equation for the change in the mass of an object as its speed approached the speed of light. The prediction was precisely the prediction of both Einstein and Lorentz. Lewis was aware of the measurements which Kaufmann had made and was also aware that Kaufmann's technique was being severely criticized. The discrepancy between the prediction and Kaufmann's experimental result, which Lewis put between six and eight percent, did not, he said, bother him.

Lewis speculated about the nature of "something" in light which possesses mass and moves at the speed of light. Identifying it with the ether, he cautiously suggested that absolute space might be identifiable with the following criterion: should the acceleration by a given force be the same in any direction, then perhaps one could say that a body was absolutely at rest. Within a year he was to withdraw the suggestion.

There are obvious difficulties with Lewis's approach. Although the mathematics is correct, and the analysis leads to results in concert with Einstein's special theory of relativity, the analysis itself is almost devoid of physical meaning. It is similar to a logical syllogism which would be correct, but not true. For example:

No person is mortal.
Einstein is a person.
Einstein is not mortal.

And in fact, Lewis's paper met stiff opposition in this country. It was heavily criticized for its lack of physical meaning. On the other hand, many of the conclusions which were formally in accord with the conclusions of Einstein's special theory of relativity or Lorentz's theory were ridiculed for being nonsensical and contrary to common sense.

According to L. T. More, Professor of Physics at the University of Cincinnati, Lewis's chief blunder was to assume that light possesses not only momentum and energy, but mass. Professor More could not resist ridiculing the implication that energy and mass are equivalent. It would mean, More asserted, that the hand of a man who stopped a ball hurtling toward him would be increased in mass. Such ludicrous conclusions came about, More said, because Lewis arbitrarily fixed the velocity of light as a constant, thereby disregarding the physical facts of life. Furthermore, he went on, if Lewis were right, it would mean that the sun was losing about 10^{17} grams of mass per year—obviously absurd. Certainly "such a quantity should have an influence on cosmic problems," he concluded.*

The tone of More's critique of Lewis dripped with sarcasm and scorn. More

* C. T. More, "On Theories of Matter and Mass," *Philosophical Magazine,* 1909, *18*:17–26.

seemed exasperated at the lack of common sense and practicality in Lewis's paper and he had no hesitation in identifying this with theoretical physics, imported from Europe, and of little use in dealing with the world. Regarding the physical meaning of Lewis's analysis, More had made some telling points. But his objections hinged on an appeal to common sense and examples which seemed to draw the conclusions that he considered absurd. More considered the entire enterprise "metaphysical" and therefore, he implied, worthless.

C. L. Speyer, Professor of Chemistry at Rutgers, had many of the same objections to Lewis's paper. In addition, he said that he simply "could not understand it." For example, how could radiant energy stick to a body when the body radiated as much as it absorbed? How could Lewis justify his assertion that all forms of energy are equivalent to radiant energy? Speyer also had grave doubts about *assuming* that mass increased with velocity.*

Lewis did not allow such critiques to go unanswered. His reply focused first on the point that energy, mass and momentum are conserved at each instant regardless of the physical processes in any physical system. And as to the special character of the speed of light, Lewis held to the position that it is nature which holds that the speed of light is constant. Here, Lewis was at one with many of his American colleagues, who also saw both the principle of relativity and the constancy of the velocity of light as empirical generalizations.

It is not clear how the next step came about. The private papers of those involved do not give a clear picture. In any event, at the December, 1908 meeting of the American Physical Society, Lewis and R. C. Tolman, who was a graduate student in physical chemistry at M.I.T., read a joint paper on Einstein's special theory of relativity. It is a remarkable paper on several counts. First, it was the first exposition of the theory in the United States. Second, it was the first paper to approach the subject without reference to electrodynamics. Third, it contained a unique thought experiment for explaining the relationship between rest mass and mass in motion. Although there is one major analytical error in the paper and it reveals weaknesses in the authors' grasp of the basic premises of the theory, it is to the two physical chemists that we owe the introduction of the theory of relativity in the United States.†

According to Lewis and Tolman, the impossibility of making an absolute measurement of a uniform velocity is a law of nature. It is an empirical generalization. Another empirical generalization is the constancy of the velocity of

* C. L. Speyer, "The Fundamental Laws of Matter and Energy" *Science*, 1909, *29*:656–659.

† G. N. Lewis and R. C. Tolman, "The Principle of Relativity and non Newtonian Mechanics," *Philosophical Magazine*, 1909, *18*:510–523.

light. These two empirical generalizations, taken together, constitute the principle of relativity. This point of view is consistent with the pragmatic experimental emphasis prevalent within the American scientific community.

Furthermore, in the thinking of Lewis and Tolman, the special theory of relativity is not a matter of rigid rods, perfect clocks, and light signals as it was for Einstein. Rather, it was a theory in which one resolved conflicts between the expectations of the observer and the dictates of the postulates. In that sense it was a phenomenological, psychologistic theory. They maintained, for example, that *changes* in the magnitude of force, mass, energy, time, and length have a fictitious significance. Distortions in the dimensions of a moving body, for example, were termed a "scientific fiction." Each observer naively considered himself to be at rest, whereas the physical condition of the object could not possibly depend upon the state of mind of the observers. Lewis and Tolman saw no choice, given that there is no meaning to absolute space, but to recast the entire foundation of physics or to adopt certain peculiarities in the units that are used to make measurements.

Their attitude is no better illustrated than in one of the innovative features of the Lewis and Tolman paper—the derivation of the relationship between the mass of an object when measured at rest and when measured in motion, relative to an observer. That derivation is now a standard element of almost every elementary treatise on special relativity. A variant of it was used in Chapter 3. In 1909, it was all new. It is a common sense solution.

However, this kind of analysis resulted in trouble for the two men. They attempted the same kind of argument for analysis of the forces on a right angle lever in equilibrium, at rest in relation to one observer and in uniform motion to a second observer in a direction parallel to one of the arms of the lever. Lewis and Tolman derived an expression for the relativistic transformation of forces for this case that is incorrect. In a German publication, Max von Laue quickly pointed out the error and noted that it was important to explain the null result of certain second order ether drift experiments in which one expected to measure the effects of static electrical changes on each other as they moved through the ether. To Lewis and Tolman this thought experiment had represented something quite different. Their concern had been to salvage the first postulate of relativity by a "fiction."

For all of its defects, the paper by Lewis and Tolman was very important. It brought the special theory of relativity to the attention of many American physicists. Lewis was invited to give a series of twelve lectures to the Harvard physics department by the chairman, Theodore Lyman, who professed complete ignorance of the subject. One lecture, a four-dimensional analysis in which a new and simplified notation system was introduced, was later published in a major German journal which reviewed innovations in electromagnetic theory and atomic physics.

Although Lewis apparently lost interest in relativity, Tolman's work on the subject continued. From 1922 to 1948 he was Professor of Physical Chemistry and Mathematical Physics at the California Institute of Technology. At the time of his death he was recognized as one of the world's leading cosmologists and one of Einstein's peers in relativity and field theory.

After the publication of the joint paper on special relativity with Lewis, Tolman concentrated on the status of the second postulate of relativity. He published several papers on the subject and defended the proposition that the second postulate is a combination of the first postulate and the experimental result that the velocity of light is independent of the velocity of the source of the light. That being the case, the second postulate, the invariance of the speed of light, is, Tolman argued, an experimental fact. This view prevails to the present day in the writings of American physicists.

REACTION TO LEWIS AND TOLMAN

The work of Lewis and of Tolman did not go unchallenged. Some physicists argued that, whereas the first postulate was experimentally verified, not only was the second postulate not verified, it was contradictory to the first. This was the point of view of Professor O. M. Stewart at Cornell University. Unlike Tolman, Stewart recognized that the second postulate, the invariance of the speed of light, is beyond experimental test because it is impossible to make a one-way measurement of the speed of light.

> However the assumption [that the velocity of light is independent of the source] has been generally accepted on account of our concept of the aether as a fixed medium filling all space. But as has been shown by Einstein and others, the first postulate of relativity leads to a rejection of this concept of the aether. In fact, in all experiments where it has been possible to test this conflict between the stationary aether theory and the first postulate, the results have been in favor of the latter. . . . Thus we have the principle of relativity destroying a concept which is used in one of its postulates.*

Of course, contrary to Stewart's assumption, Einstein never rejected the ether. He chose, he said, to perform his analysis without making use of the concept. It was part of the parsimony of his theory of measurement. Nor is the concept of the ether necessary if one assumes the invariance of the speed of light. That assumption recognizes the special role of the ultimate signal speed in any physics which uses the principle of relativity.

* O. M. Stewart, "The Second Postulate of Relativity and the Electromagnetism of Light," *Physical Review*, 1911, *32*:418–428.

L. T. More, who had leveled some very trenchant criticism at Lewis's earlier work, remained unalterably opposed to the theory of relativity. He called for a return to the fundamentals of Newtonian physics. He held that the theory of relativity as explicated by Lewis and Tolman contained an "overburdening amount of metaphysical speculation." Physics, he said, must be "cleansed" of the stifling influence of metaphysics. The chief metaphysical villain in recent physics that More objected to was the assumption of an electromagnetic world view — the attempt to account for matter and mechanical interactions in terms of the laws of electromagnetism.

> Professor Abraham and the modern school of German physicists are frankly endeavouring to give a purely electromagnetic foundation of the mechanism of the electron and to mechanical actions in general.
> Now to me and I believe to many men of science, the chief and indeed the only value of an atomic theory is to give a concrete, though crude image of matter reduced to its simplest conditions. The word electricity gives me no such image of matter; it conveys absolutely no idea of matter nor even of space or time relations.*

Of course, More's opinions about the images evoked by electricity or by the atomic theory and their relationship to space and time relationships, or to reality, constitute a metaphysic. More had definite opinions about the postulates from which one begins to construct a picture of the world. More's reasoning apparently was based on the following assumptions:

1. A physical theory is a crutch by which one may build a sensible picture of the workings of natural phenomena.
2. The theory must conform, not only to any postulate, but to postulates that are reasonable, that is, mechanical.
3. Physicists should not worry about the status of postulates; in fact, the physicist is not explaining anything; he is merely describing.
4. If postulates other than mechanical ones are used, or if the theory is not one which readily leads to a mechanical picture, we are trying to *explain* and we have converted physics to metaphysics.

More admitted that it would be necessary to modify his earlier views about the need for a mechanical ether. In the face of recent experiments More urged that no hypotheses be made about the nature of the ether. It would still be proper, however, to distinguish space through which radiant energy passed by the name "ether." Despite recent experimental work, invariable inertia would be ascribed to the electron. This could be done by splitting the mass of the electron into two

* L. T. More, "Recent Theories of Electricity," *Philosophical Magazine,* 1911, *21*:196–218.

parts, ponderable and electromagnetic. If the electron were rigid, we could expect positive results from experiments such as those done by Michelson and Morley. If the electron were deformable it would have subparts, otherwise it must get energy from the ether.

> Nor can the principle of relativity aid us in obtaining positive knowledge on such questions; at best it is a principle of negation stating . . . that all our knowledge is relative, and must be so from the fact of the finitude of our minds.

More's opinions were echoed by other physicists. In the physics section of the meetings of the American Association for the Advancement of Science in 1911, a symposium was held about the ether. The summary of this session, which appeared in *Science,* stated that:

> The significance and place of the principle of relativity was of course given principal attention and some difficulty was experienced in finding common ground on which to stand.*

The participants included Michelson, Morley, G. N. Lewis and two recent converts to the theory of relativity, D. F. Comstock and W. S. Franklin, Professor of Physics at Lehigh. He had written a popular exposition of the theory for the *Journal of the Franklin Institution.* There can be little doubt that Michelson and Morley would have found little in common with Lewis, Franklin and Comstock. The latter were committed to exclusion of the substance which the former had spent their lives in search of. According to the physicist and philosopher Philipp Frank, Michelson had confided in him that a theory of electrodynamics without an ether was not physics.†

Michelson, Morley and all the men in the United States whom we have cited in opposition to the special theory of relativity, held that acceptable theories had to conform to acceptable notions of common sense. This is precisely the thesis of L. T. More's critique. According to him there must be an ether. How else could light get from one place to another in interstellar space? It made no sense to More to talk about the increase of energy corresponding to an increase of mass. The gross examples of experience showed the absurdity of that position.

THE APPEAL TO COMMON SENSE

Common sense had been the platform from which Lewis and Tolman argued for relativity. If we must conserve energy, momentum, and mass at every instant, as

* A. D. Cole, "The American Association for the Advancement of Science, Section B—Physics," *Science,* 1912, *35*:510–516.

† Philipp Frank, Personal Communication, December 1964.

we obviously must, and we grant the experimental verification of the postulates of the theory of relativity, the results associated with the theory follow. Since the theory is based on the results of experience (that is, the postulates have been experimentally verified) discrepancies in the reports of observers in different frames of reference are the results of experience. Common sense dictated to Lewis and Tolman a resolution of the conflict. The appeal to common sense by these American authors is most beautifully typified by Lewis's ability to mold nonsense into recognizable, useful results as in his first independent publication on the subject. He assumed that "something in light has mass when it moves at the speed of light."

The kind of common sense I refer to, which is recognizable in the American response to relativity, is palpably different from the response to the theory from British scientists. Whereas the British maintained the need for a concrete medium, they participated in the Continental analysis of abstractions. Most American scientists seemed to have no patience for such activity. In the United States the appeal to common sense was most often an appeal to explanations which maintain a one-to-one correspondence between the reports of crude sense data and the basic tenets of scientific theories, and show a contempt for abstraction.

I know of no better example of this attitude than the 1911 presidential address to Section B (Physics) of the American Association for the Advancement of Science given by W. F. Magie.* Magie, Professor of Physics at Princeton, held that the theory of relativity was based on the need to explain the Michelson-Morley experiment and on the inconvenience of using fundamental equations of electromagnetic theory in unaltered form in various inertial frames of reference. He made a long list of the things the theory was *not* needed for. This list included all first order ether drift experiments and the experimental results of Kaufmann. According to Magie, if the principle of relativity were an inductive principle, there was a chance of overturning it in the future. If it were a fundamental postulate, the reason would seem to be simplicity. Magie quoted Max von Laue, who had cited the importance of simplicity as a criterion. Magie rejected that criterion on the grounds that he knew of no basis in the history of science for using it. If simplicity were ruled out as a criterion, then the deductive view was no more compelling than the inductive view. Most of the results of relativity had been predicted by other theories—for example, the theories of Lorentz and J. J. Thomson. But most important, Magie asserted, the theory of relativity completely abandoned the ether.

* W. F. Magie, "The Primary Concepts of Physics," *Science* 1912, *25*:281–292.

> In my opinion the abandonment of the hypothesis of an aether at the present time is a great and serious retrograde step in the development of speculative physics. . . . Without an aether how do we account for the interference phenomena which made [the Michelson-Morley] experiment possible?

Without the ether, it seemed impossible to Magie to maintain a wave theory of light. If light were the propagation of a disturbance a medium was needed. Did the adherents of the theory of relativity really intend, Magie asked, to return to the particle theory of Newton? Without an explanation for the propagation of electromagnetic radiation, the theory of relativity should be rejected.

> I submit that it is incumbent upon the advocates of the new views to propose and develop an explanation of the transmission of light and of the phenomena which have been interpreted for so long as demonstrating its periodicity. Otherwise, they are asking us to abandon what has furnished a sound basis for the interpretation of phenomena and for constructive work in order to preserve the universality of a metaphysical postulate.

Magie pleaded for a theory that did more than describe. A model, however imperfect, was needed. The ether provided such a model. Magie rejected Planck's comparison of the controversy about the theory of relativity to that about the acceptance of antipodes on earth:

> Many men in the Middle Ages believed that there were no antipodes but their belief was based on reasons and so far were they from being able to conceive of antipodes and to believe in their existence [that those that did believe] were pursued as heretics. I do not believe that there is any man now living who can assert with truth that he can conceive of time which is a function of velocity or is willing to go to the stake for the conviction that his "now" is another man's "future" or still another man's "past."

Magie concluded that the theory of relativity could not be fundamental because a fundamental theory must be intelligible to all people.

Of course, we know that the theory of relativity has nothing to do with claims about one man's present, another man's future or yet another man's past. And Magie was not an aberrant example of American scientific opinion. He was in the mainstream of American physics during the first decade of this century.

THE POPULAR RESPONSE, 1905–1911

If there was little notice of Einstein's special theory of relativity in the American physics literature, there was even less in the American popular literature. In 1909, *The Scientific American* announced a contest in which a $500.00 prize would be awarded for the best

> . . . simple explanation of the fourth dimension, the object being to set forth in an essay [not longer than 2500 words] the meaning of the term so that the ordinary lay reader can understand it.*

(At the time *The Scientific American* was a weekly news magazine devoted to science and engineering. Whereas there were, from time to time, feature articles on new developments in the scientific literature, a large proportion of the space was devoted to practical engineering and patent news.)

The contest was judged by two professors of mathematics, one from Brown University and one from Columbia, neither of whom appear to have published in physics. The winner of the contest was Graham D. Fitch, a member of the Army Corps of Engineers. His prize winning essay was published in *The Scientific American* in July of 1909. The exposition was purely geometric. However, Fitch raised the possibility of the reality of the fourth dimension. He discussed the question in terms of its implications for structural chemistry and suggested that the possibility of the existence of the fourth dimension "had not yet been shown to be inconsistent with any scientific fact and the limitation of space to three dimensions, though probably correct, is strictly empirical."†

In subsequent issues of the journal, three honorable-mention essays were published. The first two, by E. H. Cutler and F. C. Ferry, dissented from the view that the fourth dimension could ever be relevant to experience. The third, written by Carl Redmond, took a different view:

> The assumption of a fourth dimension has not yet led to any noteworthy useful results, but it is by no means impossible that the science of four dimensional geometry may come to have useful applications. It has been suggested . . . that an atom may be a place where ether is flowing into our space from a space of four dimensions. It can be shown mathematically that this would explain many of the phenomena of matter.

There is no clear evidence either in the call to contestants by the editors of *The Scientific American* or in the content of the winning essays of an awareness of the theory of relativity. It was not a competition which attracted professional scientists. Yet it was one of the few events between the years 1905 and 1919 which reflected popular interest or lay concern with intellectual issues that are now associated with the theory of relativity.

---

* *Scientific American,* (Jan. 9) 1909, *100*:26. The announcement was repeated in the issues of January 30, February 27 and March 20, 1909. The contest was closed on March 27, 1909, when it was announced that there had been a large number of entrants and that entries had been received from all parts of the world.

† *Scientific American,* (July 3) 1909, *101* pp. 6, 15.

CONCLUSIONS

The response reported in this chapter chronicles the American response to the theory of relativity through 1911. On the one hand, the work of Lewis and Tolman and a few others who commented favorably on that work, led the way to the open, competitive and active physics community found in the United States later in this century. The work of Lewis and Tolman was unique. Although the theory of relativity was, as we have seen, elaborated on in Germany, not even German scholars produced the kind of innovative, imaginative approach to explaining the basis of the theory that the two American chemists employed.

On the other hand, the stubborn resistance to the theory and the continued adherence to ether theories which prevailed, represented a common sense approach easily distinguishable from the British emphasis on models and utility. As Magie wrote, a useful theory must be intelligible to all people. This kind of demand of scientific theories is also reflected in the critiques by other American physicists, More's ridicule of "European electronicists," for example. The notion that scientific theories should be intelligible to all people was unique to America. It is reflected in the general view, shared by both proponents and opponents of the theory that the postulates of a theory are testable and that experiment dictates the choice of those postulates. Such attitudes are consistent with the general cultural attitude and the American fascination with technology and gadgets as well as the emphasis on the practical and experimental as opposed to the theoretical and abstract.

It might be thought that perhaps the work reported on in this chapter was not serious and is not representative of American physics. I can only point out that almost all the physicists we referred to were highly respected and were considered by their peers to be in the first ranks of their profession.* And the attitudes that we have chronicled about emphasis on the experimental and practical persisted even as the theory of relativity became assimilated into the American physics community in later decades.

* In the year 1903, with the support of the Carnegie Institution James M. Cattell began gathering data for the publication of a register of American scientists. *American Men of Science* was published in 1906 and has gone through fourteen editions. It is now titled *American Men and Women of Science.* At the time that he collected the data, Cattell used polling devices to determine who among those in the registry were the most productive and respected. Those names were denoted in the registry by a star; they became known as "scientists starred." Almost all the individuals in America who responded to the theory of relativity up to 1911, including Lewis, Tolman, More, Magie, Cole, and Franklin were starred scientists. Cf. S. S. Visher, *Scientists Starred—1903–1943—in American Men of Science* (New York, 1975).

·III·

FROM RESPONSE TO ASSIMILATION

In Part II we examined in some detail how the ideas associated with Einstein's special theory of relativity were received by the four major scientific communities in Europe and the United States in the years immediately following their publication. Almost no one in Europe and America understood or appreciated the kinematical character of Einstein's theory. And even though, by the time of the first Solvay Conference in 1911, the formalism had been established as "accepted" in fact in a majority of cases, the ideas which individual physicists associated with that formalism were the ideas embodied in Lorentz's second order ether theory.

One of the important implications of our examination of the early response to the theory of relativity is that the nature of the explanation which individuals find acceptable is more related to prior metaphysical commitment than to agreement with the evidence. This is possible because any evidence can be explained on the basis of any one of innumerable theoretical structures.

The earlier metaphysical commitments we have examined have been more or less related to what I have termed "National Styles." That is not to say that national origin is the sole determining factor. We have not tried to deemphasise the fact that there were differences in how individuals with the same national cultural backgrounds responded to Einstein's innovation. What we have shown is that over and above such individual differences, there was what can be termed a group response to the publication of the special theory of relativity which is understandable as the product of formative and controlling influences of local social institutions.

The different responses may be summarized as follows: In France there was a curious and pervasive silence—almost as if the theory of relativity had not been published. In England, while consideration of Einstein's ideas took some time, when the theory was considered, it was frequently in the context of making it compatible with traditional British empiricist concerns. At the same time, some British physicists undertook the reorganization of the theory of relativity so that it might be acceptable within the tradition of the British commitment to the notions of a mechanical, luminiferous ether. In Germany, while support for the theory of relativity was obvious and influential, it was restricted to a handful of individuals. What set the German response to relativity apart from the response in other countries was the degree to which the theory was debated and elaborated upon by various research schools. This is clearly related to the unique social structure of German higher education and its relationship to national and local scientific enterprises. In the United States, where the theory of relativity was largely ignored in much the same manner it had been largely ignored in England, the emphasis had been not only on the lack of empirical content to the theory, but on its impracticality. American physicists, more than most others, were not only skeptical, they were derisive and outraged that such utter nonsense could get a serious hearing in some quarters.

Few physicists around the world supported Einstein's ideas. Most of those who did support the formalism were actually committed to a traditional theory rooted in the Newtonian theory of measurement. In the light of these facts the question that immediately suggests itself is: Why is it the Einstein theory (as opposed to the Lorentz theory) to which people now refer? Have Einstein's ideas been accepted or are most physicists still committed to Lorentzian ideas behind the formal mask of the equations of special relativity? Furthermore, how have the ideas we have associated with the theory of relativity been propagated within and outside the scientific community? Is there a connection between popular understanding and the understanding of the physics community? In short, given the early response to the theory, the next question is, how has the theory been assimilated?

In Part III we restrict our investigation to the assimilation of the theory of relativity in the United States. It is proposed that this examination can serve as a model for how an idea such as the theory of relativity becomes assimilated in a culture over a long period of time. The American culture is particularly interesting for such a study because of the rapid growth of the scientific community in the period between 1900 and the present. We will use that investigation to raise the general question of how new ideas are introduced and how they successfully struggle for survival in what must be termed a hostile environment.

·10·

Relativity in America, 1912–1980

In Chapter 9 we discussed the premises which underlay the views of several segments of American culture near the turn of the century about science. Outside of the scientific community, the view was widely held that support of science was needed for industrial growth and that ideas should have practical import. This was a constant theme among industrialists. Within the scientific and physics communities the criterion for good science was not practical outcomes, but empirical foundations.

These two demands, while different, are compatible and harmonious. It is easy to move very smoothly from one to the other. The demand that theories be empirical meant that both the conclusions and the premises of theories be directly tested. That is a demand for practicality in another guise. Outside the scientific community, practicality meant the possibility of profit.

In both cases there was a general eschewal of metaphysics, which was identified with European culture and science. Whereas in the industrial community, the attack on metaphysics might oppose theory altogether and smack of anti-intellectualism, within science, the demand was often only that theory have an empirical base.

At the turn of the century the American scientific community was in a constant state of flux, as was American society. Concomitant with a burgeoning population and rapid industrial growth, the number of professionally oriented centers for higher education was growing rapidly and the number of scientific societies and the membership in those societies was expanding. A case in point is the growth of the American physics community. As mentioned in Chapter 9, between 1900 and 1910 there was a fivefold increase in the membership of the American Physical Society from about 100 members to about 500.

This kind of growth continued until very recently. One of the early spurs was World War I when, as folklore has it, science was enlisted in the war effort. During the period 1919–1932, the physics community in the United States became a power in world physics on a par with Europe. One commentator has said that at the turn of the century, such a development would have been unthinkable.* By 1932 the physics community was three times as large as it had been in 1919. The American Physical Society had 2500 members. Many of the new members did not come from the economically elite segment of American society as they had earlier. A significant number came from middle and lower class backgrounds. According to one historian, American physics developed such power and strength in part because of size. It was a matter of statistical probability, given the size of the community and that its membership was well trained, that many important findings were made on the American side of the Atlantic. By 1932, opportunities for good work in America were so great that European physicists were coming to America for advanced training.†

The growth of the American physics community was no accident. More and more the physicists were being trained in the United States by Americans. According to Daniel Kevles the great centers for physics before World War II were Harvard, Princeton, Chicago, Berkeley and The California Institute of Technology. But splendid contributions were being made at dozens of other universities —land-grant universities like Michigan and Ohio State, and large, private institutions like Yale, Cornell, Northwestern and the Case School of Applied Science (now part of Case-Western Reserve University).

Most of the graduates of these programs did not remain within the university setting. The opportunities for research in the industrial world were significant. Not only was there a demand for trained physicists in independent research laboratories, but most major industrial concerns like American Telephone and Telegraph or General Electric supported research labs that were superbly equipped and offered opportunities for an unfettered organization of research activities, widely described as "basic" or "pure."

The motivations for creating such opportunities are not hard to understand. At

* Nathan Reingold (ed), *Science in Nineteenth Century America* (New York, 1964), p. 251. Cf. Nathan and Ida Reingold (eds), *Science in America: A Documentary History, 1900–1939* (Chicago, 1981) pp. 97–109.

† D. J. Kevles, *The Physicists . . .* , pp. 200–221.

AT&T, for example, one of the lessons that was learned in the first two decades of this century was that research carried out by brilliant scientists paid off in ideas and led to important patents and devices. Individuals with doctorates in physics like H. D. Arnold and Frank B. Jewett led a larger and larger research group and in 1925 evolved into the Bell Telephone Laboratories. Very often such institutions maintained close alliances with research in the universities. This was especially prevalent during and after World War II.

In the 1930s and 1940s a great number of European intellectuals, including physicists like Albert Einstein, Enrico Fermi, Eugene Wigner, Edward Teller, John von Neumann, and countless others went to the United States to escape the Nazi menace and indeed, for the opportunity to fight against it. After the war, the immigration of intellectuals from Europe continued for other reasons—the United States was one of the few countries in the world with wealth and the will to support research. By the end of World War II, even the most skeptical public representatives had no doubt that the marvels like antibiotics, atomic weapons, radar, sonar and other electronic miracles, new fertilizers, plastics, etc. were the direct result of basic scientific research.*

The import of this is that during this century, American universities were transformed from a very small group of institutions producing a minuscule number of Ph.D.s into a vast number of organizations, each of them different. The best of them vied for recognized leaders in various fields, and often lured scholars to their institutions with offers of increased salaries, better research facilities, fewer teaching responsibilities, and so on. In short, more and more, the character of the structure of higher education in the United States has taken on the appearance of the structure of the German university system at the beginning of the century.

Yet there were important differences. I do not think that the social status of intellectuals in the United States has ever come close to rivaling that afforded such individuals in Europe. This is so although many Americans had themselves emigrated from Europe or had come from families who had come from Europe.

It can be argued that by the middle of the nineteenth century, there was an identifiable American culture, different from the culture of Europe. This is not to say that American culture was not affected by European views and tastes. We have already indicated the degree to which American science was influenced by British empiricism and German *Wissenschaft*. The depth of the American commitment to empiricism was unique. It is fair to say that at the time of the American Revolution, the "American culture" had already been established. Thus, while the

* For insight into these attitudes, see Harry Hall, "Scientists and Politicians," *Bulletin of the Atomic Scientists*, (Feb), 1956.

massive immigration of European scientists during the 1930s and 1940s made significant contributions to American science, the values of American science were not transformed, but perturbed. In American culture there has always lurked, not far beneath the surface, a very lively distrust of intellectuals and academicians.*

Immediately after World War II, scientists were applauded and admired for the wonders they had produced during the war, and there was an expectation that the continued production of wonders would solve all of the world's social problems. But by the middle 1960s those expectations had not been met, and there was a very active anti-intellectual and anti-scientific reaction in all segments of American society. I think that reaction accounts for the ensuing dramatic drop in enrollments in college and university science and mathematics courses, as well as a serious and widespread reawakening interest in such topics as alchemy, witchcraft, and astrology, and the identification of science with the more negative aspects of armaments, threats of environmental destruction and other concerns that are potentially threatening to civilization.

Shifting attitudes about science are reflected in the financial support for science from the public and private sectors. Before World War II, almost all support for scientific research in academic settings came from the private sector. During and after the war there was a massive influx of public money for all kinds of scientific projects and for social science research as well. The public attitude seemed to be that the lessons of the war had been that any question could be answered, any problem could be solved if enough money were spent.

With the launch of *Sputnik* a considerable amount of public funding also became available for science education. Yet since the mid 1960s support from the public sector of the economy for scientific research in the universities and for science education has been declining rapidly. It may well be that traditional American attitudes and behavior against public support for education and research will again prevail. The ready availability of public and private funding which was characteristic of the middle decades of this century for support of university research and education will then be seen as a perturbation effect of World War II.

Whereas there will not be unanimity among commentators on the meaning of the various indicators I have referred to in interpreting American attitudes toward science, the indicators themselves are a matter of public record. With that as background, we turn to an examination of how the theory of relativity was assimilated both within and outside the American scientific community in the period between 1911 and the present.

* See the essays in John Higham and Paul K. Conkin (eds), *New Directions in American Intellectual History* (Baltimore, 1979), and Hennig Cohen (ed) *The American Culture: Approaches to the Study of the United States* (New York, 1968).

1911–1919: THE LULL BEFORE THE STORM

In the period between 1911 and 1919, the literature on relativity in the United States was quite meager and represented a continuation of the trends we have already noted in Chapter 9. There was virtually no popular literature on the subject and very little technical literature. R. C. Tolman undertook to show that not only was the first postulate (the principle of relativity) proven inductively by experiment, but so was the second postulate. Tolman published a monograph on the theory of relativity in 1917.* In this monograph, he was intent on avoiding contradiction in the reasoning on which the theory was built. According to him the principle of relativity, the first postulate, was a denial of the ether. Tolman went on to note that one could construe the second postulate, the invariance of the speed of light, as a combination of the first postulate and the principle of the independence of the velocity of light from the velocity of its source. To do so, however, would lead to a strong contradiction because this latter principle had grown out of nineteenth century ether models. Ultimately, Tolman cited experimental determinations of the speed of light from both terrestrial and celestial sources as falsifying an emission theory and providing positive confirmation of the second postulate.

Tolman's 1917 convictions about the experimental confirmation of the invariance of the speed of light was based on earlier publications in which he had analyzed various measurements of the speed of light. Part of that evidence was the measurement of the speed of light from both members of binary star pairs. In Figure 1, we have shown such a pair orbiting around a common center. (The explanation for this phenomenon is precisely the same as that for the moon and earth orbiting each other. In the case of double stars, however, the two orbiting objects are likely to have comparable masses and the center around which each of the bodies orbit is not within either of the bodies as is the case in the moon-earth system. There, the disproportionately larger mass of the earth places the common center of orbit virtually at the center of the earth.)

* R. C. Tolman, *The Theory of Relativity of Motion* (Berkeley, 1917).

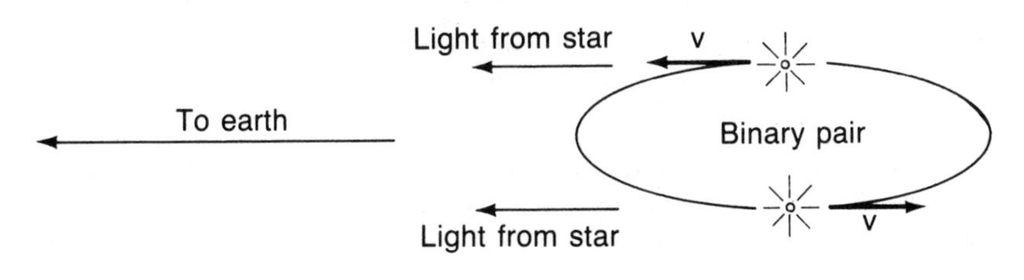

Figure 10.1

Suppose that we choose a binary pair whose orbits around each other are in the same plane as the earth's orbit around the sun. As depicted, at times, one of the pair will be moving directly away from the earth and the other toward the earth. Under those conditions one cannot detect a difference in the speed of light from either of the pairs. There are several different experimental routines that one employs in making such determinations but the fundamental problem with which Einstein wrestled in the creation of the theory of relativity has not been overcome. Either one must use two clocks which have been synchronized in order to measure the speed of light from each of the stars, or one must use mirrors to reflect the light from the stars back and forth in an apparatus. In either case, the one-way value of the speed of light cannot be measured without assumptions about how light behaves in moving from one place to another. As we have seen, it is isotropy of space with respect to the speed of light that is stipulated.

Yet Tolman insisted that the invariance of the speed of light had been experimentally determined. He remained convinced throughout his career. For example, in 1934, in a graduate text that he wrote, a classic that is still in print and widely used,* Tolman again stated that the second postulate of relativity was experimentally determined. And in the same period, in a speech he gave in honor of Albert Michelson, he said,

> . . . I do not feel that Professor Michelson approves of all the strange and bizarre conclusions which have resulted directly and indirectly from his experiment. . . . I do not quite approve of it all myself.†

In 1912 in what appears to have been the first monograph about relativity in the English language, favorable account of the theory of relativity was given by R. D. Carmichael, Professor of Mathematics at the University of Indiana.‡ Carmichael wrote that he was motivated to write the book for a course that he was teaching about relativity at a time when most people "ridiculed the theory." The treatment by Carmichael contains almost all the features that are emphasized in the early American response to the theory of relativity. The text began with a review of first and second order ether drift experiments. The postulates of relativity were introduced as experimentally confirmed statements. The theory itself, according to Carmichael, had been verified experimentally by determinations of the variation in the mass of the electron as a function of electron speed. (In fact, at the

---

* R. C. Tolman, *Relativity, Thermodynamics and Cosmology* (Cambridge, England, 1934).

† The Papers of R. C. Tolman, California Institute of Technology Archives, quoted in Jeffrey M. Crelinsten, *The Reception of Einstein's General Theory of Relativity Among American Astronomers, 1910–1930* (University of Montreal: Unpublished Ph.D. Dissertation, 1981), p. 133.

‡ R. D. Carmichael, *The Theory of Relativity* (New York, 1912).

time, all such results, some confirmatory and some disconfirmatory of the theory of relativity, were the subject of intense criticism in Europe.)

Also in 1912, Professor Leigh Page, a young instructor at Yale, published an important paper in which he showed that it is possible to use only "the principle of relativity" and the laws of electrostatics to derive both Ampere's law for electric currents and Faraday's law of magnetic fields—both laws which depend on the *motion* of charges relative to the observer. Page did not claim to have shown anything new, but it was a novel approach. Page's treatment of the *structure* of Einstein's theory was not novel. While recognizing that the premises from which Einstein derived the Lorentz transformations were different from the premises used by Lorentz, the invariance of the velocity of light was, to Page, a consequence of the principle of relativity and not itself a premise. And later, Page explained the relativity of simultaneity as a consequence of the fact that the *measured* velocity of light is the same in all directions for all inertial observers. In short, Einstein's postulates and stipulations were a subject of experimental tests for Page as they had been for earlier American commentators.*

The contributions of both Page and Carmichael will be taken up again in a later section of this chapter. During the period 1911–1919, there was little activity on problems associated with special or general relativity within the American scientific community. Part of the reason might have been the distracting influence of the first World War and a very negative opinion in this country of anything German.

THE ASSIMILATION OF SPECIAL RELATIVITY WITHIN THE SCIENTIFIC COMMUNITY, 1920–1980

One effect of the tests of the *general* theory of relativity in 1919 was to draw communities of scientists closely related to the physics community into the debate about the merits of Einstein's theory. Jeffrey M. Crelinsten has recently studied the response of the American astronomical community to Einstein's general theory of relativity between the years 1910 and 1930.† That response appears to have been similar to the early response of the physics community to the special theory of relativity (Chapter 9). There was considerable resistance to the theory on the grounds that it was metaphysical, counterintuitive, and/or destructive of the

* Leigh, Page, "A Derivation of the Fundamental Relations of Electrodynamics from those of Electrostatics," *American Journal of Science,* 1912, *34*:57–68. Page's contributions have been discussed by Edward Purcell in Harry Woolf (ed) *Some Strangeness in the Proportion: A Centennial Symposium to Celebrate the Achievements of Albert Einstein* (Princeton, 1981) p. 109. Purcell's contribution to the American literature on relativity seems to have been significantly influenced by Page's treatment. See E. M. Purcell, *Electricity and Magnetism: Berkeley Physics Course, Vol 2* (New York, 1963), esp. Chap. 5.

† Jeffrey M. Crelinsten, *The Reception of Einstein's . . . Theory . . .*

Seated: R. A. Millikan. Einstein. A. A. Michelson, and W. W. Campbell: Standing: C. E. St John, W. Mayer, E. Hubble, W. Munro. R. C. Tollman, A. Balch, W. Adams, and R. Ballard. The photo was taken at the Antheneum, California Institute of Technology, January, 1931. Courtesy of the Archives of the California Institute of Technology.

mechanical world view. Support for the general theory of relativity among astronomers was largely because of the grudging admission that experiment had confirmed its accuracy.

The evidence suggests that for many American astronomers, even the experimental results were not trusted. For just as the Michelson-Morley experiment has been repeated many times, so, too, has the test of the bending of starlight as it passes near the vicinity of the sun during a solar eclipse. Each time a solar eclipse of the sun has been reported in the popular press since 1919, there has been a report of yet another expedition of astronomers to measure the apparent positions of stars in the vicinity of the sun. Each time the Einstein prediction has been confirmed. On April 12, 1924, *The New York Times* reported that William W. Campbell, Director of the Lick Observatory in California, announced that the results of the analysis of the photographic plates exposed during the eclipse of September 21, 1922 had again confirmed Einstein's predictions, and that the experiment would never have to be done again.* Yet, each eclipse brought more field expeditions to the scene to make the same measurements.

THE EVIDENCE

Although it was a test of the general theory of relativity which brought Einstein to the attention of the world, the struggle over the assimilation of the special theory of relativity within the American scientific community continued, virtually unaffected. There is no evidence that there was a sudden rush of interest for either the special or the general theory. The evidence suggests that the same criteria which American scientists employed in their early assessments of special relativity persist to the present time.

It might be thought that one problem in assessing the assimilation of relativity by American scientists during the sixty years between 1920 and 1980 is that there has been an enormous growth in the size of the American physics group. There are many more physicists who are doing research and publishing more papers in more journals. Keeping track of it all, as well as assessing how the theory of relativity is interpreted seems impossible. But a case can be made that the primary literature in physics is not the appropriate data bank to use to assess the assimilation of relativity in the United States. A much better source is the physics textbook literature.

There are at least two reasons for this. First, the format and content of published papers in physics is standardized to transmit the *results* of research in the most efficient manner. Earlier, editors of journals might have insisted on discursive inquiries into the interpretation of the formal systems used in the analysis of

* *The New York Times,* April 12, 1923, p. 1.

theoretical and experimental results. That is no longer the case. There is a widely accepted pattern of reporting which suppresses such excursions.

Very little of the primary research literature concentrated on the theory of relativity. Rather, the formal aspects of the theory were used for other purposes—for example, to predict the mass of subatomic particles or their expected half-life in experiments in nuclear physics when the particles have a very high speed relative to the laboratory frame of reference. But using the equations of the theory of relativity says little about the terms under which the theory was understood since the same formalism might represent commitment to a totally different theory.

Textbook accounts of the theory of relativity are much more useful for examining the way the special theory of relativity was assimilated by American scientists. In the nineteenth century, the distinguishing mark of American education was the textbook. Europeans called the use of textbooks, "The American System." It has been argued by T. S. Kuhn* that when new ideas are introduced in science, not only are there few converts, but it takes long periods for the new point of view to take hold. According to Kuhn, after the transition from the old theory to the new, it is the role of scientific textbooks to transmit the new standards about the class of physical problems.

On the other hand, consistent with the theme of this book, the role that textbooks play will differ from culture to culture. Given the importance of textbooks within the structure of the American educational system, it would not be surprising to find that they have played a significant role in shaping American attitudes about innovations.

Kuhn's view about the dynamics of scientific revolutions will be examined in more detail later. Regardless of whether or not one subscribes to his view or to another view, a sensible model for the introduction of an idea like the special theory of relativity into the text literature predicts that the initial treatment of the theory occurs at the most advanced, graduate levels and filters down to intermediate courses and finally, to beginning physics courses.

It might be thought that explanations of the theory of relativity in textbooks might be misleading because, regardless of the author's convictions about the meaning, his or her convictions would give way to pedagogic expediency. But textbooks prove to be one of the few places where physicists are willing to discuss, if only implicitly, the meaning of theories and their concepts of how evidence supports theories. That makes textbooks a useful resource to judge how the theory of relativity is assimilated by scientists. Physics textbooks help established physicists indoctrinate future scientists in the meaning of the discipline.

Of course, there is the problem of selection. As the size of the American physics

---

* T. S. Kuhn, *The Structure of Scientific Revolutions* (Chicago, 1st ed., 1962; 2nd ed., 1970).

community increases, there is a parallel increase in the number of textbooks for the training of physicists and students in fields allied to physics. And so at all levels—graduate, advanced undergraduate, and introductory—the number of textbooks to evaluate is enormous. In making a selection the following criteria have been used: Attention has been paid to those texts which have been published in several editions because they are more widely used and more influential. In addition, attention was paid to other books for which independent evidence suggests that an important point of view is represented. Other books were also sampled to insure that the ranges of points of view are representative.

GRADUATE TEXTS, 1920–1945

R. D. Carmichael, who published the first English language monograph on the theory of relativity in 1912, offered a second edition of the work in 1920.* The text was identical to the earlier version except a new chapter was added about the general theory of relativity. He asserted that experiment had shown the correctness of the postulates of special relativity again and extended that argument to the general theory of relativity.

Although Carmichael was trained as a mathematician and not as a physicist, his viewpoint is important for our consideration. In late 1919 and early 1920 he published two articles in *The New York Times* interpreting Einstein's theory for a lay audience.† In those articles he again stressed that the premises of the theory had been well established by experiment. Later, in 1926, he participated in a debate sponsored by the Indiana Chapter of Sigma Xi, a national academic scientific honorary society, about the worth of the theory of relativity.‡ Before discussing that debate, let us examine in more detail the principles underlying Carmichael's position about the fundamental structure of the theory of relativity.

In addition to his other interests, Carmichael was actively interested in the philosophy of science, and published a number of articles on that subject.§ He recognized that much contemporary writing on the problem of creativity was wide of the mark. He argued that those inner verbalizations or inner images which are constantly before us, are as much the end product of "thinking" as public verbalizations and depictions. Thinking must occur at a level beyond our direct apprehension of it. Even the greatest minds in the history of Western thought,

* R. D. Carmichael, *The Theory of Relativity* (2nd ed.; New York, 1920).

† *The New York Times* December 7, 1919, Sec. I, p. 18; *The New York Times* March 28, 1920, Sec. V, p. 11.

‡ R. D. Carmichael et al, *A Debate on the Theory of Relativity* (Chicago, 1927).

§ Many of these articles were collected and published in R. D. Carmichael, *The Logic of Discovery* (Chicago, 1930).

Bacon, Descartes, and Aristotle had been at a loss when it came to understanding how hypotheses come to mind. Carmichael, like Einstein, emphasized the importance of a sensitive, finely tuned intuition.

But if we cannot understand the source of hypotheses about nature, we can understand how we choose between competing theories. According to Carmichael, there are crucial experiments by which it is possible to make these decisions. With Newton's gravitational theory and Einstein's theory of relativity, observation has helped us to discard the former in favor of the latter.

Carmichael was more self-conscious and introspective about the relationship between the postulates of theories and experimental evidence than was almost any other person in this study. About deductive systems, Carmichael understood that postulates cannot be proved. But he also made a very sharp distinction between the role of postulates in deductive systems like those found in mathematics and those in natural science. There are, he wrote, two parts to natural science. An empirical part in which fundamental laws secured by intuition are subject to the verification of a preliminary test as to their validity. In a second stage,

> . . . an attempt is made to set apart a few of the empirical laws as basic and fundamental and to derive all the remaining laws from these by strict deduction. Furthermore the logical consequences of the fundamental laws often point to the existence of new empirical laws not yet recognized. If these predictions are not verified some modification of the basic fundamental laws will be necessary: They are often corrected and extended in this way.*

But Carmichael recognized that many of the principles of science are beyond empirical testing and, in that regard he cited without amplification the theory of relativity. While this may seem paradoxical or contradictory, Carmichael seems to have had in mind that when a theory is first proposed, the postulates are often arrived at by intuition, and not supported directly or indirectly by experiment. Such was the case, he said, with the atomic theory. But he continued, there is now "almost direct empirical evidence for the basic hypotheses."†

As we saw in Chapter 3, given that the speed of light is the ultimate signal speed, and that we have stipulated that space is isotropic with respect to the speed of light, it is impossible to make a one-way measurement of that signal speed. Similarly, given that the atomic theory begins with the assumption that matter is particulate, it is also impossible to prove the existence of atoms by an experiment, even with the evidence that is now available.

Carmichael understood the structure of deductive arguments and the relation

* *Ibid.*, p. 49.

† *Ibid* pp. 58, 88. Cf. p. 161.

between postulates and evidence very well. Yet he was unwilling to eradicate the artificial distinction that he had erected between deductive systems in physical science and those for other purposes.

Earlier, we referred to a debate on the theory of relativity held at the University of Indiana in 1926 Carmichael was joined by Harold Davis, Assistant Professor of Mathematics at that university; they were opposed by Professor William D. MacMillan, Professor of Astronomy at the University of Chicago, and Professor Mason E. Hofford, Assistant Professor of Physics at Indiana.

The strategy of the affirmative position was in keeping with Carmichael's earlier position: namely, that the postulates of the theory of relativity have been established by experiment and that the conclusions have been confirmed. The opposition by McMillan and Hofford also had a familiar ring. MacMillan centered his opposition on the grounds that Einstein's postulates were counterintuitive. He interpreted Einstein's argument that the relativity of simultaneity meant that the concept of simultaneity was not acceptable and he proceeded to ridicule *that* idea. His colleague argued that, contrary to the position of Carmichael and Davis, the evidence in favor of the theory of relativity was not conclusive. He pointed to the recent work of Dayton C. Miller, Professor of Physics at Case School of Applied Science, which indicated that the results of the Michelson-Morley experiment were not null, but that the earth is moving relative to the ether in the direction of the constellation Draco at about 200 miles per second. There is, Hofford asserted, no possibility of error in this result.* On the other hand, the tests of the principle of equivalence and the general theory of relativity were not to be believed. And McMillan added that in the last hundred years science had seen four great doctrines introduced: 1) Conservation of energy, 2) Evolution of the universe, 3) The second law of thermodynamics which, while an outrage to our aesthetic senses, has a reputable experimental standing, and 4) The principle of relativity, which is actually in the domain of geometry. This principle is also an outrage to our aesthetic sense but in addition, it adds nothing to our knowledge of the world.

After the work of Carmichael and Tolman, the next treatment of the theory was by Leigh Page. In 1922 he published a graduate text on electromagnetic theory, *Introduction to Electrodynamics.* Page used this book to further his earlier approach, deriving the laws of electrodynamics from those of electrostatics by using the theory of relativity:

> The object of this book is to present a logical development of electromagnetic theory founded upon the principle of relativity. As far as the author

* Miller's results, often cited through the years by opponents of the theory of relativity, were eventually shown to be an artifact of defective experimental design. See R. S. Shankland, et al., "A New Analysis of the Interferometer Observations of Dayton C. Miller," *Reviews of Modern Physics,* 1955, *27*:167–178.

> is aware, the universal procedure has been to base the electrodynamic equations on the experimental conclusions of Coulomb, Ampere, Faraday, even books on the principle of relativity going no further than to show that these equations are covariant for the Lorentz-Einstein transformations. As the dependence of electromagnetism on the relativity principle is far more intimate than is suggested by this covariance, it has seemed more logical to derive the electrodynamic equations directly for the principle.*

As in his earlier work, Page maintained that the principle of relativity requires that the speed of light be the same in all directions in all inertial frames of reference. But then he laid out the postulates in detail, making explicit in a way that had never been done before, all of the measurement premises when one speaks, for example, of measuring a speed or velocity. In the process Page showed the role that the assumption of isotropy plays in synchronizing clocks in any frame of reference. He did not, however, link the need to stipulate isotropy with the second postulate.

It is noteworthy that in an introductory graduate text on theoretical physics which Page published in 1928 and which was subsequently published in three more editions and twelve printings,† Page argued as follows: The principle of relativity requires that the laws of physics be the same in all inertial frames of reference. When this principle is applied to the laws of electrodynamics, the speed of light is invariant. This suggests that Page did not see, in his earlier work, the need to stipulate isotropy for one signal speed. Treating the second postulate of relativity as a special case of the first postulate continued to be a very popular idea in the American text literature.

Until well after the end of World War II, the major source for introduction of the theory of relativity in the text literature was books on electricity and magnetism. At advanced levels, few such books were published. Typical of these was a book by J. C. Slater in 1941. Slater's treatment represents the most common view in American textbooks of the nature and status of the theory of relativity:

> The most decisive [of a series of experiments], . . . the celebrated investigations of Michelson and Morley have led to the establishment of two fundamental postulates as highly probable if not certain. . . . ‡

---

* Leigh Page, *Introduction to Electrodynamics: From the Standpoint of the Electron Theory* (Boston, 1922).

† Leigh Page, *Introduction to Theoretical Physics* (New York, 1st ed., 1928; 2nd ed., 1935, 3rd ed., 1952).

‡ J. C. Slater, *Electromagnetic Theory* (New York, 1941) p. 74.

Those two postulates were the principle of relativity and the invariance of the speed of light.

In 1942, P. G. Bergmann published a book entitled *Introduction to the Theory of Relativity* which had some remarkable features.* In the first place, most of the book is intended for the advanced student. However, in introducing the special theory of relativity, Bergmann not only paid homage to Einstein's popularization of relativity theory,† he replicated that treatment in detail without mathematics other than elementary algebra. Bergmann's book had little effect, as far as I can tell, on other authors until the late 1960s. Even then his influence did not change, in any essential way, the epistemological outlook of the Americans. It should be noted that Bergmann was *not* trained in the American system.

ADVANCED UNDERGRADUATE TEXTBOOKS, 1920–1945

When one turns to texts intended for advanced undergraduates, one finds that almost all treatments of relativity that appeared before the end of World War II were in texts in electricity or surveys of atomic or modern physics. Overwhelmingly the postulates of relativity were perceived as confirmed or proved by experiment. Consider, for example, the book that was until recently perhaps the most influential and widely used text in modern physics, commonly referred to as "Richtmyer and Kennard." This book was first published in 1928, by F. K. Richtmyer, Professor of Physics at Cornell University. The only reference to the theory of relativity in that edition is as follows:

> There has been much controversy and much weighing of both theoretical and experimental evidence to decide whether the relative law of variation of mass is the correct one. For our purposes, the important point is that *qualitatively* a variation of mass with velocity is to be expected, irrespective of the principle of relativity. The expression "relativity change of mass" should therefore, refer to the particular formulae used to compute masses at various velocities rather than to the origin of the phenomena.‡

The second edition was published in 1934 and contained a new appendix, entitled "Concerning Relativity." In that appendix, the Lorentz transformation equations were derived by invoking, on the authority of D. C. Miller, an opponent of the theory of relativity, the Lorentz-Fitzgerald contraction hypothesis to explain the results of the Michelson-Morley experiment. Then the Michelson-Morley

* P. G. Bergmann, *Introduction to the Theory of Relativity* (Englewood Cliffs, 1st ed., 1942; 2nd ed., 1964).

† A. Einstein, *Relativity: The Special, The General Theory.*

‡ F. K. Richtmyer, *Introduction to Modern Physics* (New York, 1928) p. 373.

experiment is cited as justification for assuming the invariance of the speed of light. The only other reference in the appendix is to Tolman's 1917 treatise for a "correct" treatment of the concept of relativistic mass. There is no discussion of the meaning of the theory or how it applies to physical systems. Einstein's name is not mentioned.

In the absence of any external evidence it is difficult to account for this state of affairs. It seems clear from Richtmyer's remarks in the first edition that he was skeptical of the theory. This is reinforced by his treatment of the theory in the second edition, but the failure to mention Einstein in the context of the theory of relativity is noteworthy and mysterious. It might represent anti-Semitic attitudes toward Einstein, although I know of no documentary evidence to support such a claim. Yet there is ample evidence of anti-Semitism in the attitudes of some twentieth century American academicians toward the work and qualifications of their Jewish colleagues.*

The third edition of Richtmyer's book appeared in 1942, three years after his death. E. H. Kennard, a colleague at Cornell University was the second author. Special relativity was the subject of Chapter 4 of this edition. After a discussion of Galilean relativity and a review of several of the more popular first and second order ether drift experiments, Kennard introduced special relativity as resting on two postulates — the principle of relativity and the invariance of the speed of light. Kennard characterized the espistemological status of these two postulates as follows:

> Of these two "postulates" the second is believed to represent a rather simple experimental fact, whereas the first is a generalization from a wide range of physical experience. There is no implication that the first postulate, which contains the new principle of relativity, is in any way self-evident; like the assumptions made in all physical theories, it is intended as a hypothesis to be tested by comparing deductions from it with experimental observations.†

Kennard's account of the epistemological status of the postulates is remarkable because it was *followed* by a detailed discussion of the problem of synchronizing clocks, the impossibility of a one-way measurement of the speed of light and recognition that " . . . measuring the velocity of light in one direction turns

* On this point see Nathan Reingold, "Refugee Mathematicians in the United States, 1933–1941," *Annals of Science,* 1981, *38*:313–338. And see the discussion of the work of Herbert E. Ives, below.

† Richtmyer and Kennard, *Introduction to Modern Physics* (New York, 3rd ed., 1942) pp. 131–132.

out . . . to involve some assumption that cannot be tested in advance." Kennard continued:

> Thus the simultaneity of events and in part, the time order in general of events at different places, are relative and not absolute concepts. At least this is true so far as physics is concerned. . . . The distinction between relations of space and those of time is thus in part a *relative* matter depending upon which frame of reference is used in somewhat the same way as relations of "right" and "left" depend on the position of the observer.*

At this point, Kennard broke off the discussion on the grounds that, while it was fascinating, his main concern was to derive the formulas for the application of relativity to physical law.

The treatment of relativity in the fourth (1947) edition and the fifth (1955) edition remained essentially the same as in the third. Whereas the third edition emphasized that the order of two events occurring in a time less than the time required for light to travel from one event to the other is relative to the frame of reference making measurements on the events, in the fourth and fifth editions, he emphasized the fact that distant simultaneity is a relative rather than an absolute concept.

The fifth edition had a new "junior" author, Tom Lauritsen, Professor of Physics at the California Institute of Technology. Correspondence between Lauristen, Kennard, and the publisher, indicates that Lauristen's contributions were concentrated in the chapters on cosmic rays.† When their collaboration began, Kennard had already revised Chapters 1–10. Yet, there was intense discussion concerning the treatment of the theory of relativity (now Chapter 2) as an example of the style of the book and the audience to which it was directed.

In a letter to Kennard, drafted on Feb. 2 [1954], Lauristen said that he had shown some of the manuscript to his colleague R. B. Leighton:

> He thinks it is ok, but he wants no part of it for his students! Too much history, too much conversation, not enough facts. I'm afraid that it is getting to be a common attitude among students. They are impatient with the historical approach and want a clean statement of current knowledge to write down in their notebooks. I'm sticking to my guns of course because I think this attitude simply produces handbook engineers, but the going is tough and I need comfort.

---

* *Ibid.* pp. 132–133.

† This correspondence is in the Kennard Papers (Folder I-2) at the Center for the History of Physics at the American Institute of Physics in New York City, and in the T. Lauritsen Papers at the Archives of the California Institute of Technology in Pasadena.

Later in the same letter, Lauritsen suggested that any discussion of the second postulate was, "of course redundant as it is merely a special case of the first," and then went on to suggest that one did not simply *assume* that the velocity of light was the same in both directions. He asked if that was not a requirement of relativity and might have even been proven by experiments measuring the speed of light.

Kennard's reply to Lauritsen, dated February 8, 1954, began as follows:

> The pedagogical question raised by Prof. Leighton is one on which I don't feel very competent. I firmly believe in a certain acquaintance with the history of physics, i.e. with the pathway trod by prior physicists in arriving at our present fund of knowledge. It may be necessary to feed it to them in rather small doses, tho [sic] my idea was always to select *significant* items, not to mention false starts unless they seemed instructive, and not to describe details that any student would naturally fill in correctly.

Kennard rejected Lauritsen's suggestion that the second postulate was a special case of the first. He did not agree that the equality of the velocity of light in both directions in one frame of reference followed from the theory of relativity. Kennard explained that the isotropy of space for the speed of light required an arbitrary convention to make it so and then said:

> Operationally, one way velocity is meaningless until clocks at different places can be synchronized; and how to do this without a new *pure* assumption? [sic]

In his correspondence, Kennard showed that he understood the relationship between one way velocity measurements, synchronization of clocks and the fact that nothing can keep up with a beam of light. In spite of this, in the text, Kennard continued to insist that the invariance of the speed of light was a simple experimental fact.

The attitudes of Leighton, Lauritsen and Kennard concerning the use of historical materials in physics textbooks is also illuminating. Though they disagreed on using historical examples, Leighton and Kennard agreed on there being correct and incorrect explanations. Current belief is the "truth"; it represents "the facts." It is ironic that Lauritsen found precisely the same biases among students so discouraging.*

* Several years later, Leighton published his own text on modern physics. His treatment of the basis of the theory of relativity was to state that Einstein had proposed a radical new approach to the problem posed by the Michelson-Morley experiment by asserting the principle of relativity and by radically altering our notions of space and time. Leighton then derived the relativity of simultaneity by implicitly assuming that the speed of light is invariant. See R. B. Leighton, *The Principles of Modern Physics* (New York, 1959) p. 5.

The sixth edition of Richtmyer and Kennard appeared in 1968. J. N. Cooper, Professor of Physics at Ohio State University was now an author. In earlier correspondence with Kennard, he revealed that he was an enthusiastic patron of the book. Kennard had kept Cooper informed of proposed changes for the 1955 edition with a request for suggestions on how the size of the book could be reduced. Cooper did not think that a serious reduction in the number of pages was possible.* Later, the publisher, McGraw-Hill, requested Cooper to write a critical review of the fifth edition to prepare for bringing out a modernized sixth edition. The authors had to decide if the book was to be a handbook, a textbook, or a history. In that context Cooper noted that,

> The sections on the philosophy of physics and related materials are wonderful for the teacher and the very mature student, but they often leave the majority of the students confused and apathetic. In such cases, these sections seem to undermine all the ideas that have been introduced.

Cooper recommended that the treatment of relativity (Chapter 2) be moved on the grounds that it is a poor topic for the first "business" chapter because modern physics courses do not cover this subject except to "take over a couple of key relations from Einstein."†

Kennard defended the idea that the book would serve a variety of purposes. He thought it could be used both as a handbook and a textbook. That, after all, he wrote, is one of the reasons the book continued to sell. He defended the inclusion of history on the following grounds:

> Sometimes history starts out in the wrong direction; then a short description may be worthwhile to illustrate the difficulties encountered in aiming research. Where history moves in the right way, give a few details that help to illuminate the subject. I rather think Richtmyer would agree with such principles.‡

* Cooper to Kennard, 16 February, 1948; Cooper to Kennard, 22 January, 1952; Kennard papers, Folder I-2.

† A typescript of the review is contained in folder I.3 in the Kennard Papers. Cooper was also associated with McGraw-Hill in another text project. He was co-author of the seventh (1964) edition of the McGraw-Hill elementary college text, *Elements of Physics,* previously published by Alpheus W. Smith, who, like Cooper, was a Professor of Physics at Ohio State University. In the first few editions the title of the book was *Elements of Applied Physics.* Through the sixth edition, published in 1957, there was no mention of the theory of relativity. The seventh edition contained a short discussion of the theory of relativity which, it said, ultimately resolved the dilemma of the results of the Michelson-Morley experiment. After Smith's death in 1968, two more editions (1972 and 1979) were published.

‡ Handwritten draft of "Comments by E. H. K. on Review by Professor John Cooper of Introduction to Modern Physics," Folder III.2, Kennard Papers. The top of the first page has the marginal note, "Sent to McG-Hill July 3 [1962]."

But in fact, when the sixth edition of *Introduction to Modern Physics* was published in 1968, the year of Kennard's death, the treatment of relativity remained in the second chapter and it was essentially the same as it had been in the third, fourth, and fifth editions. According to revisions made by the new co-author, Cooper, the second postulate was no longer described as "a simple experimental fact," but as an "experimental fact." This assertion was followed by a discussion of the need to stipulate the isotropy of space with respect to the speed of light in order to synchronize clocks.*

The treatment of the theory of relativity by Richtmyer and Kennard is in the American empiricist tradition. The postulates are "experimental facts." It is significant that none of the authors, including Kennard, Lauritsen and Cooper, or any of the editors at McGraw-Hill, saw the philosophical contradiction between the belief that the invariance of the speed of light is an experimental fact and the recognized need to stipulate isotropy of space for the speed of light. Perhaps this reflects the view that the important thing about discussing the theory of relativity was to "take over a few key relations from Einstein." It is *results,* after all, which the physicist, *qua* physicist, is interested in. This is reflected in Leighton's disdain for use of history, Kennard's distinction between history moving in the "right" or in the "wrong" direction, or in Cooper's remonstrance that philosophical discussions in the book often leave students confused.

Yet we note that virtually no other intermediate text before 1955 did more than assert that the postulates of relativity are experimental; most of them claimed that it was the Michelson-Morley experiment that provided the confirmation.

INTRODUCTORY TEXTBOOKS 1920–1945

There was even less attention to the theory of relativity in elementary textbooks. Until long after World War II, it was widely believed that it is impossible to discuss the theory of relativity without the use of sophisticated mathematical skills.

A typical example of how the theory of relativity was treated in undergraduate texts is Harvard Physics Professor F. A. Saunders' *A Survey of Physics for College Students.* The first edition was published in 1930. Through the third (1943) edition, the treatment of the theory of relativity was the same. At the end of the book, after a complex discussion of radioactivity and nuclear physics, Saunders presented a two-page summary of the results of the theory of relativity without analysis. But an earlier section of the book raised the issue of the existence of the

* Richtmyer, Kennard, and Cooper *Introduction to Modern Physics* (New York, sixth edition, 1978) pp. 54–56. A paste-up version of Chapter 2 of this edition with Cooper's modifications is in the Kennard papers, Folder IV.1.

ether, necessary if one were to make the radiant theory of heat emission appear to be reasonable.

> If the radiation of heat is a wave process, we must imagine a medium in which these waves occur. Filmy bodies such as comets appear to pass through this medium without friction; the planets revolve about the sun without being retarded in their regular paths; hence the ether must be a frictionless medium quite unlike any known material. Not only does the radiation of heat occur in it, but electric and magnetic effects also, and light. While it seems at first sight as though the ether must have very complex properties to be able to do so much, in reality only one kind of action may be required of it. All the phenomena just mentioned are closely related. This does not, of course make the action comprehensible, and in the present state of physics we cannot explain it. Because of this or for other reasons, the ether is regarded by some scientists as non-existent. Perhaps they are right but it is so great a convenience in helping us to correlate a variety of phenomena that it will be treated in this book as real, at least whenever it is useful.*

In 1953, a fourth edition of the work was published. Paul Kirkpatrick, Professor of Physics at Stanford University, was a second author. The text had been considerably rewritten and modernized. The discussion of radiant energy noted that it "can travel across empty space without loss." A chapter on the theory of relativity was added in which the second postulate of relativity, the invariance of the speed of light, was shown to have been determined by experiment. The Lorentz transformations were used to derive several kinematical relations. A virtually identical teatment was presented in the fifth, 1960 edition of this book.†

POST WORLD WAR II

The kinds of treatments thus far described are typical of those found before World War II and after the war for those books whose history began before the war. We will now examine some of the texts that first appeared after World War II.

In many ways the watershed for consideration of the theory of relativity in the American textbook literature was the publication of *Classical Electricity and Magnetism* by Wolfgang Panofsky and Melba Phillips. The first edition appeared

* F. A. Saunders, *A Survey of Physics for College Students* (New York, 1st ed., 1930; 2nd ed., 1936; 3rd ed., 1943) 2nd ed., p. 192.

† F. A. Saunders and P. Kirkpatrick, *College Physics* (New York, 4th ed., 1953; 5th ed., 1960). The quotation is from the 4th edition, p. 172.

in 1955.* It was immensely popular and widely adopted. The treatment of special relativity in this book apparently caught the imagination of later writers. Although earlier Bergmann had treated the basis of the theory of relativity in extraordinarily simple terms, it had not caught the American fancy any more than had Einstein's popularization.

The Panofsky and Phillips book is a graduate text and assumes a sophisticated understanding of calculus, linear algebra and differential geometry. Yet when the authors discussed the theory of relativity, the language level was that of high school algebra. More than that, Panofsky and Phillips examined the experimental evidence that pertained to the domain of the special theory of relativity and competing theories, including the Lorentz theory. Their conclusion is that Einstein's theory is the only one to account for all the evidence.

The emphasis that Panofsky and Phillips placed on the experimental basis of relativity was unique. Earlier, we cited works which claimed that experiment dictated the necessity of an ether based theory. Now, graduate students in physics were confronted with a text that said that ether theories did *not* account for the evidence.

For example, in one table (Table 15-1), the authors summarized fourteen trials of the Michelson-Morley experiment done over a fifty year period, showing that they gave null results. In another table (Table 15-2), later widely cited and reproduced, they concluded that only the theory of relativity satisfactorily accounts for the results of all first and second order ether drift experiments and that it also accounts for experiments in high energy physics to which the kinematical relationships of special relativity apply. Emphasizing that no single experiment proves the theory, they wrote that the experiments provide evidence for the claim that the existence of the ether is undemonstrable and that of all the choices only the Einstein theory is "plausible."

But any set of experiments can be explained from any metaphysical position. In the particular case of the special theory of relativity there are two theories which yield identical formal results—any conclusion of the Einstein theory must also be derivable in the Lorentz theory. Nevertheless, Panofsky and Phillips claimed (Table 15-2) that the Lorentz theory could not account for the results of the Kennedy-Thorndyke experiment, a modification of the Michelson-Morley experiment in which the arms of the interferometer are of unequal length.† In fact, the null result of the experiment *is* compatible with the equations of the Lorentz theory.

* Wolfgang K. H. Panofsky and Melba Phillips, *Classical Electricity and Magnetism* (Reading, Mass., 1955).

† The belief that the Lorentz theory cannot account for the results of the Kennedy-Thorndyke experiment is a common one. Cf. R. C. Tolman, *Relativity, Thermodynamics and Cosmology* p. 14.

Having established to their satisfaction that only the theory of relativity accounts for the experimental evidence, Panofsky and Phillips summarized the argument as follows:

> In 1905, compatible with the experimental facts known at that time, Einstein proposed the following postulates as a solution: (1) The laws of electrodynamics (including, of course, the propagation of light with the velocity c in free space), as well as the laws of mechanics, are the same in all inertial frames. (2) It is impossible to devise an experiment defining a state of absolute motion, or to determine for any physical phenomena a preferred inertial frame having special properties.*

This formulation of the postulates of special relativity is different from the usual one. Here, according to the first postulate, the laws of physics are the same in all inertial frames of reference and, according to the second, there is no possibility of devising an experiment to differentiate between inertial frames of reference. At first it might appear that these are simply two sides of the same statement. However, it was intended, in the first instance, to provide a criterion for what may or may not be considered a law of physics, and in the second instance, to exclude the possibility of absolute motion.

It is clear that to Panofsky and Phillips, although the invariance of the speed of light was a special case of the first postulate of relativity, it was established by experiment. If that be the case, and if the propagation of light waves is independent of the motion of the observers' frame of reference, how can observers in two frames of reference moving with respect to each other observe a spherical envelope of light spreading out from one event, say, the explosion of a flashbulb? The only way to resolve the issue, it was argued, is to use the velocity of light to define the criteria for "simultaneous events" and to drop the older custom of beginning the analysis of measurement by assuming the existence of universal time and distance scales. Panofsky and Phillips then carefully derived the transformation equations by route of derivations of length contraction, time dilation and a quantitative analysis of the relativity of simultaneity.†

The outlook and strategy of the Panofsky and Phillips treatment were consistent and clearly laid out. By implication the theory of relativity is an empirically justified theory. The motivations for the development of the theory of relativity were the lack of agreement by any other theory with the range of experimental evidence. The premises of the theory provided the criteria for acceptable laws and evidence and lead directly to the transformation equations. In effect, Panofsky and Phillips had taken over the strategy of most opponents to the theory of relativity.

* Panofsky and Phillips, *Classical Electricity and Magnetism,* p. 283.

† *Ibid.* Chapter 16.

Which theory does experiment suggest is practical? The answer was the theory of relativity.

Several years after the publication of Panofsky and Phillips's book, between 1960 and 1970, a flurry of monographs on the special theory of relativity for undergraduates was published.* It had become fashionable to teach the special theory of relativity to undergraduates. Almost all of these books used little more than algebra for the analysis. While it would be difficult to establish a direct correspondence, there can be little doubt of the importance of the treatment of the theory of relativity by Panofsky and Phillips; some of the authors explicitly cited Panofsky and Phillips. Several of them reproduced the table that Panofsky and Phillips had constructed to show that only the special theory of relativity accounted for all first and second order ether drift experiments. Almost all of these books proclaim the postulates to be generalizations from experience and experimental facts. Almost all of them develop the relativity of simultaneity as a consequence of the invariance of the speed of light.

The treatments in these books differ. Unlike almost any of the others, Taylor and Wheeler clearly state that the results of experiments like the Michelson-Morley interferometer ones show that the *round trip* speed of light in different directions is the same, but, on the following page, write that the Michelson-Morley experiments show that the speed of light is isotropic.†

The exception to these generalizations was the monograph by A. P. French, Professor of Physics at MIT. In that document, intended for undergraduates "with a modest background in physics," there is a detailed discussion of the significance of the assumption of isotropy of space for the speed of light before defining the criteria for distance simultaneity and how, in conjunction with the principle of relativity, the stipulation of isotropy can be satisfied in all frames of reference by

---

* Examples are:

David Bohm, *The Special Theory of Relativity* (New York, 1963).

T. M. Helliwell, *Introduction to Special Relativity* (Boston, 1966).

Claude Kaeser, *Introduction to the Special Theory of Relativity* (Englewood Cliffs, 1967).

R. Katz, *An Introduction to the Special Theory of Relativity* (Princeton, 1964).

N. D. Mermin, *Space and Time in Special Relativity* (New York, 1968).

Robert Resnick, *Introduction to Special Relativity* (New York, 1968).

W. Rindler, *Essential Relativity* (New York, 1969).

W. Rindler, *Special Relativity* (New York, 1960).

H. M. Schwartz, *Introduction to Special Relativity* (New York, 1968).

F. W. Sears and R. W. Brehme *Introduction to Special Relativity* (Reading, 1968).

A. Shadowitz, *Special Relativity* (Philadelphia, 1968).

J. H. Smith, *Introduction to Special Relativity* (New York, 1965).

E. F. Taylor and J. A. Wheeler, *Spacetime Physics* (San Francisco, 1966).

† Taylor and Wheeler, *Spacetime Physics,* pp. 14–15.

postulating the invariance of the speed of light. Rather than asserting that the invariance of the speed of light is a fact of experience, French states that the null result of certain ether drift experiments are compatible with the second postulate.* The book is also exceptional in that the author is British and was trained at Cambridge University.

The reason for the explosion of such monographs is not difficult to understand. It is not only that it had been discovered that it was possible to communicate the ideas in the theory of relativity without the use of sophisticated mathematical analysis. When one looks at the examples and analyses in many of these books, they almost always refer to space travel or the motions of the particles of high energy physics. Given the emerging technology of "atom smashers" and space exploration, it is easy to understand why elementary treatments of special relativity are in demand.

> One good reason is that the theory (particularly the application of $E = mc^2$) does greatly affect the everyday world and forms part of its technology.†

Closely related to understanding this interest in teaching the theory of relativity at an elementary level was a movement for science curriculum reform which began in the early fifties and became a powerful force after the launch of the first Russian satellite in 1957. In one of the curricula an attempt was made to ground the theory on measurements of the variation of mass with velocity of subatomic particles in particle accelerators, or on measurements in the variation of the half-life of cosmic rays as a function of their speed.‡

It is not surprising to find that general elementary textbooks published since World War II reflect these developments and points of view. All the editions of Robert Resnick and David Halliday's *Physics* since it was first published in 1960, use a similar argument to Panofsky and Phillips's: Only Einstein's theory accounts

---

* A. P. French, *Special Relativity* (Boston, 1968), esp. Chaps. 2 & 3. The quotation is from the preface, p. x.

† C. Kaeser, *Introduction to the Special Theory of Relativity*, p. iii.

‡ The variation in mass experiment was performed by W. Bertozzi in a film entitled "The Ultimate Speed" produced for the Physical Science Study Committee's physics course. The film is described in *The American Journal of Physics*, 1964, *32*:551–555. The time dilation experiment was performed by D. H. Frisch and J. H. Smith in a film entitled "Time Dilation—an Experiment with Mu-Mesons" and described in *The American Journal of Physics*, 1963, *31*:342–355.

In this context it is interesting to note that in 1963 the National Science Foundation supported a Cornell University summer institute devoted entirely to teaching relativity to undergraduates that was attended by fifty college and university physics teachers. One of the features of the conference was the extent to which the theory of relativity was looked upon as a practical, empirically based theory.

for the experimental basis. In all editions of Richard Weidner and Robert Sells *Elementary Modern Physics,* the invariance of the speed of light is declared to be "a fact."*

The most famous, and perhaps the most widely used introductory physics texts in the United States were, and perhaps still are, those by Sears and Zemansky. The history of these books, in two versions, regarding the theory of relativity is fascinating and illuminating. The theory of relativity was not included in the first three editions of their *College Physics* which appeared between 1948 and 1960. The first two editions of the more sophisticated *University Physics* (1949 and 1955) contained no treatment of special relativity. In 1960 however, Sears and Zemansky combined their textbook with an elementary text in modern physics written by Weir and Richards. In the combined book, *Modern University Physics,* it is stated that:

> . . . [the Michelson Morley] experiment demonstrated that one velocity is invariant—the velocity of light. Einstein accepted this as a second fundamental assumption of his special theory of relativity.†

Between the publication of the eclectic *Modern University Physics* and the third edition of *University Physics,* Sears became enamored of a new, elegant and elementary technique that had been developed by R. Brehme for graphical depiction of the Lorentz transformation equations. Sears and Zemansky included "Brehme diagrams" in the third (1963) edition of *University Physics* in an effort to raise the level of sophistication of the book. Relativity became an integral part of the treatment, beginning with mechanics. Still, according to them, the invariance of the speed of light was shown to be correct by the null result of the Michelson-Morley experiment. Furthermore, the Lorentz transformation equations

> . . . developed by Lorentz for the special case of electric and magnetic fields as seen by observers in relative motion before they were generalized by Einstein.‡

The fourth edition of *University Physics* recanted this approach on the grounds that the authors had had unrealistic expectations of the sophistication of entering college students. Although Brehme diagrams were no longer used, and relativistic analysis was removed from most of the book, Sears and Zemansky did include a discussion of the basis of the theory of relativity. According to this version,

---

* Robert Resnick and David Halliday, *Physics for Students of Science and Engineering* (New York, 1st ed., 1960; 2nd ed., 1966; 3rd ed.; 1978).

Richard Weidner and Robert Sells, *Elementary Modern Physics* (Boston, 1st ed., 1963; 2nd ed., 1968; 3rd ed., 1980).

† Richards, Sears, Weir and Zemansky, *Modern University Physics* (Reading, 1960) p. 767.

‡ Sears and Zemansky, *University Physics* (Reading, 3rd ed., 1963, 3rd printing, 1965) p. 91.

> In 1905 Einstein realized that the classical equation for combining velocities . . . is only the limiting case of a more general equation and that this general equation leads to the result that the velocity of light has the same magnitude relative to all frames of reference.*

Sears and Zemansky wrote that at the time Einstein proposed the special theory of relativity he was not primarily concerned with an explanation of the null result of the Michelson-Morley experiment. The basis of Einstein's theory is that the form of Maxwell's equations is the same in all reference frames in uniform motion.

Subsequent editions of Sears and Zemansky have added H. D. Young as a third author. The book now contains a detailed discussion of the process whereby observers in different frames of reference synchronize their clocks, but it is assumed before performing the thought experiments that the speed of light is invariant. According to Young, the speed of light is known precisely, to six significant figures, and according to Einstein's principle of relativity the laws of physics, including the law stating the isotropy of the speed of light, are the same in all inertial frames of reference.†

This analysis of the American physics text literature reveals that, first, there has been little change in the basic premises that Americans have held about the relationship of evidence to belief about the theory of relativity. To this day, American physicists are largely committed to the idea that a theory like the theory of relativity is viable, not only because the conclusions of the theory conform to experience, but also because the premises of the theory have been tested and demonstrated.

More than that, the text literature reveals that rather than reifying a revolution that has already occurred, the American textbooks have insured that, despite the revolutionary views of physicists like Einstein, the new formalism can be adapted within an epistemological framework that is familiar and comfortable. As has been the case with so many transitions and swings of the pendulum in the history of ideas, the more things change, the more they remain the same.

BRIDGMAN'S OPERATIONISM AND EINSTEIN'S RELATIVITY

Earlier, we discussed a number of monographs on the theory of relativity published after 1960 and intended for undergraduates in physics. One mono-

* Sears and Zemansky, *University Physics* (Reading, 4th ed., 1970) p. 608.

† Sears, Zemansky and Young, *University Physics* (Reading, 5th ed., 1976; 6th. ed., 1981) Chapter 43.

Cf. Sears, Zemansky and Young, *College Physics* (Reading, 4th ed. 1974) Chapter 14.

graph we did not discuss was the posthumously published *A Sophisticate's Primer of Relativity* by P. W. Bridgman, which appeared in 1962.*

The character of Bridgman's book is different enough to warrant special consideration. It was the culmination of Bridgman's half century struggle to understand Einstein's theory of relativity. But more than that, Bridgman's life-long quest to make clear and unambiguous the underpinnings of physical thought was intimately connected to his struggle to understand the structure of the theory of relativity.† Detailed study of Bridgman's successive analyses of the foundations of special relativity provides insight into the process of assimilation of ideas like those in that theory by a person committed to a well articulated point of view.

Perhaps no other figure in the history of American physics is so identified with empiricism as Bridgman. He specialized in the physics of materials under high pressure, for which he received the Nobel Prize in 1946, but he is even better known as the father of *operationism,* a philosophy which holds that the meaning of physical terms is conferred on them by specifying how they are measured. According to Bridgman his goal was to develop

> . . . something approaching more to a systematic philosophy of all physics which shall cover the experimental domains already consolidated as well as those which are now making us so much trouble.

Bridgman's public expression of this was first given wide hearing in his book *The Logic of Modern Physics,* published in 1927. It is difficult to overemphasize the influence of this book on American physics and philosophy.‡

---

* P. W. Bridgman, *A Sophisticate's Primer of Relativity* (Middletown, 1st ed., 1963; 2nd ed., 1983). The first edition of the book contained a forword and afterword by Adolph Gruenbaum. The second edition of the book contains a new introduction by A. I. Miller, hereafter cited as "Miller, 'Introduction . . . '" The second edition of *Sophisticate's Primer of Relativity* was published after the text of this chapter was completed. This section of the chapter has been revised to take note of Miller's contributions and of differences in our analyses.

† Much of the information about the evolution of Bridgman's thought about the theory of relativity is in unpublished material in his papers at the Harvard University Archives. I take this opportunity to acknowledge the help of Maila Walter, Harvard University, who was generous in sharing her knowledge of the organizational structure and contents of the Bridgman Papers. I am also indebted to Ms. Walter for sharing with me her draft manuscript, "Laboratory Practice and the Realities of Physics: The Operational Interpretation of P. W. Bridgman," and for taking the time to discuss many of the issues in this section with me.

‡ The quotation is from P. W. Bridgman, *The Logic of Modern Physics* (New York, 1927), pp. ix–x. For brief accounts of Bridgman's influence in physics and philosophy see Albert E. Moyer, "Percy Williams Bridgman," *Dictionary of American Biography,* supplement Seven, 1961–1965 (New York, 1981) 74–76 and E. C. Kemble, et al, "Bridgman, Percy Williams," *Dictionary of*

The first paragraph of *The Logic of Modern Physics* indicated how influential Einstein's theory of relativity had been to Bridgman:

> Whatever may be one's opinion as to our permanent acceptance of the analytical details of Einstein's restricted and general theories of relativity, there can be no doubt that through these theories physics is permanently changed. . . . Reflection on the situation after the event shows that it should not have needed the new experimental facts which led to relativity to convince us of the inadequacy of our previous concepts but that a sufficiently shrewd analysis should have prepared us for at least the possibilities of what Einstein did.*

The passage indicates that Bridgman may have had some misgivings about the theory (an issue we will return to below) but it also indicates the central role of the theory of relativity in helping Bridgman focus on what he perceived to be the problem. According to Bridgman,

> The attitude of the physicist must . . . be one of pure empiricism. He recognizes no *a priori* principles which determine or limit the possibilities of new experience. Experience is determined only by experience.†

Bridgman then developed the notion that the meaning of terms in physics is defined solely by the operations used to measure them. This is what he meant by operationism. For example:

> What do we mean by the length of an object? We evidently know what we mean by length if we can tell what the length of any and every object is and for the physicist nothing more is required. To find the length of an object, we have to perform certain physical operations. The concept of length is therefore fixed when the operations by which length is measured are fixed: That is the concept of length involves as much as and nothing more than a set of operations by which length is determined.‡

From the problem of defining the concept of length, Bridgman moved to the problem of the concept of time (including simultaneity), mass, and eventually he raised the issue of "spreading time over space" the problem of distant simultane-

Scientific Biography (New York, XVI Vols, 1970–1980) II, pp. 457–461. I am indebted to Albert E. Moyer for sharing with me a draft of his article "P. W. Bridgman's Operational Perspective on Physics: Origins, Development and Reception."

* Bridgman, *The Logic of Modern Physics*, p. 1.

† *Ibid*, p. 3.

‡ *Ibid*, p. 5.

ity. The need, Bridgman said, was to make the second postulate of relativity operational; and without recognizing that he had done so, he implicitly assumed the possibility of making a one-way measurement of the speed of light.*

BRIDGMAN'S EARLY STUDIES OF RELATIVITY THEORY

Bridgman's study of the theory of relativity had begun in earnest not long after the death of his colleague, B. O. Pierce, in 1914. Bridgman was asked to teach Pierce's courses in electromagnetism in the Harvard Physics Department, courses which included the special theory of relativity. He has said that he found the foundations of the theory "obscure" and cause for "intellectual distress."† It was in order to relieve that distress that Bridgman undertook what was to become a lifelong struggle to come to terms with the theory of relativity.

As we quoted him earlier, in 1927 Bridgman expressed some doubts about long term viability of the special and general theories. The evidence suggests that by 1927 he had satisfied himself about the special theory of relativity. In fact, it was a key model for his early ruminations about operationism. The general theory, however, was troublesome to him. On September 1, 1921, in some notes to himself, Bridgman had described Einstein's concept of generalized coordinates as "bunk." "Einstein has been deluding himself by a metaphysical preoccupation." In order to rectify this situation, the general theory should be made over in the image of the special theory so that its bases become "a perfectly definite set of physical operations."‡ And three weeks later in a letter to his old high school friend, Richard C. Tolman, he wrote:

> I have been trying to read Einstein this summer, and am in a funk about the whole thing. The part that bothers me is that I have absolutely no grasp of the part that the generalized principle of relativity plays in yielding specific results. . . . I am still more troubled by this question after reading Einstein's mathematical paper. . . . Whenever Einstein wants a particular result he has to say either that we must take the simplest solution or else that this or that assumption seems probable. . . . I am also troubled by my failure, and as far as I can see the failure of everybody else to put the fundamental facts of the new theory in a simple form. The old theory [special theory of relativity] could be

* *Ibid.* Esp. pp. 70–74.

† P. W. Bridgman, "Remarks on the Present State of Operationism," in P. W. Bridgman, *Reflections of a Physicist* (New York, 1955) p. 161. For an account of Bridgman's education, and relationship to contemporaries in the philosophy of science, see A. E. Moyer, "Bridgman's Operational Perspectives . . . ," Cf. Miller, "Introduction. . . . " p. ix.

‡ P. W. Bridgman, "Remarks on Generalized Relativity," Sept. 1, 1921, pp. 1–2; Cf. Moyer, "P. W. Bridgman's Operational Perspective in Physics. . . . ", p. 23.

> expressed in a great many ways, and nearly everyone who had anything to do with it got up different ways of showing that simultaneity must be relative and that a beam of light has mass. But there is nothing of the sort here, and it seems to me significant.*

In his reply† Tolman agreed that the generalized principle of relativity could not lead to specific results without auxiliary hypotheses. Two were required: The principle of equivalence and the notion that the distribution of matter in the region will determine the nature of the gravitational field. This was reassuring to Bridgman.‡ As far as he was concerned,

> It merely is now a matter of experiment to find whether the hypothesis that there is no natural preferred system of coordinates is true or not.§

BEING OPERATIONAL VERSUS OPERATIONISM

Bridgman's satisfactions with special relativity were based on his perception that the fundamental entities of the theory were operationally determined and that the basic postulates had been verified by experiment. His unpublished papers show, however, that beginning in 1959, his "intellectual distress" about the basic ideas of the special theory of relativity returned. That distress focused on the problem of making operational the process of spreading time over space.

On the one hand, Bridgman insisted that the meanings of terms in physics were to be conferred on them by the operations employed to measure them. On the other hand, he recognized that there were important abstract concepts in physics which could never be measured. In order that his operationism not be so restrictive as to make it impossible to introduce these concepts, Bridgman had first suggested that some operations were mental—for example the introduction of the concept of continuity when determining "whether a given aggregate of magnitudes is continuous."‖ Later he had felt the need to expand the concept of *mental operations* into something he called *paper and pencil* operations. These pencil and paper operations would allow for the calculation of entities which, in principle, were beyond the possibility of measurement, for example the potential difference between two points inside a solid.#

Bridgman's reasons for the introduction of operationism had been to provide

* Bridgman to Tolman, September 21, 1921, Bridgman Papers HUG 4234.8.

† Tolman to Bridgman, September 27, 1921, Bridgman Papers, HUG 4234.8.

‡ Bridgman never accepted the general theory of relativity. On this point see Miller, "Introduction . . . ," pp. xl ff.

§ Bridgman to Tolman, October 8, 1921, Bridgman Papers, HUG 4234.8.

‖ Bridgman, *The Logic of Modern Physics* p. 5.

P. W. Bridgman, *The Nature of Thermodynamics* (Cambridge, 1941) pp. vii–ix.

criteria for excluding that which was meaningless from the discourse of physics and to aid the physicist in being clear about the meaning of the terms within his or her discourse. For example, as there is no way to perceive it, the question of whether or not every dimension in the universe might have suddenly doubled or halved has no meaning.* Bridgman never recognized that the moment he admitted the concept of mental or pencil and paper operations, he opened the door to admitting any concept whatsoever. It is an age old problem for all empiricist based philosophies.

From the point of view of a working physicist Bridgman's approach is quite reasonable. Among physicists, operationism was, and still is, a widely supported viewpoint. There *were* critiques and very often the criticism was directed at the narrowness of admissible discourse allowed by strict adherence to operationist tenets. Bridgman's response to such attacks was always to deny that he was a philosopher or that his was a philosophical system. Rather, he could claim, operationism was a technique for doing physics, a guide to follow in the laboratory in order to be clear about what is being measured and how those measured quantities can be used without conceptual error in deriving other, non-measured quantities. Taking Bridgman at his word here gives us an insight into the relationship between his philosophical writings and his work in physics.† However, Bridgman did not heed his own dictum.

Bridgman was cautious about the role of logic in science. In 1934, Rudolph Carnap, a leader of the logical positivists, invited Bridgman to join the Vienna Circle.‡ Bridgman's answer is worth quoting in detail:

> [There is a difference] between my general point of view and that of the Viennese Circle. I have the feeling of this difference particularly with regard to logic. Logic is a tool of human thought weileded [sic] by human beings. I do not think it is a perfect tool and I believe that there will always be conceptual situations which can never be contemplated with complete logical certainty if one pushes his analysis far enough. If the Viennese Circle said things like this out loud a little more emphatically, I think we would be in almost complete accord.§

And earlier, in 1928, Bridgman had written to Professor E. B. McGilvray at the University of Wisconsin that

---

* P. W. Bridgman, *The Nature of Physical Theory* (Princeton, 1936), p. 12.

† This is the point of view that Maila Walter has emphasized in her conversations with me about the import of Bridgman's work. Cf. Walter, *op cit.*

‡ The Vienna Circle is a name given to the twentieth century logical positivist movement because so many of its members lived and worked in Vienna.

§ Bridgman to Carnap, Sept 19, 1934. Bridgman Papers, HUG 4234.10.

> . . . it would be a great mistake to think of Einstein's theory [of relativity] as a logically consistent structure; it is lousy with implicit physical assumptions. It is these unexpressed assumptions that give it its physical content and value.*

Yet Bridgman's lifelong struggle to understand the foundations of the special and the general theory of relativity hinged on his trying to make the theory a completely logical, completely consistent model of operationism.

SPREADING TIME THROUGH SPACE 1959–1961

During the last three years of Bridgman's life, he became transfixed with the problem of spreading time through space. In an unpublished working paper entitled "Frames of Reference," dated March 10, 1959,† Bridgman proposed adjusting clocks in each frame of reference so that the speed of light was the same in all directions. He recognized that such a process would require an independent procedure for synchronizing the clocks. He returned to the problem of providing that independent synchronization six more times in the course of this forty-two page draft. At one point he proposed the following procedure:

> The sweeping search light may be used to set clocks in the two systems. The arrival of the search light beam at any locality is an event recognizable equally in both systems, as by illumination of a screen. Illumination of a stationary or moving screen is equally observable in both systems. Now let the sweeping beam move along the *X* axis with infinite velocity. We set all clocks in the stationary system to read zero when the beam arrives. In the moving system the *velocity of the sweep is also infinite,* and we set clocks in the same way.‡

A year and a half later, in the margin next to the emphasized portion of the above quoted material, Bridgman wrote,"This is not true 27/10/60." From other things he wrote at this same time, he had realized that the procedure quoted above confuses group and wave velocities. By that time he was working on the final manuscript of *Sophisticate's Primer.* In the interval he had produced a

---

* Bridgman to McGilvray, April 8, 1928, Bridgman Papers, HUG 4234.12.

† "Frames of Reference, 10/3/59," 42 pp. Bridgman Papers, HUG 4234.50. Note that Bridgman consistently indicated the date as Day/Month/Year. Dates placed in the margins at various points suggest that the document was written over a period of about two months.

‡ *Ibid* p. 8. Emphasis in original. See also, pp. 11, 17–23, 29, 32, 38.

number of other unpublished draft manuscripts,* all of which returned again and again to the problem of operationalizing the one-way velocity of light and thereby operationalizing the concept of simultaneity.

In "Reformulation of Special Relativity," Bridgman proposed two different techniques (pp. 2–4 and p. 10) for making direct measurements of the one-way speed of light using a single clock. In fact, since the first technique relies on light being reflected from semi-transparent screens to a central observer, it is a two-way determination. The second technique confuses measurements of the group velocity and the wave velocity and is not a measurement of the speed of light at all.

In "Inertial Systems" Bridgman again attempted to make a one-way measurement of the speed of light by making observations on the rate at which light moves in a direction, A to B, perpendicular to the line between the observer, O, and the path of the light beam (pp. 6–9). His method for attempting to determine if the speed of light traveling from A to B were different from that traveling from B to A implicitly assumed that clocks located at A, B and O had already been synchronized. Bridgman wrote,

> I do not see what is the matter with my specific suggestion for distinguishing between the right-left and left-right velocities.†

In "Comments on Ives' Papers on Relativity" Bridgman again took up the question of one-way velocities and the problem of the synchronization of clocks. Herbert E. Ives had been a highly respected physicist at the Bell Telephone Laboratories. Bridgman began his analysis of some of Ives' ideas by noting that with his colleague G. R. Stillwell, Ives had undertaken an experiment to measure the alterations in the spectra of gasses as a consequence of their motion. The experiment was viewed by most physicists as confirming the predictions of relativity theory. Bridgman held that the experiment was the first to give direct evidence of the change in the rate of a moving clock. Bridgman then discussed the fact that Ives never accepted the Einstein theory. Ives never relinquished his commitment to the luminiferous ether, and like Lorentz and Poincaré, cham-

---

* The manuscripts include the following:
  "Reformulation of Special Relativity 11/7/59."
  "Inertial Systems 24/7/59."
  "Causality and Relativity Effects 10/8/59."
  "Comments on Ives' Papers on Relativity 18/8/59."
  "On Velocity 15/2/60."
  "The Incredible First Law of Motion 17/8/60."
  "Two Footnotes to Special Relativity Theory 4/9/60."

† Bridgman, "Inertial Systems," p. 10. Cf., Miller, "Introduction. . . . ", p. xxxix, for a reproduction of the manuscript page containing Bridgman's sketch.

pioned an alternative to Einstein's theory of relativity which began with the premise that nature conspired to prevent the detection of absolute motion through the mechanism of the Lorentz contraction.

Ives was not alone in this. In the spring of 1954, Ives' colleague at Bell Labs, K. K. Darrow, wrote to H. P. Robertson for information which might help Darrow in writing a remembrance of the late Ives. Darrow was extremely well known and a respected physicist who wrote what are considered classic popularizations of modern physical theories. Robertson was an applied mathematician at the California Institute of Technology who had made his reputation on the application of general relativity to practical situations. In his reply, Robertson pointed out the irony of the fact that Ives' excellent experimental work had been used to support Einstein's theory, when in fact Ives remained committed to the ether as a fundamental assumption. Darrow's response is revealing:

> "How very amusing that Ives should have proved what he thought he was disproving, or fortified what he thought he was weakening! Of course, a hasty reading of your letter suggests that maybe he showed that the same phenomena are explicable in two alternative ways, one being the special theory of relativity and the other a theory involving a "preferred frame tied to the aether."*

Darrow would have been even more surprised to learn that Robertson himself did not accept Einstein's formulation of the theory of relativity. Convinced that the theory must be practical and experimentally based, Robertson reconstructed the special theory of relativity so that it began with two postulates, the principle of relativity and the Lorentz contraction, both of which he held to be the direct result of several key experiments.†

Bridgman's analysis was critical of Ives' treatment of one-way velocity. Ives had criticized Einstein for "using the convention" that the space was isotropic with respect to the speed of light, on the grounds that such a definition was not operational.‡ One only measured the "to and fro" speed of light. An operational technique for synchronizing clocks, Ives proposed, would be to move identical

---

* See Robertson to Darrow, June 1, 1954 and Darrow to Robertson, June 7, 1954, Robertson Papers, Folder 1.40, California Institute of Technology.

Miller states that Ives' critiques of relativity were constructive and scholarly. ("Introduction. . . . " xxiii). Privately, Ives held Einstein in contempt. After World War II Ives was an active participant with an industrial chemist in a scheme to undermine the reputation of "The Jew Einstein." See the Ives Papers in the Manuscript Division of the Library of Congress.

† See H. P. Robertson and T. N. Noonan, *Relativity and Cosmology* (Philadelphia, 1960). Cf. H. P. Robertson, Lecture Notes, Robertson Papers, Folder 19.12.

‡ H. E. Ives, "Extrapolation for the Michelson-Morley Experiment," *Journal of the Optical Society of America,* 1950, *40*:185–191.

clocks from one point to another at a speed approaching zero. The speed could not actually reach zero because operationally this would mean that the clock would never reach its goal. In his own analysis of the problem, Bridgman pointed out that the only way the clocks at the two points would be properly synchronized would be if the transported clock had moved at zero velocity, which would be equivalent to admitting Einstein's stipulation of the isotropy of space with respect to the speed of light. Bridgman was critical of Ives' interpretation of the "operational methodology":

> Operational methodology does not demand that the limiting process be carried out *physically*—there is no reason why the physical operation should not be carried out for several values of q [the speed of transport] different from zero and then mathematically extrapolation be made to zero.*

Whereas Miller considers such a statement to show how Bridgman had "further developed the operational point of view since 1927,"† for me it is an example of how Bridgman had confused two different points of view: the physicists' notion of being operational as a way of keeping straight the meaning of measurements and calculations, and the philosophers' notion of operationism in which all meaning is conferred by how measurements are made. The confusion was convenient for it allowed Bridgman to go around those hurdles which a strict adherence to operationism would not allow him to vault. Such philosophical opportunism is to be expected from Bridgman the physicist. It is discordant because it occurs in the context of Bridgman's philosophical search for an operational determination of the one-way velocity of light. That search continued on the very next page of his Ives paper.

Bridgman's struggles with attempts to make one-way measurements of the velocity of light continued through 1960.‡ In January of 1961, Bridgman began drafting *Sophisticate's Primer* under extraordinarily difficult conditions. He suffered from Paget's disease and, with only a few months to live, was in great pain.§ Still, at each step of the way, Bridgman took care to date not only parts of the manuscript, but all marginal notes. *Sophisticate's Primer* is a book dominated by the problem of spreading time over space.

Bridgman first asked rhetorically if it were not now possible, given improve-

* Bridgman, "Comments on Ives' Papers on Relativity," p. 3. Cf. *Sophisticate's Primer* pp. 64–66; Miller, "Introduction. . . . " pp. xxiv–xxv.

† Miller, "Introduction. . . . " p. xxiv.

‡ See Bridgman, "On Velocity," "The Incredible First Law of Motion," and "Two Footnotes to Special Relativity Theory."

§ Kemble et al, *op. cit.*

ments in measuring techniques, to make a one-way measurement of the speed of light. He answered the question by saying that the result would be irrelevant.

> It must be admitted at once that it is perfectly possible "in principle" to measure such a one-way velocity and to obtain a unique result. In this sense "one-way" velocity has physical "reality." But the reason it is irrelevant for our present purposes is that what we get by such a measurement is information about the combination "properties-of-light-plus-the-method-by-which-time-has-been-spread-through-space." [sic].

Bridgman said further that there was no way to isolate the one-way speed of light out of the combination. He insisted that the concept of "isotropic light propagation" had real physical content, that is, it is operational. But in fact, Bridgman was not then referring to the one-way velocity of light but to the round trip velocity.*

Bridgman immediately plunged back into a discussion of the possibility of determining the one-way speed of light:

> Is there not something "objective" connected with the propagation of light independent of how we spread time through space? The answer is that there is indeed something here with real physical content and that our physical intuition was correct in thinking that there must be such a thing.

Bridgman then argued that the question we should be addressing is not the isotropy of space with regard to one-way, but with regard to round trip, values of the speed of light. Again he concluded that the question of one-way isotropy is "irrelevant" because it is impossible to separate the problem of synchronizing clocks from the problem of making the one-way measurement of the speed of light.†

Still not satisfied, Bridgman once more raised the issue, except he had switched his focus from one-way measurements of the speed of light, (nonproper measurements requiring two clocks) to round trip measurements (proper measurements requiring one clock).

> As long as it is assumed that the propagation of light is characterized *only* by its two-way velocity in any direction, it would appear that an experimental check of isotropy is necessary for *every* direction. For one can imagine a medium constituted of infinitely fine threads radiating in every direction, each thread having its own characteristic velocity of propagation. . . . Hence it would logically appear that in order to verify the physical correctness of relativity theory, the velocity of light should be checked for every direction.

* Bridgman, *A Sophisticate's Primer of Relativity,* pp. 42–45. The quotation is on pp. 43–44.
† *Ibid.,* pp. 44–48.

For practical purposes, Bridgman continued, it is only necessary to check the round trip speed of light in a few directions.*

Bridgman would not let the matter rest. He pressed on and on in his attempt to discover a technique for operationalizing, either a one-way measurement of the speed of light or the concept of simultaneity independent of each other. Again he concluded that it would be impossible:

> Any apparent one-way velocity . . . must conceal somewhere an assumption about the way time is spread over space.†

The question of one-way measurement of the speed of light recurred in the text until the very end of the book. Bridgman summarized all of these arguments as follows:

> Of the various sorts of velocity, it is only the one-clock velocities, including in particular round-trip velocity and self-measured velocity, which are physically significant in the sense that they do not depend on the arbitrary way in which time has been spread through space. This means that one-way, two-clock velocities correspond to no physical "reality" and that a determination of them is irrelevant in describing the physical nature of a system. Questions with real physical content, such as the isotropy of the propagation of light, have to be formulated in terms of one-clock velocities, not one-way two clock velocities as usually implied.‡

The care and detail that Bridgman put into the question of the possibility of determining the one-way value of the speed of light is not strange. It was *the* over-riding concern of almost all his earlier studies of the theory of relativity. What was different in *Sophisticate's Primer* was his recognition that in a single frame of reference it was impossible to make a one-way measurement of the speed of light. He dismissed such a concept on the grounds that it was "irrelevant."

The casual reader would have no way of knowing that Bridgman included the following note, obviously not intended for publication, with the manuscript:

> [A] major change [in my thinking] was my realization, early in 1961, that "one-way" velocity can have no physical significance by itself but is essentially a two-clock concept and has meaning only when the method has been specified by which time is spread over space. This realization negatived my efforst [sic] to find some physical method of measuring one-way velocity, and completely stultified the point of view contained in my MS "Two Footnotes to Relativity Theory" which I was on the point of

---

* *Ibid,* p. 49.

† *Ibid,* pp. 49–69. The quoted material is on p. 68.

‡ Bridgman, *Sophisticate's Primer,* p. 147.

> offering to the American Journal of Physics. The second revision of the primer . . . begun in March 1961 is written in accordance with this new insight.*

The note indicates that after forty years of trying to operationalize one-way measurements of the speed of light, Bridgman had realized that it was an impossible task. The realization must have been a shock. But the note also indicates that Bridgman believed that he had taken that impossibility into account.

And yet in the midst of his ruminations about one-way measurements and spreading time over space, Bridgman produced a proof that the one-way speed of light is the same in all directions in systems containing more than one inertial frame of reference. The proof is flawed because he imbeds the conclusion in the premises, but it gave Bridgman confidence that the isotropy of space could be operationalized when considering more than one frame of reference. The appearance of this "proof" is strange in the context because Bridgman had made one-way measurements of the speed of light "irrelevant." Its presence gives insight into the depth of Bridgman's commitment to philosophical operationism and his continued confusion between the concept of "being operational" as a technique for doing physics and "operationism" as a guiding metaphysic for understanding the world of experience.†

PHILOSOPHY AND PHYSICS

In discussing the work of Henri Poincaré, I pointed out that when he talked *about* physics Poincaré took a philosophical position known as conventionalism. But when Poincaré worked as a physicist, his position was more akin to that of a realist. Bridgman's confusion between "being operational" and "operationism" is related to Poincaré's. It is the confusion between being logical, that is, thinking well as a physicist, and treating physics as if it were a branch of formal logic.

That confusion is not restricted to physicists who concern themselves with the

* This note was dated 7/4/61 (April 7, 1961). The draft to which Bridgman refers has the date 4/4/61 (April 4, 1961) on the first page, but contains pages with earlier dates going back to August, 1960. Cf. Miller, "Introduction. . . . ", p. xxxiv.

† *Ibid*, pp. 88–90. Cf. Miller, "Introduction. . . . " pp. xxxii–xxxv. Miller states that he is sure Bridgman would have revised the proof in a later edition of *Sophisticate's Primer*. Whether Bridgman would have revised the proof or not, is not the issue. The issue is Bridgman's unyielding commitment to operationalizing the fundamental premises of a theory like the theory of relativity. While noting that Bridgman's derivation is an error, Miller says that it is not "merely an error." He describes it as "an act of desperation." Because Bridgman said that he had written this draft of *Sophisticate's Primer* in the light of understanding that it was impossible to operationalize one-way measurements of the speed of light, I cannot agree with Miller's interpretation.

philosophical status of the field. Philosophers of science often fall prey to the same confusion. In *Sophisticate's Primer* Bridgman's discussion of the issue of one-way measurements of the speed of light is defined, in part, by his concern over the treatment of the problem by two philosophers of science, Hans Reichenbach and Adolph Gruenbaum. Both Reichenbach and Gruenbaum were interested, not only in the logical structure of the theory of relativity, but in logical reconstructions of the theory.

There is ample justification for such examinations in their own right. As we have seen, there are an infinite number of possible logical structures which might account for a body of empirical evidence. Alternatives to Einstein's formulation should not be considered better or worse. They are different. But all too often, the possibility of logical reconstructions have led philosophers to use them as a substitute for evidence in historical arguments.* The possibility of such reconstructions should not blind us to the fact that science is not logic.

Reichenbach was well aware of the dangers. In 1949 he wrote:

> The philosopher of science is not much interested in the thought processes which lead to scientific discoveries; he looks for a logical analysis of the completed theory, including the relationships establishing its validity. That is, he is not interested in the context of discovery, but in the context of justification. . . . The philosophy of physics . . . is not a product of creed but of analysis. . . . it endeavors to clarify the meaning of physical theories, independently of the interpretation by their authors, and is concerned with logical relations alone.†

It was in this spirit that Reichenbach undertook an analysis showing that it was not necessary to assume isotropy of space for the speed of light. Space could be

* Some examples are:

I. Lakatos, "Falsification and the Methodology of Scientific Research Programme," in I. Lakatos and A. Musgrave (eds), *Criticism and the Growth of Knowledge* (Cambridge, England, 1970); I. Lakatos, *The Methodology of Scientific Research Programmes* (Cambridge, England, 1978); E. Zahar, "Why did Einstein's Programme Supersede Lorentz'" *British Journal for the Philosophy of Science,* 1973, *24*:95–123, 223–262.; A. Gruenbaum, "The Genesis of the Special Theory of Relativity," in H. Feigl and G. Maxwell (eds), *Current Issues in the Philosophy of Science* (New York, 1961); A. Gruenbaum, "The Special Theory of Relativity as a Case Study in the Importance of the Philosophy of Science for the History of Science," in B. Baumrin (ed) *Philosophy of Science* (New York, 1963); A. Gruenbaum, "The Bearing of Philosophy on the History of Science," *Science,* 1964, *143*:1406–1412.

† H. Reichenbach, "The Philosophical Significance of the Theory of Relativity," in P. A. Schillp (ed), *Albert Einstein: Philosopher-Scientist* (New York, 1949) pp. 287–312, pp. 292–293. Cf. G. Holton, "Einstein, Michelson, and the Crucial Experiment," in G. Holton, *Thematic Origins of Scientific Thought* (Cambridge, 1973) esp. fn 152. Holton's discussion has important bearing on the questions raised here.

anisotropic.* In that case, decisions about distant simultaneity would require calculation. From the point of view of logic it makes no difference. However, from the point of view of our intuitions about simultaneity, stipulating anisotropy makes no sense whatsoever.

About the theory of relativity Reichenbach said,

> . . . it appears amazing to what extent the logical analysis of relativity coincides with the original interpretation by its author.†

Note that Reichenbach has already limited the possibilities to *the* logical analysis. It is a small step to argue that logical reconstruction serve as a heuristic for historical investigation.‡ Not only does such a step ignore the multiplicity of logical reconstructions, it biases historical accounts to those that follow the rules of formal logic. Just as Bridgman confused logic with physics, philosophers of science such as Gruenbaum, Reichenbach and those committed to rational reconstruction, have confused logic with history.

THE POPULAR RESPONSE TO THE THEORY OF RELATIVITY

Except for noting in Chapter 9 that between 1905 and 1911 virtually no attention was paid to Einstein or to his scientific contributions in the American popular press, we have not alluded to popularizations of the theory of relativity or to the nature of the response that Einstein's innovations received in the popular press. In this section, we summarize the kinds of responses that the theory of relativity received in the American press. Understandably, the response to the theory of relativity in the popular press immediately became entangled with impressions of Einstein, the kind of person he was thought to be, and assessments of what he thought of others.

The lack of attention paid to Einstein in the popular press continued until 1919. There is no mention of him, for example, in the *New York Times Index* before 1919.§ In early November of that year, at a joint meeting of the Royal Society and the Royal Astronomical Society of London, it was announced that the results of the analysis of photographs taken of stars in the vicinity of the sun during

* Reichenbach's treatment has been explicated by Gruenbaum in Bridgman, *Sophisticate's Primer of Relativity* (1st edition) "Afterword."

† Reichenbach, *The Philosophical Significance . . .*, p. 293.

‡ Lakatos,"Falsification . . ." and Zahar, "Why . . ." See especially Gruenbaum's "The Genesis of the Special Relativity," and the elegant response to that paper by Michael Polanyi, in Feigl and Maxwell (eds), *Current Issues . . .*, pp. 53–55.

§ See A. Pais: *Subtle is the Lord: The Science and the Life of Albert Einstein* (New York, 1982), p. 309.

Einstein in his study in his Berlin home, 1919. World Wide Photos.

a solar eclipse earlier that year had been completed and it strongly supported predictions of Albert Einstein's general theory of relativity. The world response to this announcement was a dramatic explosion of interest, both in the theory of relativity and in Einstein himself.

That the world was caught by surprise, while not paradoxical, had nothing to do with there having been no attempt to communicate the ideas contained within the theory. Einstein's papers were a matter of "public" record. And shortly after the theory reached its final form, in 1917, Einstein published his own non-mathematical account of both the special and general theories of relativity. That popularization was widely read in Europe. Some measure of the degree of insulation that existed between cultures can be gleaned from the fact that, by the time the fifth edition of Einstein's popular account was translated into English and the book made its first appearance in the English and American markets in 1920, the German-language version had gone through some ten editions.

An important factor in the lack of interest in Einstein's work in the American popular press was the First World War. What is somewhat surprising is the public explosion in 1919. This is especially true in light of the general view that the theory was not one that the public could understand.

The sources of the popular explosion of interest by Americans and the American press in Einstein and relativity are mysterious. It was like the explosion of interest in hoola hoops, or bobby socks, or deely boppers. In the case of the Einstein madness, no doubt it represented for some a relief from the dreariness and horror of World War I. But whatever the initial emergence of the fascination with Einstein and the theory of relativity represented, it was in character with traits that we have already identified as part of the American concept of ideas and practices in science and how those ideas relate to ideas and practices in other segments of the culture.

Before 1919 almost no one outside of physics knew of Albert Einstein. After 1919 almost no one did not know of him. He became the model, not only for the scientists, but for the university academic in the United States. That model included the following: a genius whose ideas are abstract and totally incomprehensible and impractical. And personally, someone who is absent-minded, slovenly in appearance and a child in all things except his or her expertise. There is some evidence that Einstein willfully contributed to that myth,* but it is a myth that has more to do with American concepts of science as a social institution than with the character of Albert Einstein.

Once Einstein came to public attention he never left it. But as with all celebrities, the attention paid to him was cyclical. After the initial interest in late 1919 and early 1920, a brief period of calm ensued which lasted until just before his first visit to the United States in 1921. If the proportion of space in the public press allotted to Einstein's comings and goings can be used as a measure of American fascination with him, then Einstein's visit here (on a fund raising tour on

* Lewis Pyenson, personal communication.

SECOND NEWS S

For News, Subscriptions
or Advertising
Call ATlantic 6100.

SATURDAY
MORNING,

Pittsburgh Po

Constantly Expressive—Einstein W.

This **WORLD** of **OURS**
-by-
Clare Boothe Brokaw

FOR MEN ONLY

THIS column is addressed exclusively to men. It is about clothes. Lady readers will gather nothing from it except perhaps the hint that even in the field of fashion, the male is still the conqueror.

* * *

FASHION is a domain of human activity which, during the past 50 years, seems to have been exclusively appropriated and dominated by the female sex. Nevertheless, I am about to prove (I think I am) that in spite of the almost universal standardization of men's dress, and the strict, unimaginative sartorial rules laid down for their observance, clothes still do make the man. More important than that, only men have been able to make their clothes fully express their personalities. In fact, such is the genius of the male sex in this respect, that there is hardly a famous man living today who can not be instantly identified by a brief description of some particular article of dress he is in the habit of wearing.

* * *

CHARLIE CHAPLIN'S shoes, for instance, are symbols of a personality that all the world recognizes as Charlie—even without Charlie in them. From Shanghai to Key West you can festoon a theater front with a pair of sloppy shoes, and, if you will, a flexible bamboo cane, and battered plug hat and you need not mention Charles' name, or show his face, for everyone will know who is play-

Professor Albert Einstein in a group of characteristic expressions, all moving slowly toward his famous smile in which his face actually lights up as he replies to the questions of his interviewers.

These portrait photographs, taken yesterday
newspapermen who crowded around him, sh
eagerly; next, getting a gleam of insight int

Clipping from *Pittsburg Post-Gazette* 29 December, 1934. Story in Column at far right is typical of press treatment of Einstein. The picture caption emphasizes that Einstein's

ION

t-Gazette

DECEMBER 29, 1934.

SPORTS, FINANCIAL, CLASSIFIED SECTION

Up to His Subject

—Post-Gazette Photos.

nswered the 20 first, listening uestion; third, explaining slowly and patiently; fourth, making his point clear; and last—the smile itself. He did not hesitate to say, "I don't know," frequently in reply to queries by his interviewers.

ATOM ENERGY HOPE IS SPIKED BY EINSTEIN

Efforts at Loosing Vast Force Is Called Fruitless.

SAVANT TALKS HERE

Now Indicates Doubt Of Relativity Theory He Made Famous.

Blind chance, or cause and effect—which ever you prefer—may run the universe!

Space may be infinite, or it may be finite, nobody knows!

It may be curved, or not curved, just as you please!

Whatever you decide, no one can contradict you, because no one has so far been able to prove any one of these contentions. Stil, you may be wrong, because some of the contentions may be proved in the future.

That is the contention of Prof. Albert Einstein.

But the "energy of the atom" is something else again. If you believe that man will someday be able to harness this boundless energy—to drive a great steamship across the ocean on a pint of water, for instance—then, according to Einstein, you are wrong now.

Energy of Atom.

The idea that man might some day utilize the atom's energy brought the only emphatic denial from the noted scientist yesterday when he was interviewed by a score of newspapermen at the

frequent reply to queries was, "I don't know." Reproduced with permission of the *Pittsburg Post-Gazette.* Courtesy of the American Institute of Physics Niels Bohr Library.

behalf of Hebrew University) was followed by the public in the manner of the arrival of the Beatles in the mid-nineteen sixties. Another wave of public attention welled up after Einstein published his first paper on unified field theory in February, 1929. He came to public attention again with his visits to the United States in 1930, in 1931, and finally when he accepted a permanent appointment to the then new Institute for Advanced Study in Princeton, New Jersey.

At first, attention was focused on Einstein's scientific work and his personal traits. Einstein, however, was a public figure who believed in using his fame for what he considered good causes. He was quite astute at using the public forum provided to him by the press to further causes in which he believed. His political views, then, were also a matter of public record, public interest and concern.

The meeting of the Royal Astronomical Society in which the eclipse results were announced took place on November 6, 1919. The first announcement in *The New York Times* was on November 9. The following day *The New York Times* published a second article under the headline "Lights All Askew in the Heavens." A facsimile of the *Times* headline is reproduced on page 313.* It is hard to imagine a set of headlines more calculated to mystify readers of that newspaper.

The article contains all those elements which characterized future images of Einstein in the American press: J. J. Thomson informed the reporters that it was not possible to explain the theory in words intelligible to the non-scientist. Other scientists were quoted as reassuring the public since it would not affect anything on this earth. And the article ended with the following paragraph, tacked on almost as an afterthought:

> When he offered his last important work to the publishers he warned them there were not more than twelve persons in the whole world who would understand it, but the publisher took the risk.

It is hard to imagine that Einstein himself was not responsible, tongue in cheek, for that remark. Several weeks later, on December 3, *The New York Times* carried a story from its Berlin correspondent who had interviewed Einstein. Einstein, obviously in his element, told the reporter that the idea for general relativity had come to him in 1915 as a result of observing from his study, a man fall from a roof on a nearby building. Luckily the man fell on a pile of rubbish and was not hurt. Einstein said that he hurried downstairs to interview the man about his experience and when the man reported that he had not experienced the sensations commonly associated with gravity, the key to the general theory of relativity fell into place. Obviously, Einstein was having a good time. (In fact, the principle of equivalence, which this story illustrates, was first suggested by Einstein in 1907.) And when the

* Cf. Pais, *Subtle is the Lord . . .* , p. 309.

LIGHTS ALL ASKEW IN THE HEAVENS

Men of Science More or Less Agog Over Results of Eclipse Observations.

EINSTEIN THEORY TRIUMPHS

Stars Not Where They Seemed or Were Calculated to be, but Nobody Need Worry.

A BOOK FOR 12 WISE MEN

No More in All the World Could Comprehend It, Said Einstein When His Daring Publishers Accepted It.

Special Cable to THE NEW YORK TIMES.

LONDON, Nov. 9.—Efforts made to put in words intelligible to the non-scientific public the Einstein theory of light proved by the eclipse expedition so far have not been very successful. The new theory was discussed at a recent meeting of the Royal Society and Royal Astronomical Society. Sir Joseph Thomson, President of the Royal Society, declares it is not possible to put Einstein's theory into really intelligible words, yet at the same time Thomson adds:

"The results of the eclipse expedition demonstrating that the rays of light from the stars are bent or deflected from their normal course by other aerial bodies acting upon them and consequently the inference that light has weight form a most important contribution to the laws of gravity given us since Newton laid down his principles."

French Government to Open Cheap National Restaurants

PARIS, Nov. 9.—"National restaurants," it is officially announced, will be opened within a month, where meals without wine will be served, at 2 francs, in wooden barracks, which it is purposed to heat. The barracks will be built and operated by the authorities.

A protest has been issued by the head of the restaurant proprietors' organization on the ground of unfair competition, as the State has access to army stocks and is relieved of many taxes.

HOLLAND UNLIKELY TO GIVE UP KAISER

Dutch Reiterate on Anniversary of His Flight Their Views on Right of Asylum.

NO DEMAND FOR HIM YET

Negotiations with Belgium Over Territorial Adjustments Are Said to be Going Well.

THE HAGUE, Nov. 8, (Associated Press.)—Former Emperor William came to Holland a year ago next Monday. During this time there has been no demand, officially or unofficially, for his extradition or delivery up to the Allies, nor has Holland at all changed its viewpoint toward him.

BALFOUR WRITING A BOOK.

ORLANDO POINTS TO ITALY'S GAINS

Treaty Assures Her Frontiers, He Says, Though Fiume Deadlock Continues.

CITES WILSON'S OPPOSITION

Sees It as "Insurmountable Obstacle" to the Acceptance of Italy's Concessions.

PALERMO, Nov. 9, (Associated Press.)—Ex-Premier Orlando, in a speech to the electors last night, reiterated the defense which he made in the Chamber in September of his actions, and declared that the situation on the Adriatic question today was the same as he left it in the beginning of June.

It was impossible to reach a solution, said Signor Orlando, because Italy could not increase her concessions, the acceptance of which found "an insurmountable obstacle in the opposition of President Wilson."

To Vote for Deputies by Groups.

ROME, Nov. 8.—Even those citizens who don't ordinarily take much interest in politics are awaiting the election of Nov. 16 with the greatest interest, tinged with a certain amount of anxiety.

Chivalric Foes in the Campaign.

Republican Won by 1,300 in Oklahoma; Favored Reservations in League Fight

OKLAHOMA CITY, Okla., Nov. 9.—Returns from fifteen scattered precincts reporting today increased to 1,300 the lead by which J. W. Harreld, Republican, was elected to Congress in the Fifth District yesterday over Claude Weaver, Democrat.

ASTOR CAMPAIGN ENTERING LAST LAP

Close of the First Week Sees All Three Candidates Claiming Victory.

LADY ASTOR CITES BIBLE

Tells Hearers That the 'One Genuine Socialist in History' Taught the Golden Rule.

PLYMOUTH, Nov. 9, (Associated Press.)—Andrew Bonar Law, Government leader in the House of Commons, sent the following message to Lady Astor today:

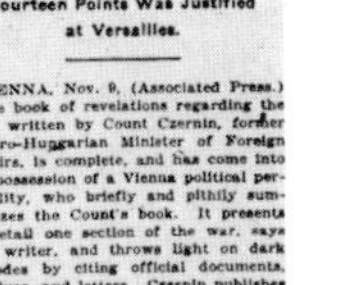

TWO FAMILIES CLAIM RETURNED SOLDIER

War Risk Agent Holds Bromley or Reddy at Norwalk—Marked Like Supposed Dead Man.

DANBURY, Conn., Nov. 9.—A remarkable case of dual identity was revealed today when a man arrested last night in Norwalk was brought here by O. J. Waite, an investigator of the War Risk Insurance Bureau.

CZERNIN'S BOOK PREDICTS MORE WAR

"In This Peace There Lie the Germs of Death," the Fallen Diplomat Says.

HAS A CHAPTER ON WILSON

Says That Austria's Suspicion of the Fourteen Points Was Justified at Versailles.

VIENNA, Nov. 9, (Associated Press.)—The book of revelations regarding the war, written by Count Czernin, former Austro-Hungarian Minister of Foreign Affairs, is complete, and has come into the possession of a Vienna political personality, who briefly and pithily summarizes the Count's book.

ALLIED MILITARY PLAN LACKS OUR SANCTION

American Delegates Fear That it May Call for an International General Staff.

PARIS, Nov. 9, (Associated Press.)—The American delegation to the Peace Conference has not yet approved the plan for a military commission to succeed the Supreme Council in carrying out the German Peace Treaty and executing its military provisions.

New York Times, November 10, 1919, p. 17; reporting the confirmation of Einstein's predictions of the apparent displacement of star positions near the sun during solar eclipse. Reproduced from the Collection of the Library of Congress.

reporter tried to confirm the accuracy of the claim that only twelve men in the world could understand his work, Einstein laughed.

Whether or not Einstein was responsible for the "only twelve men can understand" story or not, it had significant consequences in the popular literature. For example, *The New York Times* often referred to that presumed feature of Einstein's work when he was in the news:

On November 11, 1919, the newspaper editorialized as follows:

> Quite the most disturbing feature of the situation is the assumption that only men of wonderful learning have the ability and therefore the right to see what there is in the fact that light being subject to turn from a straight path by a mass of matter like the sun, must itself have of matter at least the qualities of weight. Not a few mathematicians who are merely amateurs can at least glimpse significance in that, and it is hard to avoid the suspicion that if the masters could know any more they wouldn't recoil from the tasks of putting what it means to them in words for the rest of us to at least mull over. If we give it up, no harm would be done, for we are used to that, but to have the giving up done for us is—well, just a little irritating.

And on November 13, 1919, in the light of several news stories which reported that some scientists like R. A. Millikan were skeptical of Einstein's theory and that the bending of starlight was probably due to local atmospheric diffraction effects, *The New York Times* editorialized as follows:

> People . . . who have felt a bit resentful at being told that they couldn't possibly understand the new theory, even if it were carefully explained to them ever so kindly and carefully will feel a sort of satisfaction on noting that the soundness of the Einstein deduction has been questioned by R. A. Millikan. . . .
>
> [The refraction proposal] is understandable as well as plausible, and it is hard not to hope that it is true.

Over the years such remarks were often found in that newspaper. On April 26, 1920, in response to a report that quoted J. J. Thomson as saying that the new theory of gravitation could not be understood without the theory of invariants and the calculus of variations, the reader of the editorial page found this:

> Now we know! Anybody without the equipment mentioned will waste his time on trying to understand the new relativity.

And on September 6, 1920, the *Times* was of the opinion that, while Einstein cannot be blamed for it, there is something irritating about a theory which only twelve people can understand.

On January 31, 1921, readers of the editorial page were treated to the following that appeared two days after a news story reporting that Einstein had suggested a way of measuring the size of the universe:

> . . . Professor Einstein has provided some more light reading for the few but fit thinkers who jauntily turn over the pages of Laplace's "Celestial Mechanics" at breakfast and look on Sir Isaac Newton with condescension. . . .
>
> To most of us geometry is not practical but impractical and not a tincture of Euclid lingers in our intellectuals. To fictions such as time, space, mathematics, the universe, we should be all kinder. . . .
>
> Yet whatever else be fictional in a fictional Whole, we believe implicitly in Professor Einstein because of the subtlety of his characters and his plot and the limpidity of his dialogue.

Such editorial remarks appeared regularly in *The New York Times* throughout Einstein's life.

A good illustration of the mystification process is afforded by the treatment of the publication in early 1929 of one of Einstein's first papers on unified field theory. The title of the six-page paper was "Unified Field Theory" (*"Einheitliche Feldtheorie"*). The first article on the publication of that paper appeared on page 1 of *The New York Times* on January 12, 1929. The story referred to the article as "a book of five pages" with the title "A New Field of Theory."

A week later, on January 19, 1929, Einstein was reported as being amazed at the publicity over "A New Field of Theory." He had received hundreds of cables from newspapers all over the world with provisions for prepaid answers. More than a hundred journalists camped outside his office. In one article, the *Times* reporter announced that the five-page book contained only equations and required ten years to write. Thus a measure of how difficult the work was could be gleaned from the fact that Einstein had been able to write a half a page a year. Considering that Einstein's original work on relativity had contained only three pages (the origin of that figure is unknown) and had spawned more than three thousand books, not to mention countless articles, the *Times* correspondent found it impossible to guess at the impact of the field theory paper. More incredible than the *The New York Times* stories on this development is that on February 1, 1929, the *New York Herald Tribune* published a verbatim translation of Einstein's paper complete with the tensor equations (See pages 316–317). In a side bar to that article the *Herald Tribune* let its readers in on how amazed printing experts were at the feat of cabling such a complicated story across the Atlantic, complete with mathematics.

This gives only a glimpse of the response to Einstein in the American popular

16 NEW YORK HERALD TRIBUNE, FRIDAY, FEBRUARY 1, 1929

Cabled Text of Prof. Einstein's New Theory and Comment From American Scientists

Einstein's New Theory; Text of His Treatise

Einstein's Work Held Mere Clew To World Riddle

Only Suggests That Electro-Magnetism, Gravitation Are One, Says Sheldon

Great Mathematical Feat Tends to Wave Interpretation of Physics Phenomena

By H. H. Sheldon

Professor of Physics, New York University

How Einstein Text Reached Herald Tribune Readers

Die lokalen n-Beine sind alle "parallele" gestellt. Parallele und gleiche Vektoren sind solche, welche—auf ihr lokales n-Bein bezogen—gleiche Koordinaten haben. Die Parallelverschiebung eines Vektors wird durch die Formel

$$\delta A^{\nu} = -\Delta^{\nu}_{\alpha\beta} A^{\alpha} \delta x^{\beta}$$

gegeben, wobei in

Excerpt from his "Field Theory" pamphlet as printed in German

Technical Feat Of Transmission Amazes Experts

Intricate Einstein Formulas Described in Words by Cable and Reconstructed

Navy Scientist Assails Einstein Theory of Light

Glen H. Draper, Still in His Twenties, Denies Speed of Its Rays Is Absolute

Propounds Five Reasons

Spectrum and Other Variations Cited in His Statement

New York Herald Tribune, February 1, 1929, p. 16. A cabled verbatim translation of an Einstein paper on Unified Field Theory. Another story on the same page gives account of

As "covariant differentiations" we use only that which is formed by means of the capital Delta.

It is labeled according to the custom of Italian mathematicians, by the semicolon. Accordingly

$$A_{\alpha;\sigma} = A_{\alpha,\sigma} - A_{\alpha}\Delta^{\alpha}_{\sigma}$$

and by the equation

$$A^{\alpha}_{;\sigma} = A^{\alpha}_{,\sigma} + A^{\alpha}\Delta^{\alpha}_{\sigma}$$

As the s, [illegible] as well as the

$$g_{\mu\nu}(= {}^{\alpha}h_{\mu}\,{}^{\alpha}h_{\nu})$$

and the s superscript mu v have disappearing covariant deductions, these values can be exchanged according to desire as factors with covariant differentiation signs.

From the method of labeling hitherto used I depart only by defining the tensor, capital Lambda, with omission of the factor one over two, through the equation

$$L^{\cdot}_{\cdot\cdot} = \Delta^{\cdot}_{\cdot\cdot} - \Delta^{\cdot}_{\cdot\cdot}$$

The principal difference between the current formulae of absolute differential calculus which the introduction of the unsymmetrical disarrangement law brings with it lies in the divergence formation. Take $T^{\cdot\cdot\sigma}_{\cdot\cdot}$ as tensor according to wish with upper index sigma. Its covariant derivative is expressed if we only write the supplementary member in connection with index sigma. Equation

$$T^{\cdot\cdot\sigma}_{\cdot\cdot;\tau} = \frac{\delta T^{\cdot\cdot\sigma}}{\delta x^{\tau}} + \cdots + T^{\cdot\cdot\alpha}\Delta^{\sigma}_{\alpha\tau}$$

If one multiplies the equation with determinant h after one has reduced sigma and tau so that one receives, according to my introduction, a tensor density, then Gothic T on the right side of this equation

$$hT^{\cdot\cdot\sigma}_{\cdot\cdot;\sigma} = \frac{\delta \mathfrak{T}^{\cdot\cdot\sigma}}{\delta x^{\sigma}} + \cdots + \mathfrak{T}^{\cdot\cdot\alpha}\Lambda^{\sigma}_{\alpha\sigma}$$

The last member on the right side is missing if the disarrangement law is symmetrical. It is itself a tensor density just as the other members of the right side together which we in accordance with the usual method of naming label as divergence of

$$0 = L^{\cdot}_{kl;m} + L^{\cdot}_{lm;k} + L^{\cdot}_{mk;l} + L^{\cdot}_{l\alpha}L^{\alpha}_{lm}$$

$$+ L^{\cdot}_{l\alpha}L^{\alpha}_{mk} + L^{\cdot}_{m\alpha}L^{\alpha}_{kl}$$

That is just the condition that the capital Lambda lets itself be expressed in the stated way through the h. Through a single contraction of equation No. 3 we get by substituting ϕ for $L^{\alpha}_{\cdot\alpha}$ for the sake of brevity, the following important identity:

Equation No. 3a.

$$0 = L^{\alpha}_{kl;\alpha} + \phi_{l;k} - \phi_{k;l} - \phi_{\alpha}L^{\alpha}_{kl}$$

These we transform by introducing anti-symmetrical tensor density in k and l and get the equation

Equation No. 4

$$\mathfrak{V}^{\cdot}_{kl} = hL^{\cdot}_{kl} + \phi_{l}\Delta^{\cdot}_{k} - \phi\Delta^{\cdot}_{l}$$

Equation No. 3 is then changed into the simple form

Equation No. 3b

$$(\mathfrak{V}^{\alpha}_{kl})_{;\alpha} = 0$$

The tensor density $\mathfrak{V}^{\alpha}_{kl}$ fulfills still a second identity which is of significance for the following. We are supported in this deduction by the following law of exchange of divergence formations in tensor density in any order desired

Equation No. 5

$$\mathfrak{A}^{\cdot\cdot}_{/\sigma/\tau} - \mathfrak{A}^{\cdot\cdot}_{/\tau/\sigma} = (-\mathfrak{A}^{\cdot\cdot}L^{\cdot}_{\sigma\tau})$$

The dots on the Gothic A mean indices as desired, which are the same in all three members of the equation, especially those which are unaffected by divergence formations.

The proof of equation No. 5 rests especially on equation No. 2, but also on definition formula.

Equation No. 6

$$\mathfrak{A}^{\cdot}_{\cdot\cdot} \cdots \mathfrak{A}^{\cdot}_{\cdot\cdot} + \mathfrak{A}^{\cdot}_{\cdot\cdot} \cdots - \mathfrak{A}^{\cdot}_{\cdot\cdot} \cdots$$

Equation No. 5 hangs closely to- [illegible]

deficiency could easily be remedied if in place of the vanishing of

$$\mathfrak{V}^{\alpha}_{k\gamma/\alpha}$$

we postulated the equation

$$\overline{\mathfrak{V}}^{\alpha}_{k\gamma/\alpha} = 0$$

wherein

$$\overline{\mathfrak{V}}^{\alpha}_{kl}$$

means the tensor

Equation No. 9

$$\mathfrak{V}^{\cdot}_{\cdot} - \mathfrak{V}^{\cdot}_{\cdot} - \epsilon h(\phi_{\alpha}\Delta^{\cdot}_{\cdot} - \phi_{\cdot}\delta^{\cdot}_{\cdot})$$

differing as little as one likes from

$$\mathfrak{V}^{\alpha}_{kl}$$

Then we get the very equations of Maxwell—all in first approximation—if we form divergence field equations according to index alpha. Next we get—by passing to the limit, epsilon equal to zero—now as before differing as little as one pleases from the equation

$$\mathfrak{V}^{\alpha}_{k\gamma/\alpha} = 0$$

which just gives in first approximation the correct law of gravitation.

Field equations for electricity and gravitation will accordingly be correctly supplied on first approach through the statement

$$\overline{\mathfrak{V}}^{\alpha}_{k\gamma/\alpha} = 0$$

with the additional condition that one should pass to the limit epsilon equal to zero.

Therewith it is accompanied with the existence—valid on first approach—of the identity

Equation No. 8a

$$\mathfrak{V}^{\alpha}_{k\gamma/\alpha/\gamma} = 0$$

In field equations on first approximation a distinction of laws for gravitation on one side and for electricity on the other appears, which separation of course represents such a characteristic feature of nature.

It is now a question to make useful for strict consideration that knowledge won on first approximation. It is clear that we, too, have proceeded at this point from an identity which corresponds to equation No. 8a. This is obviously the

This equation system, together with

Equation No. 10a

$$(\mathfrak{V}^{\alpha}_{kl/l} - \mathfrak{V}^{\alpha}_{k\tau}L^{\cdot}_{\sigma\tau})_{/\alpha} = 0$$

forms a complete system of field equations.

If instead of starting from equation No. 10 we proceed directly from No. 10a, so we might not have reached electromagnetic equation No. 11. Besides, we would have no guarantee that systems No. 11 and No. 10a were compatible with one another.

But it appears to be certain that these equations are consistent with each other as the original equation No. 10 has sixteen conditions for the sixteen values superscript s figure h subscript mu. Among the sixteen-equations four identities exist necessarily on account of the common covariance of these equations.

Among the twenty field equations there exist accordingly on the whole eight identical relations, whereof of course only four are explicitly stated in the text.

It has already been said that equation No. 10a on first approach contains the gravitation equations and that equation No. 11—in connection with the existence of a potential vector—contains the Maxwell equations for a vacuum. I have also been able to show that, vice versa, for every solution of these equations an h-field satisfying equation No. 10a exists.

By reduction of equation No. 10a we arrive at a divergence condition for the electrical potential

Equation No. 12

$$f^{\cdot}_{\lambda} - \tfrac{1}{2}\mathfrak{V}^{\sigma}_{k\tau}L^{k}_{\sigma\tau} = 0$$

$$Zf^{\cdot} \cdot \mathfrak{V}^{\cdot}_{\cdot 1} \cdot 2h\phi^{\cdot}$$

Together these form equation No. 12.

Deeper investigation of the consequence of field equations Nos. 11 and 10a will have to show whether Riemann's metrics in connection with distant parallelism really provide an adequate conception of the physical qualities of space. According to this investigation, it is not improbable.

[P]alestine Appeal Is Cabled by Prof. Einstein

[Du]ty of U. S. Jews to Contribute [t]o Homeland, He Declares

[illegible] appeal from Professor Albert Einstein to the Jews of America, cabled [from] Berlin just before the publication of his "New Field Theory," was made public here yesterday by Louis Lipsky, president of the Zionist Organization of America. He asks support for the United Palestine Appeal campaign.

Professor Einstein long has been actively devoted to the Zionist cause. "Since the economic ruin of east European Jewry conferred upon the Jews of America the historic task of providing the material needs of world Jewry," his cable read, "appeals for money have been made to America every year. It is your duty as men and as Jews to help our impoverished brethren; but the solution of the Jewish problem does not lie that way.

"The rebuilding of Palestine as the Jewish national home differs fundamentally from all other Jewish activities of our time. This is a movement to aid, not individuals, but an entire nation. Your help is asked, not for support of the weak but for strong, healthy pioneers eager to work for the renaissance of the Jewish people. Far too little has been contributed hitherto; even the Jews of America have given much less than they should and are able to give."

Nevertheless, Professor Einstein continues, the work of rebuilding the Jewish home in Palestine is progressing.

to interfere with each other and produce darkness. This property of mutual annihilation is one characteristic of waves. But a wave must have some medium along which to travel, and so we invent a hypothetic substance called the ether. We can neither see, feel, taste, smell or weigh this material, so we say it is wholly transparent, tasteless, orderless, weightless, etc. Michelson and others have tried to detect a relative motion of this material with respect to the earth without success. This was the starting point of Einstein's original deductions.

For many years the wave theory fell from respect. We were given in its place the quantum theory, which said that energy, including light, etc., traveled in little bundles called quanta. But we were never able wholly to dissociate them from wave characteristics. We are less inclined to try to do so than we were.

Recently the electron, which was at one and the same time both our smallest speck of electricity and our smallest particle of matter, has been shown to have the characteristics of a wave. The atom is made up of a bundle of such waves. Perhaps they are analogous in some respects to a watch with a number of balance wheels. They appear as a solid particle outside, but have a periodic motion characteristic of a wave within. From this it would appear that the entire universe, including ourselves, is made up of wave bundles. This is given further support from the fact that we now know that mass can be changed into radiant energy and the reverse. Thus our universe is frittering away into radiation to build a new one. Light has long since been shown to be bent toward the sun in passing near it, thus showing evidence of having mass. A strongly magnetized piece of metal can be lifted against the force of gravitation by another magnet. Since this earth itself is a great magnet it would repel from its surface a magnet of sufficient strength and lightness, a thing which we may some day look forward to.

Final Answer Still Far Distant

All these things tend to show a unity of all the known phenomena of physics and to show that they are in some respect inseparable from a wave interpretation. The suggestion that electromagnetism and gravitation are the same has given us one more step in the solution of the problem of our existence, the final answer to which we are still far away from. But in so far as the secret of the universe is concerned we are little if anything further along than we were before, although Einstein's work marks a great mathematical achievement.

Perhaps we cannot better express the present state of affairs than did that poet-mathematician Omar Khayyam in but two verses:

For in and out, above, about, below,
'Tis nothing but a magic shadow-show,
Played in a box whose candle is the sun
Round which we phantom figures come and go.

Myself when young did eagerly frequent
Doctor and Saint, and heard great argument
About it and about: but evermore
Came out by the same door as in I went.

British Tax Drafters Bow Low to Einstein

From the Herald Tribune London Bureau
Copyright, 1929, New York Tribune Inc.

LONDON, Jan. 31.—The most conspicuous British reaction to the publication of details of the new Einstein theory was a three-column cartoon in "The Evening News" today, showing the framers of the British income tax laws bowing low and taking off their hats before the professor as the new master of incomprehensibility.

[illegible] Einstein text, as issued in German. The Herald Tribune did not use [illegible] before the mathematicians and physicists of America.

Einstein's Work Hailed as Key to Science Riddles

(Continued from page one)

Poor, professor of celestial mechanics at Columbia, who is the outstanding opponent of Einstein in this country, both commented at some length upon the bearing of his new achievement.

Both were agreed in their explanation of just what it was that Einstein actually had done in the present paper, although Professor Poor denied vigorously that the accomplishment would be found to have anything either of value or significance. Professor Gronwall, on the other hand, considered it an advance of the utmost importance.

Professor Einstein's paper, said Professor Gronwall, had supplied "a unified mathematical treatment for electricity and magnetism and for gravitation. It does not mean that electricity and gravitation have been shown to be identical, but merely that the mathematical treatment of the two is on the same basis."

Sees Important Consequences

This, in Professor Gronwall's opinion, is the sole immediate achievement of the new theory. He explained, however, that it carried with it important consequences, and may easily lead to other consequences far more important still. To make this clearer Professor Gronwall had to go back into history.

"Newtonian gravitation," he said, "was thought of as a property of matter, and was considered as action 'at a distance' of one body upon another. Electro-magnetic phenomena, on the other hand, were thought of as properties of a 'field'—an electric field or a magnetic field. Clerk-Maxwell worked out the formulæ governing electro-magnetic fields, but it had been impossible to relate these to gravitation.

"In his generalized theory of relativity, published in 1916, Einstein showed that gravitation was not a property of matter, acting at a distance, but a property of space, and susceptible of treatment in terms of 'field' equations. In 1918 and in 1923 Einstein himself endeavored to relate his field equations for gravitation with the Clerk-Maxwell field equations for electro-magnetic phenomena, but rejected his own efforts. Both Weyl and Eddington worked toward the same end, but their work was extraordinarily difficult and made the problem so terribly much more complex as to be unsatisfactory.

Common Mathematical Basis

"Einstein has now succeeded in finding a mathematical basis common to the two sets of formulæ, and the great beauty of his work is that it is so simple."

Professor Gronwall hastened to explain that he meant the underlying ideas were simple and that even they were only simple to a mathematician. In technical language, what Einstein has now done, he said, is "to make a more complete analysis of the coefficients that enter into an expression for the square of the line element.

"These coefficients are now split up into finer parts and are chosen so as to give not only a slightly modified equation for the gravitational field, but also to give the Clerk-Maxwell equations for the electro-magnetic field. So that in this way the two fields come out of the same general mathematical reasoning."

A point which other commentators have not stressed is that in doing this Einstein modified his 1916 field equations for gravitation. However, Professor Gronwall said, the new equations for gravitation agree with the old ones within the degree of accuracy necessary.

But the consequence of this correlation between electro-magnetism and gravitation, Professor Gronwall showed, is that electro-magnetic phenomena now becomes properties of space, as graviation already had become under Einstein's treatment. They must be explained upon the basis of the non-Euclidean geometry as the result of a "warping" or "curvature" in space itself and not as the result of causes recognized by the older physics.

The first thing that will be necessary with the new work, Professor Gronwall continued, will be to test it out against the known facts of nature—"to apply it to special cases and see if it agrees with experimental observation." If it does, it may prove to hold the key to a number of science's most difficult mysteries.

To say what might ultimately be done with such a theory, Professor Gronwall went on, would be "pure guess-work." But he hazarded one or two guesses. "It is quite probable," he said, "that we could confidently expect to find with its aid some important information as to the nature of light, and that we will at last find a general access to the phenomena treated by the quantum theory."

Professor Gronwall is a research mathematician in the quantum theory. This is a theory worked out to account for certain observed phenomena of energy; the difficulty is that it has never been related to other theories of physics and seems incredible upon the basis of everything else that is known about the universe. To relate it to the main body of physics would mean a great step forward in our knowledge of the universe.

'Mathematical Handybox,' Says Skeptic

In attacking the new work, Professor Poor explained that the mathematical part of the paper was so incomplete as published in the earlier editions of yesterday's Herald Tribune that he had no had an opportunity to check the equations or follow the mathematical methods adopted. "But from the opening paragraphs of the paper itself," he went on, "and the explanations cabled from Berlin, it is possible to form a reasonably accurate idea as to what Einstein has accomplished.

Professor Poor agreed that it was a purely mathematical construction, but differed from those who have hailed the paper in his view of its ultimate value. "From all these reports," he said, "it would seem that Einstein has evolved some intricate mathematical formula, from which he can derive, by proper substitutions and adjustments, the ordinary formulas of gravitation and, by other substitutions and adjustments, he can equally well find the formulas of electricity or electro-magnetism. He has simply devised some combination of mathematical symbols, meaningless in themselves, but which may represent certain effects of gravitation when proper values are assigned to these symbols; and which may represent certain effects of electrical action when certain other values are assigned to them.

"The word 'theory' in his statement does not mean the law of gravitation; it means a collection of formulas. He claims, and claims only, to have found some sort of mathematical handybox from which he extracts the formulas of gravitation or the formulas of electricity at will."

Even supposing, however, that this established any valid identity between Einstein's gravitational field equations and the electrical field equations, Professor Poor would deny the significance of the result because he denies the value of Einstein's work in 1916 on gravitation.

[illegible] exceptions, namely [illegible] Fizeau's experiment on light [illegible] cepheid variable stars, the Michelson-Morley experiment, and the spectral line-shift due to the receiver moving in the direction of the ray of light.

"From these five I made certain assumptions as to the real causes of the effects obtained. The first assumption naturally was that light is a real phenomenon, and at any given time light was at a fixed place in space relative to its source. In sound the wave-front is spheric relative to the place in the medium of the emission.

"The second followed immediately. It is possible to have light pass an observer at different velocities by the motion of the observer. This is a direct analogy to sound, because when the receiver moves toward the place of emission in the medium with a velocity 'V' the sound-waves pass the receiver with greater velocity than they would if the receiver had remained stationary, by an amount equal to 'V.'

"The third assumption was naturally that light was itself the medium in which the energy is propagated.

"This led to the fourth assumption that light media, which I call quanta, have all those properties commonly assigned to ether and that the source is the generator."

Scientists Asked to Test It

Captain C. S. Freeman, U. S. N., superintendent of the observatory, has become interested in the Draper theory and is trying to interest distinguished scientists of the nation in it. All the minor details of the theory may not yet be correct, Mr. Draper admitted, but there has not yet been any experimental evidence to prove that any part of the theory is at fault. He has been working on the theory since his graduation from the University of Utah in 1923.

"Einstein's theory of relativity," he said, "assumes that light travels at a maximum velocity attainable and relative to the receiver. The velocity is constant in all directions, irrespective of the relative motions of the source and receiver and the medium on which light travels.

Would Measure Effects Only

"With this view the Einstein theory not only fails to explain aberration, but also the spectral line shifts, due to the observer moving toward the source of light, and as far as I am able to understand his theory there is no explanation given for determining the direction of the motion of a star relative to the observer.

"I assume in my theory that effects only are measurable and that [the] causes are deductions made from the observations, but that these deductions are true only when the effect will follow the sequence of the causes given by the theory, that is, as soon as a cause does not produce the expected effect, it is due to a fault of the deductions.

Hagen Sues for Divorce; Golfer Charges Desertion

Wife Quit Him in 1927, [He] Says in Coast Petition

LOS ANGELES, Calif., Jan. 31 [illegible] Charging that his wife had deserted him nearly two years ago, Walter Hagen, the golfer, filed suit for divorce here to-day.

The British open champion declared he had been "kind and indulgent," [that] he had always provided an excellent home and used every care for her happiness and welfare," but that his wife, Edna C. Hagen, left him March 3, [1927,] at St. Petersburg, Fla., and had refused to return.

The Hagens were married in [New] York on April 6, 1923. They have [no] children. Hagen is passing the winter in Southern California following [his] petition in numerous golf tournaments in this section.

how equations were sent by cable. Reproduced from the Collections of the Library of Congress.

press. Reporters were usually at sea about the meaning and significance of Einstein's work and they dwelt on the professorial image—sloppy, absent-minded, henpecked. When they tried to understand the technical issues, they were almost always assured by scientists that the theory was practical but that it was impossible to understand without a knowledge of very sophisticated mathematics. It is not surprising that science is held at such a distance in American culture.

It was not until the day that Einstein died, April 19, 1955, that the editorial page of *The New York Times* relented. Describing Einstein as a hard headed realist, they went on to write:

> What distinguished Einstein both as a scientist and as a man was his passionate devotion to the truth. He hated arbitrary laws, conventions, coercions of one type or another, any form of pedantry. Thomas Hardy once prayed: "God give me strength to find a fact though it slay me." Einstein had that strength. It is this that set him apart as a scientist and the mathematician who saw relationships in the outer world to which other men were blind.

Anyone who had followed the opinion of that newspaper or most other American newspapers over the years would have found such an assessment strangely out of tune. And so it would be today.

CONCLUSIONS

In this chapter we have surveyed the assimilation of Einstein's theory of relativity in the United States. Within the scientific community the formalism one associates with special relativity has become almost second nature during the last seventy-five years. That formalism is an essential tool in the design of high energy physics technology, and an important ingredient in nuclear technology. But we have also seen that the meaning of the formalism has remained ambiguous. Even now, in most physics books, the interpretation of the formalism is more representative of Lorentz's or Poincaré's outlook than Einstein's. The American interpretation of the meaning of the theory of relativity is based on the belief that the theory is correct because both the postulates and the predictions of the theory are in agreement with measurement and observation. Such an interpretation is more easily integrated into traditional American views about the relationship between evidence and theory than is Einstein's view that theories are the free creation of the human spirit.

From the beginnings of this country, faith in the achievements of science has existed side by side with a distrust of the theoretical and the intellectual. These contrasting attitudes are reflected in the public response we have sampled to the theory of relativity. Such attitudes are reflected as well in the public support which science receives when it is seen as the precursor of industrial growth and in the public antagonism directed toward science when the studies of university scientists who work with public funds are characterized as impractical, or when science is vilified as the source of the materials and practices of environmental pollution. Science is also often charged with an amorality that allows for the dispassionate consideration of societies run by computers, fortress America bristling with nuclear arms capable of destroying everyone in the world ten times over and the gradual substitution of technological efficiency for the confrontations between people that make us uniquely human.

The character of the response to the theory of relativity by scientists in the United States mirrored the social environment of American science. This should give us perspective on the role of science as a social institution in a culture. Science does not find the truth, does not create technological tools or decide on how they are used. Men and women in science, as we have attempted to show in this chapter, and in the rest of this book, struggle to understand how the world works. That struggle never ends. While we keep changing our minds about acceptable premises, less frequently do we change our minds about the relationship between those premises and the evidence we gather from the world in support of our views. The creation of technological tools of any sort is not the sole responsibility of the scientific community. Those tools are the result of political consensus for which all segments of the culture must share equal responsibility.

·11·

Relativity and Revolutions in Science

IN THE COURSE OF THIS BOOK we have examined the intellectual and social context out of which the theory of relativity emerged, how the theory was initially received in four different scientific cultures and how it has been assimilated since in one of those cultures. The development and assimilation of the theory of relativity have been widely acclaimed as one of two major revolutions in physics in the twentieth century. (The other is the development of quantum physics.) We now examine briefly the notion of *scientific revolution* and how it might be applied to the theory of relativity.

SCIENTIFIC REVOLUTIONS

The concept of *revolution* in science is recent.* Widespread discussion of the possibility of such phenomena did not occur until the publication of T. S. Kuhn's *Structure of Scientific Revolutions,* which first appeared in 1962.†

Before the introduction of Kuhn's book, the general view that distinguished science from all other disciplines and their relationship to the world of technology and commerce was that it was progressive, cumulative, certain and the source of industrial and technological innovation.

* One of the first, if not the first, book length studies that explicitly made reference to scientific revolution was A. R. Hall's *The Scientific Revolution, 1500-1800* (London, 1954; 2nd ed., 1962).
† T. S. Kuhn, *The Structure of Scientific Revolutions* (Chicago, 1962; 2nd ed. 1970).

The storm that ensued after the publication of *Structure of Scientific Revolutions* is hard to describe. Some sense of it can be had from the "Postscript-1969" that Kuhn added to the second edition of the book. The breadth of interest in that book can be understood when we realize that in twenty years, more than 550,000 copies have been sold. For a book by an academician, published by an academic press, about the nature of revolutions in science, that is a phenomenal record.

In *Structure of Scientific Revolutions* Kuhn attempted to provide a theory for how change periodically occurs in our theories about the nature of the physical world. These changes almost always are accompanied by intensive battles between conservatives who wish to maintain the *status quo* and those who wish to promulgate the revolution. Kuhn sought to understand in a general way, the dynamics shaping such battles and how new ideas are assimilated within the affected scientific community. In doing so, Kuhn felt the need to introduce a specialized set of words and phrases such as "paradigm," "paradigm shift," "normal science," "problem solving," "anomaly," etc. It is a measure of the degree to which the book took the intellectual world by storm that such phrases will never again have quite the same meaning that they had before the publication of *The Structure of Scientific Revolutions.*

Everyone read *The Structure of Scientific Revolutions.* Everyone. It has become a kind of cult book and Kuhn a kind of cult figure. Social scientists have flocked to the book. They have used it, not so much to study the concept of revolution in science, but as a warrant for claiming their disciplines and their works to be scientific. I am still called upon to give lectures on *The Structure of Scientific Revolutions* in social science courses where the pertinent question on everyone's lips is where this particular discipline is located on the "Kuhnian" periodization of scientific revolution: "Are we in the paradigm, the pre-paradigm, or post-paradigm state?" Scientists, social scientists, and humanists were all puzzled and intrigued by Kuhn's claim that science does not proceed in a linear, continuous, progressive development but is punctuated by discontinuous, precipitous, revolutionary breaks. To many people in these disciplines, such views are now labelled "Kuhnian."

Philosophers and philosophers of science responded in a somewhat different manner. *The Structure of Scientific Revolutions* is a monograph that was originally published in a series which bears the title, "The International Encyclopedia of Unified Science." The editor-in-chief of that series, Otto Neurath, and the associate editors, Rudolph Carnap and Charles Morris, were leaders in the founding of logical empiricism, one of the more radical empiricist camps. From the time of the publication of *The Structure of Scientific Revolutions,* philosophers have analyzed the book as a tract in epistemology and the status of knowledge. Kuhn has been attacked as an anti-rationalist, or a relativist, or both. Analytic philosophers and those who have struggled with various theories of verification or falsification, that is, with the question of the criterion for meaningful statements, have been particularly attentive and hostile to Kuhn's work. Some have accused him of total solipsistic subjectivity.* As with the social scientists and physical scientists, the element of Kuhn's work which seems to have caught the attention and imagination of these scholars is Kuhn's claim that science does not proceed by careful step by step accretion, by making a unique logical connection between data and theory to reveal the true nature of the physical world.

In replying to the many responses to his thesis, Kuhn himself has directed most of his efforts toward the criticisms of philosophers. While struggling against the charge of relativism and solipsism, he has been an enthusiastic participant in the general analysis that philosophers of science have devoted to his work.

Forgotten in all of this is that, in the first instance, *The Structure of Scientific Revolutions* was not a treatise on epistemology and it does not provide a check list for social scientists. Kuhn was attempting to provide a theory of how ideas are born, struggle to exist, and are then able to replace other, well established ideas about the way the world works. One of the reasons that consideration of this aspect of the book was lost is that there was very little direct critical response to Kuhn's work by historians of science. There was good reason for this. It was the claim of revolutionary, discontinuous change that captured the imagination of most scholars outside the history of science. Within the history of science, this was no surprise at all. The concept of *scientific revolution* was a well established idea within the discipline. If it had not been made explicit, at least the idea was in the air.

There were two surprises about Kuhn's thesis for historians of science. The first was that the notion of revolutionary change was itself so revolutionary to most other scholars. The second was that as a result of his position, Kuhn was being charged with relativism, solipsism, anti-rationalism, and even anti-intellectualism. In fact, a careful reading of Kuhn's book supports his contention that he did not hold such views. If anything, he shares a great deal with the empiricist tradition of the editors of the monograph series within which *The Structure of Scientific Revolutions* was originally published. While most philosophers, social scientists, and natural scientists view him as an externalist looking to factors outside the substance of science for shifts in belief within science, it is my view that Kuhn is an internalist committed to looking almost exclusively at internal substantive influences.

* For entrée into the literature on Kuhn's work see, T. S. Kuhn, "Reflection on My Critics," in I. Lakatos and A. Musgrave (eds.), *Criticism and the Growth of Knowledge* (Cambridge, England, 1970) and "Second Thoughts on Paradigms," in F. Suppe (ed.), *The Structure of Scientific Theories* (Urbana, 1971).

This is not a pejorative judgment. It is simply a recognition that, whereas Kuhn sometimes refers to the probable importance of social, cultural, and political changes as factors in determining the shifts of belief within a scientific community, in the end, it is measurement and data, and only measurement and data, that are the instruments of change within his theory. Indeed, the driving engine for change within Kuhn's theory is the building up of a collection of empirical anomalies at odds with predictions of currently acceptable theories, which finally can no longer be ignored or swept under the rug. In attempting to resolve these anomalies scientists turn to and embrace hitherto unacceptable theories. The theories themselves gain acceptance because of the agreement of prediction with measurement in a wider range of instances than was previously possible.

In that regard, Kuhn's thesis is a relatively conservative one. It might even be called an empiricist theory of change. In fact, Kuhn has always been explicit and emphatic in denying that he would entertain any kind of irrational factor as a possible instrument of change of belief within a scientific community about the nature of the physical world. He shares much more in common with his critics in the philosophical community who are concerned with establishing criteria of confirmation or falsification than he recognizes or admits to.

RELATIVITY AND SCIENTIFIC REVOLUTIONS

While I do not share Kuhn's assumptions about science or about history, I think his work is extraordinarily powerful and important. Many features of his analysis are illustrated by the factors surrounding the introduction of the theory of relativity. There was considerable disquietude near the turn of the century among scientists concerning the failure of either the mechanical or the electromagnetic world view to provide a comprehensive account of first and second order ether drift experiments. On the other hand, it is noteworthy that there was not complete satisfaction with the theory which Lorentz (and Poincaré) produced in 1904 although that theory was formally identical to the theory that Einstein introduced a year later.

Part of the reason for that lack of satisfaction was that the premises of the electromagnetic world view could not be easily satisfied by the Lorentz theory. On the other hand, a theory such as Abraham's which followed electromagnetic assumptions very closely, was itself incapable of explaining much more than that the mass of electrons appeared to increase as the velocity of electrons increased. There seemed to be no possiblity of developing a theory which both satisfied the assumptions of the electromagnetic world view and accounted for all of the available evidence.

Recall also that the amount of dissatisfaction with Lorentz's work varied from community to community. Americans were, initially, quite satisfied with Lorentz's theory. It seemed to explain the evidence. German physicists, who paid attention more to the underlying philosophy of theories, were less satisfied. It was dissatisfaction with the underpinnings of the different theories that motivated Germans to explore and elaborate different theories and it was that process of elaboration that more and more revealed to German physicists the difference between Einstein's and Lorentz's theories and convinced them of the heuristic power of Einstein's theory.

It would be a mistake to dismiss the role of experimental evidence in the process whereby the theory of relativity first gained recognition or in the process of its assimilation. After all, it was experimental evidence that created the disquietude in the first place in much the manner that Kuhn has suggested. And it was the successes of the predictive power of both the special and general theories of relativity that eventually were so convincing to scientists in the United States and elsewhere.

But, as we have pointed out, any set of experimental evidence can be explained from any theoretical point of view one wants to adopt. In the case of special relativity, at the time that the decisions about rival theories were being made, there was no convincing evidence one way or the other.

In Chapter 10 it was pointed out that Kuhn's suggestion for the role that textbooks play in his scenario of the course of scientific revolutions was inadequate to account for the American case. Rather than convincing Americans of the new point of view, textbooks appear to have shielded the scientific community by placing the new point of view squarely within a metaphysical framework that was comfortable and comprehensible. In fact, when one examines the details of the American case, it is not clear that any kind of revolution has taken place at all.

But is that so different from what occurs in political and social revolutions? What *was* so different in the United States before and after 1776? Or in France after 1789? Or in Russia after 1917? Whereas the forms and framework of the political structure of these countries had changed, those changes themselves have little to do with being an American or a Frenchman or a Russian.

This book has been dedicated to the notion that that which is the case for social, political and educational institutions, is also the case for scientific institutions. The fate of ideas like the theory of relativity is as much a function of culture as is the fate of any other product of the human intellect.

Appendix ·1·

Trigonometry

THE PURPOSE OF THIS APPENDIX is to introduce the rudiments of trigonometry to those who are not familiar with the subject. Our goal is not to teach it in a rigorous or thoroughgoing way. The reader is urged to study the examples. Practice translating formal equations into ordinary language and vice versa. There are many fine texts available on the subject of trigonometry in a variety of styles. You are encouraged to continue the exploration of this fascinating subject in one of them.

Trigonometry has a reputation among secondary school students for being very difficult. While it is true that one can prove very beautiful and subtle relationships, the basic part of the theory is straightforward. Utilizing the logic of Euclidean geometry, trigonometry relates angles to distances. This is why it is important in land survey—it allows us to calculate a distance when all that we can measure are other distances of our choosing, and the angles between the distances we have chosen and the distance we want to measure.

We will make use of two basic premises or postulates, both theorems in Euclidean geometry.

1. Corresponding parts of similar triangles are in proportion to each other.
2. The Pythagorean theorem: In a right triangle the square of the hypothenuse is equal to the sum of the squares of the two other sides.

As we are using these two Euclidean theorems as postulates it is not possible to say more about them within the trigonometric logical system. However, we have

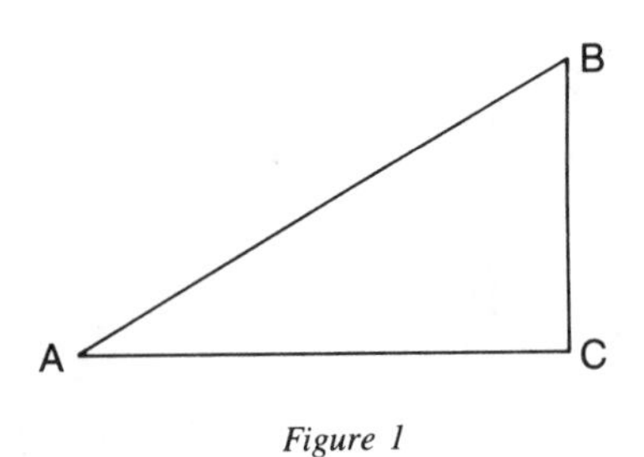

Figure 1

discussed the meaning of the similarity theorem in Chapter 1. Here is a non-standard proof of the Pythagorean theorem which requires no knowledge of geometry.

Figure 1 is a right triangle. Since one of the angles is 90° and the sum of all three angles must be 180°, the sum of the angles at *A* and *B* is 90°. It is correct that for any right triangle the sum of the other two angles is 90°. The side opposite the right angle, *C*, side *AB*, is the longest side of the triangle. It is given the special name, ***hypothenuse***. For a given angle, say angle *A*, the side *BC* is called the side ***opposite*** the angle for obvious reasons (note that the side opposite angle *B* must be *AC*). *AC* is called "the side *adjacent*" to angle *A* (*BC* is the side adjacent angle *B*). The origins of the Pythagorean theorem are lost in the murkiness of prehistory. It is named after the Greek mathematician and cosmologist Pythagoras, but there is evidence to suggest that he probably brought it back from Egypt where it had been used in land survey. If we label the sides of the triangle, *a*, *b*, and *c* as in Figure 2, the algebraic statement of the theorem is

$$c^2 = a^2 + b^2 \tag{1}$$

Note that one interpretation of this algebraic statement is that the area of the square *built* on the side we call the hypotenuse is equal to the sum of the areas of the square built on the other two sides (Figure 2). This is a physical interpretation of a mathematical statement. The abstract mathematical relationship has been translated into the measurement of area, for example, of land measure.

The *proof* of the theorem is presented in Figure 3. Actually, no words are needed. Study and compare the drawings for a few minutes to see that the theorem stated in (1) is correct. In Figure 3 we have composed two equal squares, each with sides of length $a + b$. Algebraically, squaring $a + b$ gives the following result:

$$(a + b)^2 = a^2 + 2\,ab + b^2 \tag{2}$$

Examining the figure at the left we see the areas a^2, b^2 and the two remaining rectangles of area *ab*. Those two rectangles have been divided equally into four right triangles whose sides have been labeled *abc*.

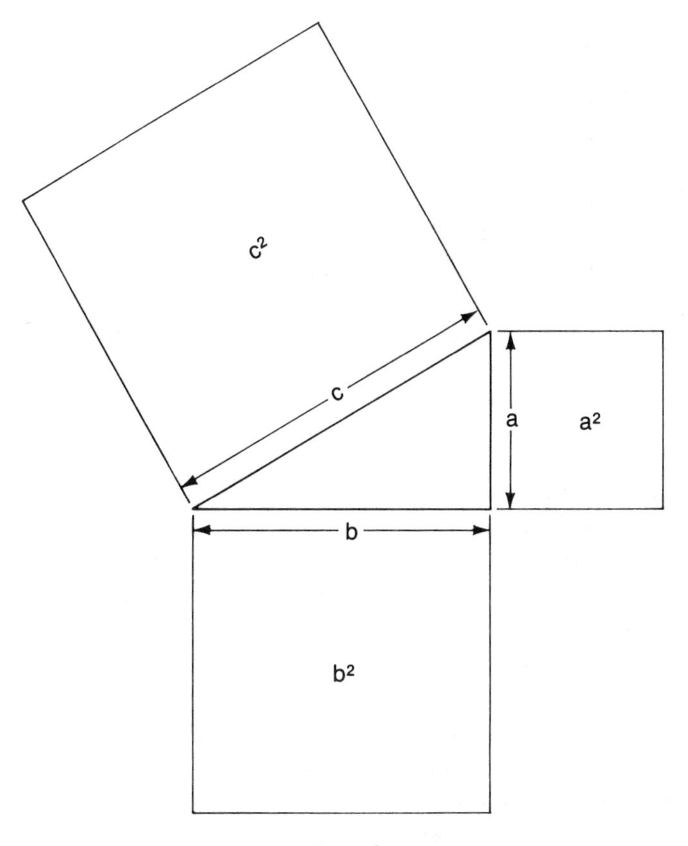

Figure 2

The square figure at the right in Figure 3 has the same area as the one at the left as it is also $a + b$ on a side. It has been divided differently. The central part of the figure has the area c^2 and what remains are the *same* four right triangles having sides *abc*. Therefore, the area $a^2 + b^2$ in the figure at the left will equal the area c^2 in the figure at the right. Whereas this proof of the Pythagorean theorem is not developed within the classical framework of Euclidean geometry, there is no question that the theorem is correct. To say that the Pythagorean theorem is true has a different meaning. If space is Euclidean, that is, if it satisfies the assumptions of Euclidean geometry, then the Pythagorean theorem is true in the context of comparing areas in space. Since the earth itself is not a plane, the Pythagorean theorem is clearly *not* true for areas on the surface of the earth. However, since the curvature over short distances cannot be easily detected, within the limits of error

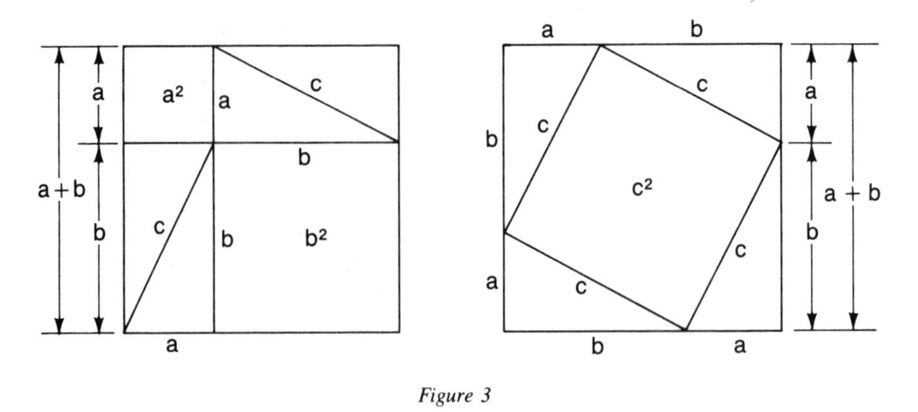

Figure 3

expectation, the Pythagorean theorem gives acceptable results. On the other hand, if space had a subtle curvature, the theory would not apply to the world.

Earlier it was stated that the subject of trigonometry relates angles to lengths. We will develop the subject by examining the very limited universe of right triangles. Eventually, it can be generalized to other triangles and then to situations in which no triangle, as such, is involved. We will not explore more general cases.

We begin with three *definitions:* sine, cosine and tangent. The definitions give convenient names to ratios of two of the sides of any triangle. Since there are three sides to a triangle, a, b, and c, and we are going to be considering two of them at a time, there are six possible combinations, a/b, b/a, a/c, c/a, b/c, c/b. As three of those ratios are the inverse of the other three, they are superfluous. The name "sine" is a medieval rendition of the equivalent word in Arabic. "Cosine" means, "related to sine" and the relationship is unyielding. "Tangent," a word introduced into the Western mathematical vocabulary at the end of the sixteenth century, is related to another meaning of tangent: a line drawn perpendicular to the radius of the circle which meets the circumference.

The sine, cosine and tangent (abbreviated, sin, cos and tan) are operators like $+$, $\times$, $-$, and $\div$. They tell you to perform certain operations.

In Figure 4 we have drawn four right triangles. Three have different shapes, two of them, *ABC* and *AKL* are similar, as drawn the way they are, one inside the other, the angles are equal.

The sine of an angle (for now restricted to the angles in right triangles) is *defined* as the ratio of the side opposite the angle to the hypothenuse. Referring to the triangles in Figure 4:

$$\sin B = \frac{\text{side opposite } B}{\text{hypothenuse}} = \frac{b}{c}$$

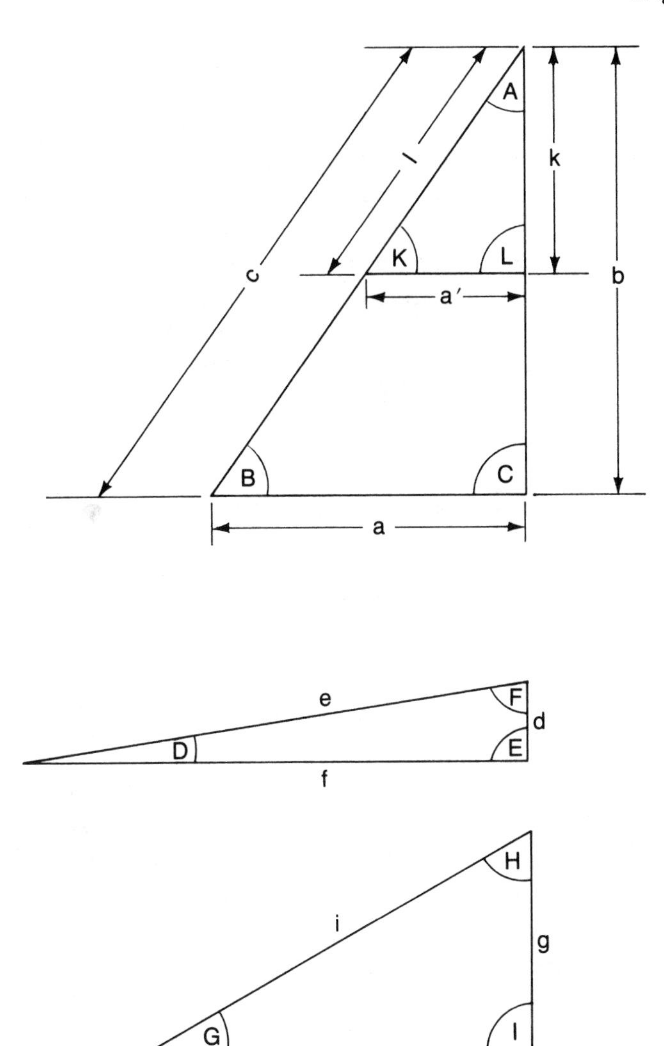

Figure 4

Similarly:

$$\sin D = \frac{d}{e}, \quad \sin G = \frac{g}{i}, \quad \sin A = \frac{a}{c},$$

$$\sin A = \frac{a'}{l}, \quad \sin K = \frac{k}{l}, \quad \sin F = \frac{f}{e}, \quad \sin H = \frac{h}{i}$$

In each case the sin of the angle is a set of directions to find the ratio of the side opposite the angle to the hypothenuse. It is important to note that once you have specified an angle, for example, 30°, you have fixed the shape of the triangle and thus fixed the size of the sin. This is the case since the sum of the two angles that are not right angles in the triangle are 90°. Once you have specified one of the angles as 30°, the other angle is 60°. Since triangles having the same angles are similar, that is, corresponding sides must be in proportion to each other, the sin of 30° is the same number regardless of the right triangle in which it is found.

Consider the two triangles we have drawn, one inside the other. These triangles are similar, and $a'/l = a/c$. In other words, regardless of the size of the triangle, the *ratio* of the side opposite the angle to the hypothenuse is the same for a given angle. In any right triangle in which one of the angles is 30°, the side opposite the 30° angle is half the length of the hypothenuse. Therefore, the sin of 30° will *always* be 0.5.

If the side opposite the 30° angle is one foot, the hypothenuse in that 30-60-90 triangle is two feet. If the side opposite the 30° angle is 3 cm, the hypothenuse in that 30-60-90 triangle is 6 cm.

The cosine of an angle in a right triangle is defined as the ratio of the side adjacent to the hypothenuse. In Figure 4:

$$\cos B = a/c$$

Similarly:

$$\cos D = \frac{f}{e}, \cos G = \frac{h}{i}, \cos A = \frac{b}{c} = \frac{k}{l}, \cos K = \frac{a'}{l}, \cos F = \frac{d}{e}, \cos H = \frac{g}{i}$$

For the angle 30° the cosine has the value 0.866, regardless of the triangle in which it is found. If the length of the side adjacent the 30° angle is 0.62 meters that represents 0.866 the length of the hypothenuse which then is 0.72 meters.

The only other combination of sides that are left to compare are the sides adjacent and the side opposite. The ratio is called the tangent and the definition is the ratio of the side opposite to the side adjacent.

$$\tan B = b/a$$

Similarly:

$$\tan D = \frac{d}{f}, \quad \tan G = \frac{g}{h}, \tan A = \frac{a}{b} = \frac{k}{a'},$$
$$\tan K = \frac{k}{a'}, \tan F = \frac{f}{d}, \tan H = \frac{h}{g}$$

For a 30° angle the tangent is 0.577. If the side opposite a 30° angle is 0.5 yards, the length of that side must be 0.577 the length of the side adjacent which is

0.866 yards. Once we define the sin and the cos, the definition of tangent cannot be independent. It is simple to prove that the tangent is equal to the ratio of the sin to the cos.

$$\sin B = b/c, \cos B = a/c$$

$$\frac{\sin B}{\cos B} = \frac{b/c}{a/c} = b/c \cdot c/a = b/a = \tan B$$

An equivalent relationship will be true for any angle. Note that the sin of one angle of a right triangle is the same ratio as the cos of the other angle. In a 30-60-90 triangle, sin 30° equals cos 60° and vice versa. In a 21-69-90 triangle sin 21° equals cos 69°, etc. Given the definitions and the postulate about similar triangles this is correct.

Earlier I stated that there are six possible combinations of ratios when three things are taken two at a time, as we do when we look at the ratios of the sides of triangles.

For completeness, we mention the other three ratios. The secant is defined as the ratio of the hypothenuse to the side adjacent, that is, the inverse of the cosine. The cosecant is defined as the ratio of the hypothenuse to the side opposite, that is, the inverse of the sine. The cotangent is the ratio of the side adjacent to the side opposite, that is, it is the inverse of the tangent. These three functions are the inverse of the three functions we have defined. Table 1 summarizes the postulates and the definitions.

As we have stated, specifying one of the angles other than the right angle in a right triangle fixes the other angle and thereby determines the shape of the triangle. Therefore, to specify one of the angles of a right triangle determines the *ratios* of the lengths of the sides regardless of the size of the triangle. This raises the possibility of publishing tabulations of the three trignometric functions. Table 2, at the end of this appendix, is such a table.

Choose any angle in the table and the ratio of the sides can be read. Thus, the table tells us that for 31° the sine is 0.515, the cosine is 0.857, and the tangent is 0.601. What does this mean? It means that in a right triangle having an angle of 31° (the other angle of the triangle will be 59°) such as in Figure 5, $a/c = 0.515$, $b/c = 0.857$, and $a/b = 0.601$. We specify one length, for example, the hypothenuse of the triangle is 2 meters: It follows that side a is .515 of 2 meters and side b is .857 of two meters, or 1.03 meters and 1.71 meters, respectively. The methods for construction of such tables go beyond this discussion. Let us take several simple cases to see that the table makes sense. If the angle is 0, the side opposite is 0 and the sin is 0. At the same time, the hypothenuse and the side adjacent lie on top of each other and the cos is 1. This is confirmed by the table.

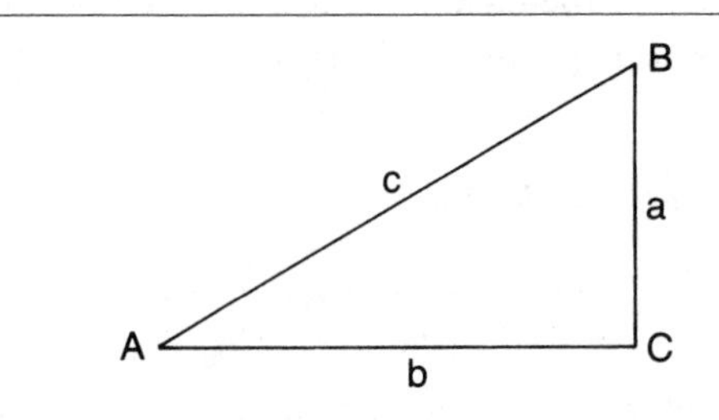

Postulates:

I. Corresponding parts of similar triangles are in proportion to each other.

II. The Pythagorean theorem: The square of the side of the hypothenuse of a right triangle equals the sum of the squares of the other two sides of the triangle.

Definitions:

$$\text{sine } A = \frac{\text{length of side opposite } A}{\text{hypothenuse}} = a/c = \cos B$$

$$\text{cosine } A = \frac{\text{length of side adjacent } A}{\text{hypothenuse}} = b/c = \sin B$$

$$\text{tangent } A = \frac{\text{length of side opposite } A}{\text{length of side adjacent } A} = b/a = \sin A/\cos A$$

Table 1: Postulates and Definitions of Plane Trigonometry

Similarly, for $90°$ the sin is 1 and the cos is 0. This is also confirmed by the table. Suppose that the triangle is an isosceles right triangle, that is, a triangle with two sides of the same length. Each of the angles is $45°$ and if each of the other sides has a unit length, the hypothenuse is $\sqrt{2}$ (Apply the Pythagorean theorem to Figure 6):

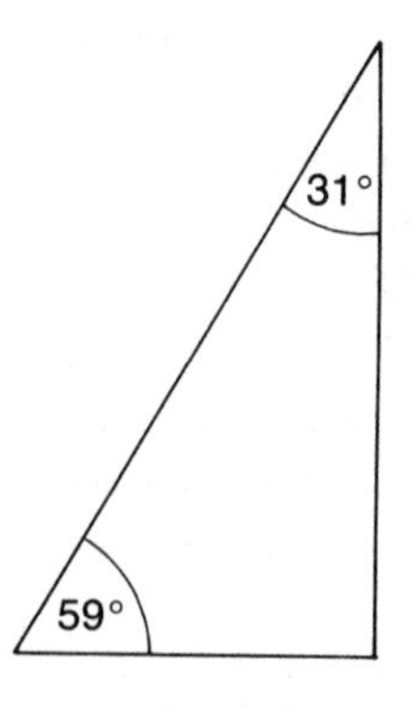

Figure 5

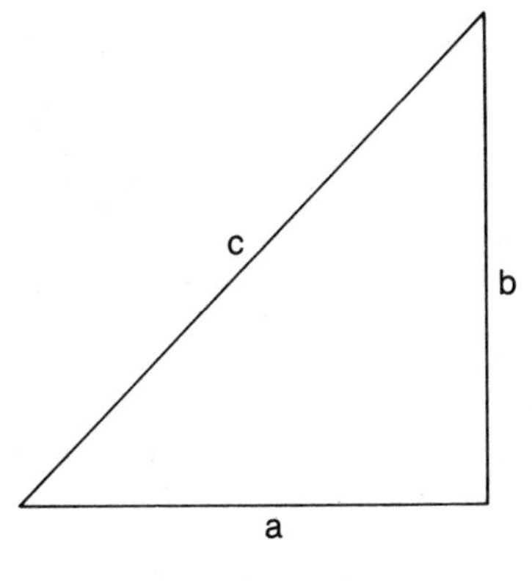

Figure 6

$$c^2 = a^2 + b^2$$

Both a and b are one unit long and thus:

$$c^2 = 1^2 + 1^2 = 2$$
$$c = \sqrt{2} = 1.414. \; . \; .$$

As both of the angles are 45° and the sides a and b are equal to each other, the sin 45° equals the cos 45° and the value of these functions is sin 45° = cos 45° = 1/1.414 = 0.707. Examination reveals that the sin of any angle will be the same as the cos of 90° minus that angle: sin x = cos (90° − x). Also the tan x = sin x/cos x for any angle in the table. For very small angles, as depicted in Figure 7, the length of the side adjacent will be very close to the length of the hypothenuse. It follows that the sin of small angles will have approximately the same value as the tan of those angles. The two functions agree to one part in a thousand from 0° to 7° and to one part in a hundred from 0° to 12.° Therefore for very small angles one could substitute the sin function for the tan function with little error.

Let us see how to use trigonometry in several practical situations. A common one used in land survey finds a distance that is inconvenient to measure with a ruler or a tape. For example, in Figure 8, a person stands on the shore of a lake and wishes to know the distance across the lake to a tree on the other side. For that purpose he or she can make use of an instrument known as a transit:

Figure 7

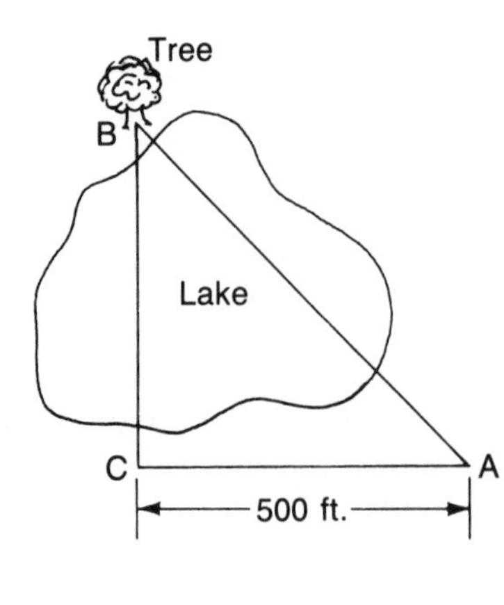

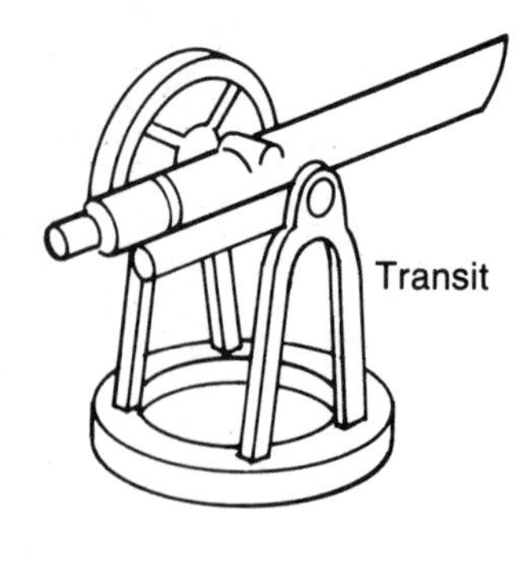

Figure 8

A transit is a small telescope mounted on a base calibrated in degrees. As one looks through the telescope, the image has superimposed on it a set of cross hairs corresponding to the pointer, which is attached to the body of the telescope lined up along the base. It is possible to adjust the position of the scale so that, no matter what direction the telescope is pointed in, the operator can arbitrarily set the calibrated scale to zero. Generally, two people are required to do land survey. To start, one person sets up the transit at C on a tripod, levels the bubble-level in the base and sights the tree. It is the distance between C and the tree (labeled B) which is to be determined. The scale is set to read $0°$. The transit is rotated through an angle of $90°$ and the second person holds up a stick or sign so that it can be seen by the operator of the transit. The stick is placed a convenient distance from the transit, say 500 feet, shown as point A in Figure 8. Note that the right angle C has been established and the distance CA has been determined. It is easy for the transit to be moved to point A, and for the other person to place the stick at C in place of the transit. The operator of the transit sights back to the stick and resets his scale to read $0°$. Without moving anything else, he or she rotates the transit and sights the tree. The angle A is read from the scale on the transit. The distance BC is determined since

$$\tan A = BC/AC; \; BC = AC \cdot \tan A$$

We have measured AC and angle A and the problem is solved.

In this appendix we have only made use of the definitions of trigonometric functions and shown that they allow us, with the postulates, to make a general table of the ratios of the sides for given right triangles. In order to generalize the notion of trigonometric function to angles greater than $90°$ and to angles considered independent of triangles, we would have to examine other areas of mathematical analysis.

Table 2: Trigonometric Table

Angle (deg)	*Sin*	*Cos*	*Tan*	*Angle (deg)*	*Sin*	*Cos*	*Tan*
0	0	1.000	0				
1	.017	1.000	.017	46	.719	.695	1.036
2	.035	.999	.035	47	.731	.682	1.072
3	.052	.999	.052	48	.743	.669	1.111
4	.070	.998	.070	49	.755	.656	1.150
5	.087	.996	.087	50	.766	.643	1.192
6	.105	.995	.105	51	.777	.629	1.235
7	.122	.993	.123	52	.788	.616	1.280
8	.139	.990	.141	53	.799	.602	1.327
9	.156	.988	.158	54	.809	.588	1.376
10	.174	.985	.176	55	.819	.574	1.428
11	.191	.982	.194	56	.829	.559	1.483
12	.208	.978	.213	57	.839	.545	1.540
13	.225	.974	.231	58	.848	.530	1.600
14	.242	.970	.249	59	.857	.515	1.664
15	.259	.966	.268	60	.866	.500	1.732
16	.276	.961	.287	61	.875	.485	1.804
17	.292	.956	.306	62	.883	.469	1.881
18	.309	.951	.325	63	.891	.454	1.963
19	.326	.946	.344	64	.899	.438	2.050
20	.342	.940	.364	65	.906	.423	2.145
21	.358	.934	.384	66	.914	.407	2.246
22	.375	.927	.404	67	.921	.391	2.356
23	.391	.921	.424	68	.927	.375	2.475
24	.407	.914	.445	69	.934	.358	2.605
25	.423	.906	.466	70	.940	.342	2.747
26	.438	.899	.488	71	.946	.326	2.904
27	.454	.891	.510	72	.951	.309	3.078
28	.469	.883	.532	73	.956	.292	3.271
29	.485	.875	.554	74	.961	.276	3.487
30	.500	.866	.577	75	.966	.259	3.732
31	.515	.857	.601	76	.970	.242	4.011
32	.530	.848	.625	77	.974	.225	4.331
33	.545	.839	.649	78	.978	.208	4.705
34	.559	.829	.675	79	.982	.191	5.145
35	.574	.819	.700	80	.985	.174	5.671
36	.588	.809	.727	81	.988	.156	6.314
37	.602	.799	.754	82	.990	.139	7.115
38	.616	.788	.781	83	.993	.122	8.144
39	.629	.777	.810	84	.995	.105	9.514
40	.643	.766	.839	85	.996	.0870	11.430
41	.656	.755	.869	86	.998	.0700	14.300
42	.669	.743	.900	87	.999	.052	19.081
43	.682	.731	.933	88	.999	.035	28.635
44	.695	.719	.966	89	1.000	.017	57.286
45	.707	.707	1.000	90	1.000	0	∞

Appendix ·2·

Kinematics: The Galilean Description of Motion

THE WORD "KINEMATICS" LITERALLY MEANS "a description of motion." It does not matter if anything in the world actually moves that way. All that is of interest when one describes a motion is to learn how a thing will move under the specified conditions. If the description of the motion is identical to a motion that occurs in nature, it is tempting to conclude that the premises in the analytical description represent the natural situation in which that motion occurs. Not all initial conditions will do. In order to be satisfying, the premises have to conform to established beliefs about the world.

It might be possible to construct other theories which predict the circumstances under which the premises in the kinematic argument come about. Such theories would be *dynamical,* and would specify under which conditions the motion would take place. Galileo's description of motion, in which the speed of an object increases in proportion to the time elapsed, is a kinematical argument. He identified the result of that analysis with the motion of objects in free fall. He never seriously attempted to account for the reason the motion occurred. Fifty years later, Newton provided a dynamical theory which explained why one can expect the speed of objects in free fall near the surface of the earth to increase in proportion to the time elapsed. That theory is the subject of a later appendix. In this appendix we will examine Galileo's description in some detail, following almost point by point his argument as it appeared in his 1638 publication, *Dialogue Concerning Two New Sciences.**

* Galileo Gallilei, *Dialogue Concerning Two New Sciences,* translated by H. Crew and A. De Salvio (New York, 1954), "The Third Day," "The Fourth Day."

The book was written near the end of Galileo's life. He was under house arrest as part of his punishment for having been found guilty in 1632 of heresy for teaching belief in the Copernican theory of the heavens. There were personal and political issues in his trial and conviction, as well.

The book was dedicated to his daughter, who was caring for him. He needed such care not only because of his age (he was in his seventies), but because he had lost his sight. The book was written in Italian, rather than Latin, the universal language of scholarship. In effect, Galileo was going over the head of the Church directly to the educated lay person.

Galileo was a masterful writer. The *Dialogue* is between three characters, Salviati, the learned, modern, natural philosopher; Sagredo, an intelligent, interested, lay person, open and questioning, seeking wisdom about the natural world; Simplicio, an Aristotelian whose theories and explanations are challenged by Salviati, almost always successfully, and with uncompromising wit and sarcasm.

In his earlier book, *Dialogue Concerning Two Chief Systems of the World,* Galileo made use of the same cast of characters. It is widely believed that one of the factors that turned Pope Urban VIII against Galileo and almost assured his conviction for heresy was that some of Galileo's enemies convinced Urban that Simplicio was a buffoon who was a caricature of the Pope.

Galileo had a pen which could be used like a deadly stiletto. He orchestrated these dialogues in order to emphasize the truth, while destroying with wit, sarcasm, and sometimes glee, theories and viewpoints he considered to be uninformed or untenable.

The *Dialogue* contains four "days" or chapters. The first two are about the first of the two new sciences: problems which we associate with "strength of materials." The third and fourth days are devoted to an analysis of motion, the second of the two new sciences. The Third Day deals with two kinds of motion: uniform motion and uniformly accelerated motion. In the Fourth Day it is the motion of projectiles that is analyzed.

The mathematical language that Galileo used was geometry. We will not follow him in this because it is clumsy and unfamiliar. We will use simple algebra and trigonometry. In Galileo's defense, it should be pointed out that algebra was not widely available in his time and he had been trained in geometry. But more important, almost no one could have read his work had he used the language of algebra.

The Third Day of the *Dialogue* begins with a summary of what will be revealed. First, Galileo tells us that, although there have been previous superficial analyses of falling bodies, as far as he knows, he is the first to point out that when a body falls from rest, the distances traversed in equal time intervals stand to each other as the odd numbers, beginning with 1. Second, whereas it has long been known that projectiles follow a curved path, Galileo claimed to be the first to say that the path is a *parabola*.

Galileo's choice of words here is important. From his other writings, it is clear that he held the role of number and geometric shape to be extremely important. He has been identified as a Pythagorean, Platonic mystic about the role of number and shape. The book of nature, he once wrote, is written in mathematical characters. Again he stated that God made all things in number, weight and measure. It is characteristic that Galileo chose to label only two of his results, one related to pure whole number and the other to perfect, geometric shape. The implication of his statement about successive equal intervals of time is that when an object falls, the distance traversed is proportional to the square of the *total* elapsed time. Galileo had emphasized the role of pure whole number in a most dramatic and startling way.

It is left as an exercise for you to prove that the sum of successive odd numbers generates the squares of all the numbers. For example, the sum of the first two odd numbers is four, the square of two. The sum of the first three odd numbers is nine, the square of three. You are asked to prove that this will be the case for all the odd numbers.

The Third Day of the *Dialogue* is divided into two parts. The first part discusses uniform motion and the second, uniformly accelerated motion. The analysis makes use of mathematical machinery developed during the fourteenth century. Although Galileo made use of an older analysis, there are crucial differences. The older analysis had been developed by scholars at Merton College, Oxford, and at the University of Paris. Two kinds of motion had been analyzed: Uniform and Difform (Accelerated) Motion. There are two types of difform motion: Uniform Difform Motion and Difform, Difform Motion. And of the Difform, Difform Motion, there are two types: Uniform Difform, Difform Motion, and Difform, Difform Difform Motion. Of the two types of Difform, Difform, Difform Difform Motion . . . and so on. Never was the question raised about the physical significance of these kinematical manipulations. Galileo, on the other hand, talked about the kinematics of motion that occurs in the world of experience and restricted himself to consideration of two kinds of motion—uniform motion and uniform accelerated motion.

UNIFORM MOTION

Galileo began by defining uniform motion: "Uniform motion is motion in which the distances traversed by a moving object during any equal time intervals are themselves equal." There follows a caution that Galileo means *any* time interval, no matter how small, for as he points out, it might be that between any two points *between* the measurements, the object might not be moving at a uniform rate and we would have measured the *average* speed of a non-uniformly moving object. Galileo appears to have been struggling with one of the core ideas of the differential calculus, a subject that was developed fifty years later by Newton and by Leibnitz. Galileo did not have the language to express these ideas easily and they never emerge from his work explicitly.

Following the definition, there are four postulates:

I. In the case of one and the same uniform motion the distance traversed during a longer time interval is greater than the distance traversed during a shorter time interval.
II. During one and the same uniform motion, the time required to traverse a greater distance is longer than the time required for a lesser distance.
III. In one and the same time interval, the distance traversed at a greater speed is larger than the distance traversed at a lesser speed.
IV. The speed required to traverse a longer distance is greater than the speed required to traverse a shorter distance during the same time interval.

You might wonder why Galileo spelled out four such obvious postulates. The reason is that Galileo's proofs were geometric and he was forced, when considering motion, to compare lengths to lengths, times to times and distances to distances. These four postulates were required for Galileo to proceed with his general proofs of uniform motion. In discussing uniform motion, there are three variables, the distance traversed, the speed (constant for any given motion), and the time elapsed. In order to demonstrate how these variables relate one to the other, Galileo treated each as fixed and then compared the ratio of the other two. Three variables taken, two at a time, can be arranged in six different ways and in this section of the Third Day there are six theorems. We will not follow Galileo in those proofs. Mathematically, the definition for uniform motion can be stated as

$$\Delta x = v(\Delta \mathrm{t}) \quad (v \text{ constant}) \tag{1}$$

We see (Figure 1) that this is the equation of a straight line passing through the origin. Equation (1) embodies all six of Galileo's theorems on the subject of uniform motion. It is almost assured that had algebra been the language of discourse for these problems in Galileo's time, he would not have bothered to prove his six theorems.

It is important to realize that in equation (1) *all* of the properties of the equation of a straight line can be given a physical interpretation. Thus each point on the abscissa represents a particular time interval (Δt), and each point on the ordinate represents the distances traversed (Δx). The speed v is not represented in the graph

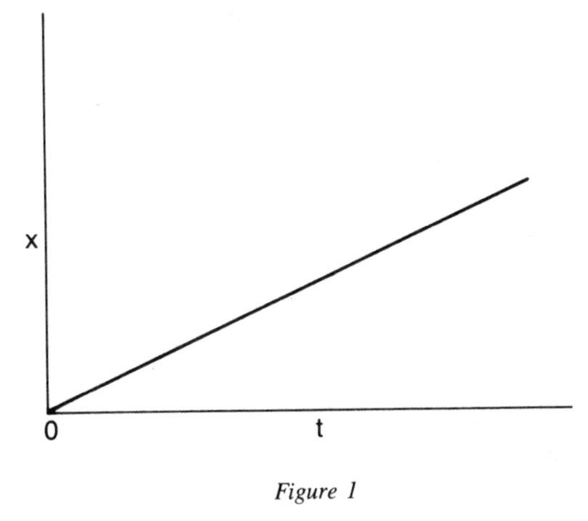

Figure 1

in Figure 1, however, examining the equation shows that v plays the role of the slope of the line.

Equation (1) and the graph in Figure 1 are not the most useful ways of depicting uniform motion. Δx represents some distance interval $x_1 - x_0$; Δt represents some time interval, $t_1 - t_0$, where x_1 is the position of the object at some time t_1 and x_0 is the initial position of the object at the time t_0, the start of the measured interval. Very often it is convenient to set t_0 equal to zero, and even to set x_0 equal to zero. Substituting into (1):

$$x_1 - x_0 = v(t_1 - t_0); \; x = x_0 + vt \tag{2}$$

The subscript has been dropped from the final position because we have implicitly set $t_0 = 0$.

Term by term, the physical meaning of equation (2) can be obtained by comparing it to the equation of a straight line, $y = mx + b$. x is the final position of the moving object. x_0 is the initial position and x is represented by the y intercept. v plays the role of the slope m. The graph for two different motions is represented in Figure 2. In the first case (I), at $t = 0$, the object was at the origin. In II, at $t = 0$, the object was not at the origin but at some other point x_0 from whence it moved to x_1 in the time t. (It is important to bear in mind that we have set $t = 0$ and that actually the variable t is $\Delta t = t - 0 = t$.)

There is another way to graphically represent the distance traversed by an object at constant speed. It will be extremely useful later. Rather than graphing the distance traversed against the time, in Figure 3 we have graphed the speed versus the time. As the speed is constant, the graph will be a straight line parallel to the t axis. What if we were interested in how far the object had traveled between two

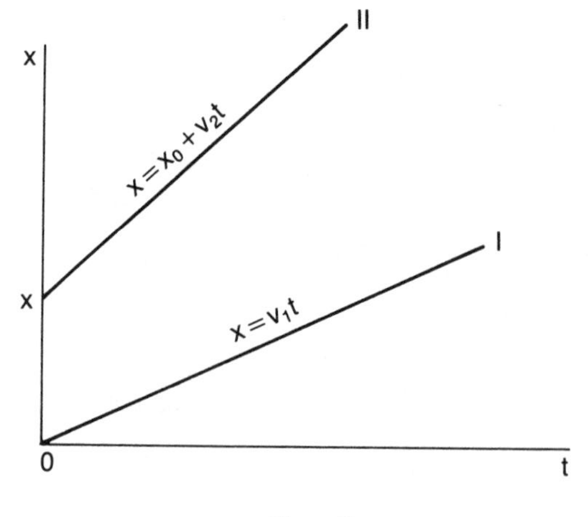

Figure 2

times, say t_1 and t_2? Equation (1) can be used and the distance traversed would be $\Delta x = v(t_2 - t_1)$.

In Figure 3 we have erected verticals at t_2 and t_1 to intercept the straight line representing the speed. Examining the equation and comparing it to the figure we see that the product $v(t_2 - t_1)$ is the shaded area in the figure. That area is the rectangle bounded by the times and a portion of the straight line representing the speed. The dimensions of that product are $[L]/[T] \cdot [T] = [L]$. It seems reasonable to allow the *area* of the rectangle under the curve to represent the distance traversed over a specified time interval. This is an abstract use of the concept of area, but in principle it is no more abstract than allowing the symbol "t" to stand for time or "x" to represent a position, or the symbols "Susan" to stand for a person.

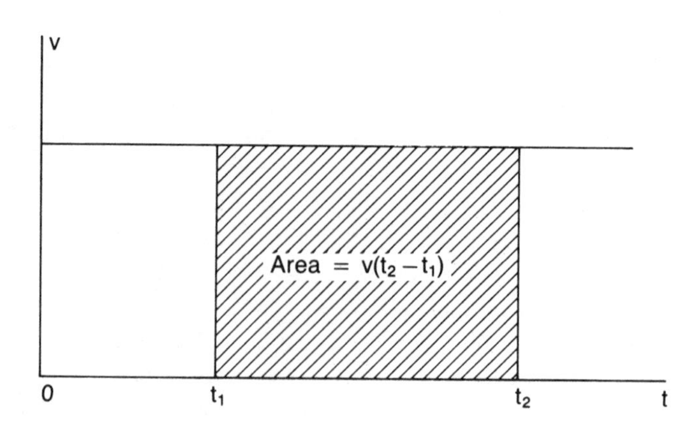

Figure 3

UNIFORMLY ACCELERATED MOTION

Galileo's title for this section of his work is "Naturally Accelerated Motion" by which he means the motion of freely falling bodies. The section opens with a general preface in which Galileo writes that, if the speed of an object in free fall increased in proportion to the time, it would be the simplest way for nature to arrange things because in equal increments, the same effects would occur.

Objections are immediately raised by the Aristotelian, Simplicio, and the novice, Sagredo, that should the postulate be invoked then in starting from rest, an object would move through all values of speed, however small. At the start, as the speed will be proportional to the time elapsed, were the object to move at such a slow speed, it would require a very long time to traverse any distance. Yet, when we actually observe falling bodies, they seem to acquire great speed in virtually no time at all.

In his response, Galileo proposed an experiment that he never actually performed. Reading the description of it, one concludes that it would have had the outcome claimed for it, and would have given credence to Galileo's claim. Such descriptions have been given the name "thought experiments." Galileo loved to propose them. But more to the point, thought experiments are crucial to physical reasoning.

Galileo's thought experiment is based on the premise that the effects of a falling body are due, not only to the body's weight, but to the speed acquired by the body as it falls. Thus, if a large weight like a boulder falls from a great height, it can force a large spike into a wooden block. Should the boulder fall from only half the height, its speed will be less and the spike will not penetrate as far. Should the boulder fall from a height half as far again, the effects on the spike will be still less. But since the weight of the boulder has not changed, the diminishing effects on the spike must be the result of the diminishing speed of the boulder. Thus, if the height fallen is infinitely small, the speed will be infinitely slow. This is the result that Galileo wanted. The argument appears to be flawless and convincing.

Galileo turned to an even more convincing argument: the symmetry between bodies falling and bodies being thrown up to high heights. He argued that when an object is thrown upward against the force of gravity, the object gradually slows down and moves slower and slower until finally it comes to rest. Simplicio objected to this argument on the grounds that near the end of the motion an object moving infinitely slowly would never arrive at its destination and would never come to rest. In his response Galileo, through the voice of Salviati, made the crucial point that Simplicio's argument would be true if the object continued to move at a very, very slow speed, but that the object never has any speed in the sense of retaining it. Rather the object *passes through* all values of speed from its initial high speed when it leaves the hand of the thrower to zero when it comes to rest in the instant it reverses itself and begins to fall back to earth. The argument anticipates the mathematical limit so crucial to the development of differential calculus.

When Sagredo suggests that perhaps this would help to understand the cause of gravity, Simplicio objects because to throw an object is violent motion and that is different from the natural motion of free fall. Sagredo's attack on the traditional distinction between natural and forced motion is interrupted by Salviati who says flatly that it really doesn't matter at this point what the cause of gravity is. For now, he says, let us be satisfied to investigate and demonstrate some of the properties of uniformly accelerated motion and show that this is the way things fall in nature; that the speed increases in proportion to the time elapsed.

The informed layman, Sagredo, says that it would have been clearer to him had Salviati suggested that the speed of an object increases in proportion to the distance traversed. This is more intuitive. Salviati, always sympathetic and understanding to Sagredo, responds first by saying that it is comforting to have such a companion in error because that is what he once thought. But the notion can be discarded after a moment's thought.

Salviati's argument is disarmingly simple. If the speed of an object is proportional to the distance traversed, an object must traverse these different distances in the same amount of time. For example, if the speed with which an object traverses eight feet were twice that with which it traversed the first four feet, then the time intervals to traverse the two distances would be the same. But for a body to fall four and eight feet in the same time is only possible if there is a sudden and instantaneous increase in the motion. We have seen that that cannot be true.

At this point, Galileo could not resist giving a dig to those who held him under house arrest. Sagredo remarks ruefully that Salviati seems to be able to handle these difficult questions with remarkable ease. Such easily acquired knowledge is not often held in high esteem. Salviati's reply is worth quoting:

> . . . It is very unpleasant and annoying to see men who claim to be peers of anyone in a certain field of study, take for granted certain conclusions which later are quickly and easily shown by another to be false. I do not describe such a feeling as one of envy, which usually degenerates into hatred and anger against those who discover such fallacies; I would call it a strong desire to maintain old errors, rather than accept newly discovered truths. The desire at times induces them to unite against these truths, although at heart believing them, merely for the purpose of lowering the esteem in which certain others are held by the unthinking crowd. Indeed I have heard from our Academician many such fallacies held as true but easily refutable. . . .

Modern commentors often point to that passage when speaking of the claim that Galileo was the first experimental scientist, or the first modern scientist. As we have said, several others were working along the technical lines developed by Galileo, and furthermore, earlier scholastics had been more insistent on literal readings of the evidence of the senses than Galileo and most others who have followed him. Passages like this, while pointing up Galileo's sagacity and courage, demonstrate how people who hold different values interact in power situations. The modernity of the passage can be used as an indication of the degree to which people of the sixteenth century were like people today. They knew about the same *amount* of things that we know, they knew *different* things because different things were important to them. They argued in the same way people argue today. The forces available to them, especially on an organized national level were different, and this difference is in no way trivial. But that is a matter of technology, not of change in human character or human intellectual capacity.

Salviati then began the quantitative study of uniformly accelerated motion with the following definition:

> An object is said to be uniformly accelerated when its speed receives equal increments in equal times.

In other words, if v is the speed of the object and t represents the time,

$$\Delta v \propto \Delta t; \qquad \text{or} \qquad \Delta v = a(\Delta t) \tag{3}$$

where a is a constant of proportionality which we identify as the acceleration.

Salviati answers that in addition to this definition, our Author makes one assumption, namely:

> The speeds acquired by one and the same body moving down planes of different inclinations are equal when the heights of these planes are equal.

By height of the plane, Galileo meant the vertical height of the plane. For example, in Figure 4, there are two inclines (generally, the inclines will be the hypothenuses of right triangles), AC and DC; nevertheless, the height of these inclines is the same, the vertical distance CB. The assumption states that the speed of an object, say a metal sphere, rolling down the incline will be the same at the bottom, whether the ball rolls down AC or DC. Furthermore, this would be the speed that the object acquired would it have fallen from C to B.

When Sagredo indicated that should one be allowed to consider friction non-existent, the assumption would appear reasonable and probable, Salviati replied that "I hope by experiment to increase the probability to an extent which shall be little short of a rigid demonstration."

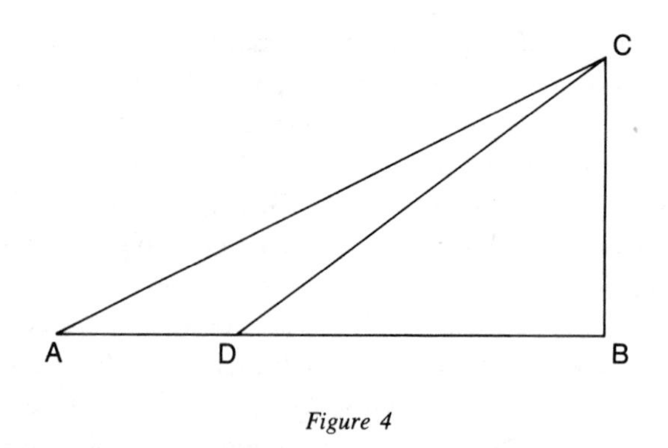

Figure 4

That is a most curious statement. First, if the claim that the speed of objects descending different inclines from the same vertical height is the same, is to be demonstrated by experiment, why was it necessary for Galileo to state it as an assumption? The answer is that the statement cannot be verified by experiment. The speed of an object at the bottom of the plane cannot be measured. At best, one might make distance and time measurements which would give information on the *average* speed at the bottom of the plane. Galileo warned us about confusing average speed with instantaneous speed. Furthermore, the experiment that Galileo proposes to make the assumption reasonable is a *thought* experiment. It is not a thought experiment about the behavior of objects on inclines. Rather it is a thought experiment about the probable behavior of a pendulum.

In Figure 5, we have shown a pendulum swinging around the point A in the plane of the page. Suppose that we draw the pendulum out to C and release it.

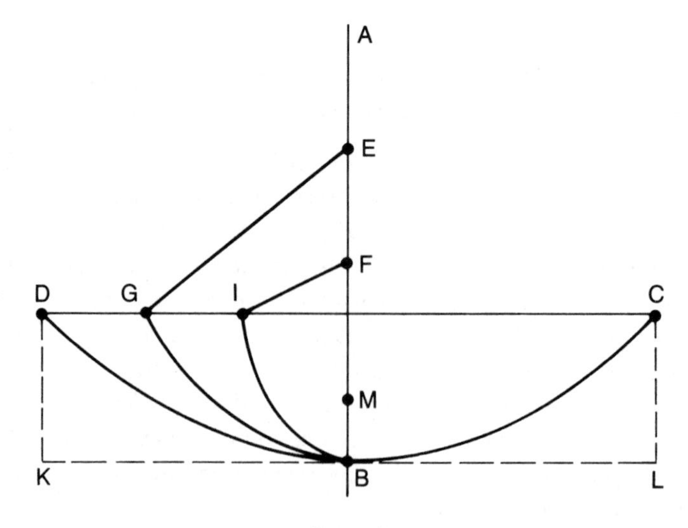

Figure 5

Barring frictional effects and air resistance, the pendulum will gain enough speed in falling from C to B to climb to D on the other side. The vertical height DK equals the height CL. And barring interference from outside influences, the pendulum will continue on this path forever. The bob of the pendulum is a falling body acted upon only by the gravitational force and constrained to move along the arc of a circle with radius of length AC = AB = AD. Suppose that we place a pin at point E along the line AB. When we draw the bob up to point C and release it, the path of swing is the same as it was until it reaches point B. When the cord encounters the pin at E, however, the pendulum bob is no longer able to follow path BD. It now acts as a pendulum with cord of length EB and moves along the arc of a circle defined by BG. Similarly, if the pin is moved to F, the pendulum moves along the path BI. In each case, barring frictional or resistance losses, the points G and I lie along the line DC. If one places the pin so low that the cord and bob cannot reach the height defined by line DC, the cord begins to wrap itself around the pin.

There is nothing to prevent us from running this thought experiment backwards. Rather than drawing the bob of the pendulum to C, we can, initially, draw it to D. It will climb to C on the other side. If we place the pin at E and draw the pendulum bob up to G, as the pendulum falls the pin will release the cord when the bob arrives at B and the pendulum will climb on the other side to C again. Similarly should the pin be placed at F, and the bob drawn out to I, the bob would still climb to C on the other side.

Imagine now that we replace the pendulum *arcs* with the chords of those arcs. This is shown in Figure 6. Moving along the inclines DB, GB, IB, the end object acquires the same speed as evidenced by the fact that it climbs to the same vertical height at C on the other side.

We have made an enormous inferential chain of reasoning. We began by considering objects falling. We moved to the motion of objects descending along inclines and from there to objects swinging at the end of flexible cords in order to conclude that in all cases, the same speed is acquired in descending from the same height regardless of the path if friction and resistive losses are ignored. It is a clever,

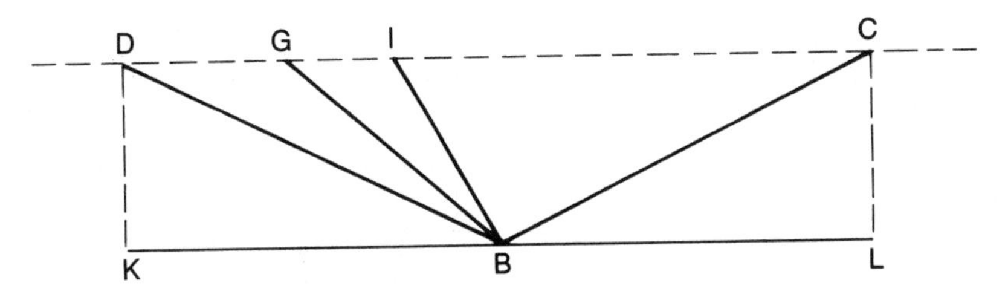

Figure 6

even brilliant, argument and one that is characteristic of the relationship between experiment and belief. In high energy physics, for example, the concept of "detecting" a particle has become enormously abstract. Very often two detectors located as a theory dictates, are connected to an electronic device which emits a signal if signals arrive from the detectors simultaneously. One of the detectors, however, has been connected to the device with a delay line that retards the arrival of the signal. Simultaneous arrival of the signals at the device means that the events occurred a certain time interval apart. When that happens, theory dictates that a certain particle must have decayed or broken apart and created other particles. "Detection" of these so-called daughter particles in geometric and temporal relationship to each other is signaled by the simultaneous arrival of information.

Galileo could show that defining uniform acceleration as in equation (3) led to the conclusion that the distance traversed is proportional to the square of the time elapsed. He argued that that is the relationship between distance and time found in the descent of objects on the incline.

As stated earlier, the mathematical proof (if A then B) makes use of mathematics developed during the fourteenth century. The analysis allows one to treat uniformly accelerated motion—as defined in equation (3)—as if it were uniform motion. This allows a solution of the problem of dealing with instantaneous speed without using the calculus that was not introduced until fifty years after Galileo's death. It is sometimes called the mean speed theorem or the Mertonian rule. The proof we provide below is adapted from that given by Oresme (d.1382) and is essentially the proof Galileo provided in his first theorem of the section on accelerated motion. Figure 7 is a graphical representation of equation (3) in which the object starts from rest at time $t = 0$.

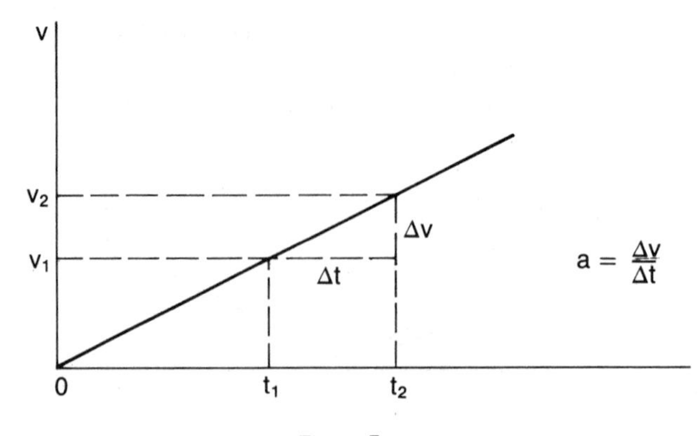

Figure 7

$$v = at \qquad t_0 = v_0 = 0 \tag{4}$$

Note that we are plotting speed on the ordinate and time on the abscissa. This is the equation of a straight line. The speed increases in proportion to the time and the slope of the line, a, is the acceleration,

$$\frac{\Delta v}{\Delta t}$$

Earlier, we found the distance traversed under the assumption of constant speed by taking the product of the speed and the time elapsed. That is not possible in this case because the speed is constantly changing. However, the graphical concept employed earlier, that the distance traversed be represented by the area under the curve bounded by the time interval and the speed, still has validity as the space has not changed. Comparing Figure 3 to Figure 7, we see that the abscissas and ordinates are the same.

The area under the curve in Figure 8 is given by the area of the right triangle $0,v_f,t$, where v_f represents the final velocity at the end of the time period. But the area of a triangle is no more than half the product of the base and the height. Therefore, the area under the curve is given by $v_f t/2$. The space can be called "velocity-time space", or, v-t space for short. The area under the curve in Figure 8 can represent the distance traversed, but it is the distance traversed when the speed is changing at a constant rate.

$$x = ½\, v_f t. \qquad v_0 = t_0 = 0 \tag{5}$$

It is here that the mean speed theorem is useful. In Figure 9 we have restated the situation for uniform acceleration in v-t space. Let us consider the speed the object has at a point midway during its motion. In other words, t_m is half the value of t. At that point the speed of the object will be v_m as shown in Figure 9. Two triangles

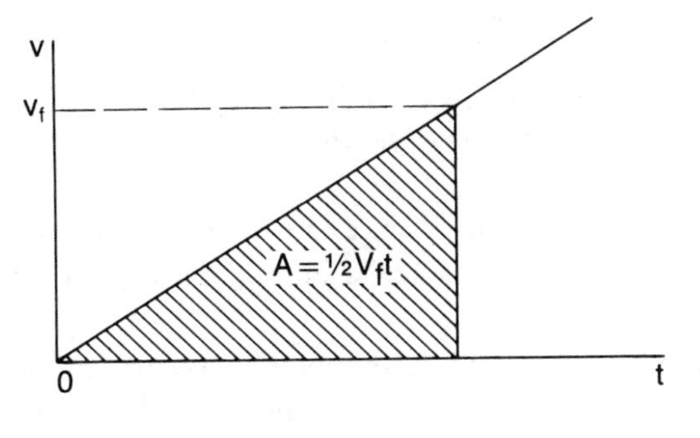

Figure 8

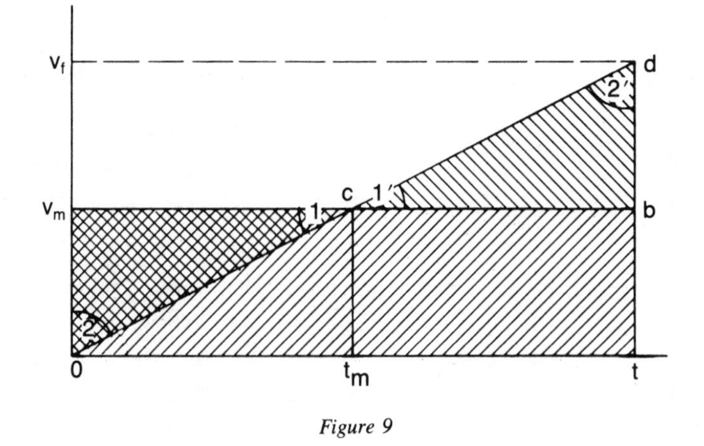

Figure 9

have been created. Triangle Ocv_m is identical to triangle cbd for the reasons: Each has a right angle (by construction), angle 1 = angle 1′ (they are formed by the intersection of two straight lines) and angle 2 = angle 2′ (alternate interiors, or because the sum of the angles of the two triangles is 180°). But, as we have chosen t_m to be the midpoint between the origin and t, the length $v_m c = cb$, so the triangles are identical. That being the case the line Ov_m must be half the line Ov_f. In other words, for uniformly changing motion, the speed of the object at the midpoint in *any* time interval is halfway between the initial and final speed of the interval. Since the change in the speed is constant, the speed at the midpoint is the mean or average speed for the interval. Since triangle cdb is identical to triangle $cv_m O$ they have the same area. Imagine that we begin with the triangle Odt. By removing the area cdb and placing it at Ocv_m, we transform the triangle into a rectangle $v_m btO$ of the same area. The area of triangle Odt is identical to the area of the rectangle $Ov_m bt$. The physical interpretation of this analysis is as follows:

The Distance an Object Traverses During Any Period of Constant Acceleration is the Same as the Distance it Would Have Traveled Had Its Speed Been Constant and Equal to the Average or Mean Speed for the Time Intervl.

This is the result in equation (5) since ½ v_f is the mean speed of the interval in which the object begins from rest and acquires a final speed v_f in a certain time interval.

We can analyze the situation just a bit more to express the distance traversed in terms of the acceleration. Since the change in speed, Δv (in this case $v_f - 0$) is nothing more than the product of the acceleration and the time elapsed—the defining equation for consant acceleration, equation (3)—equation (5) can be manipulated in the following way:

$$x = \tfrac{1}{2} v_f t$$

But

$$v_f = at$$

therefore:

$$x = \tfrac{1}{2}at^2 \qquad v_0 = t_0 = 0 \tag{5a}$$

What we have shown is that if one assumes that the speed changes in proportion to the time elapsed (constant acceleration) then the distance traversed should be proportional to the square of the time elapsed. This is what Galileo set out to prove.

Let us derive the law geometrically and algebraically in a general fashion. In Figure 10, we have shown the line describing equation (3),

$$\Delta v = a\Delta t; \; v = v_0 + a(t - t_0).$$

The y intercept is v_0. The area bounded by t and t_0 is the area of a trapezoid. To find the magnitude of that area is easy.

As shown in Figure 11, it is the sum of the area of a rectangle of area $v_0(\Delta t)$ and a triangle of area $\tfrac{1}{2}(\Delta v)(\Delta t)$. The area of the triangle is the same as the area of

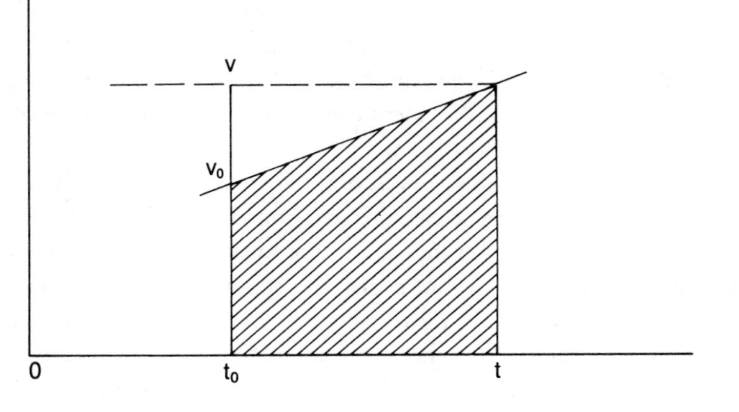

Figure 10

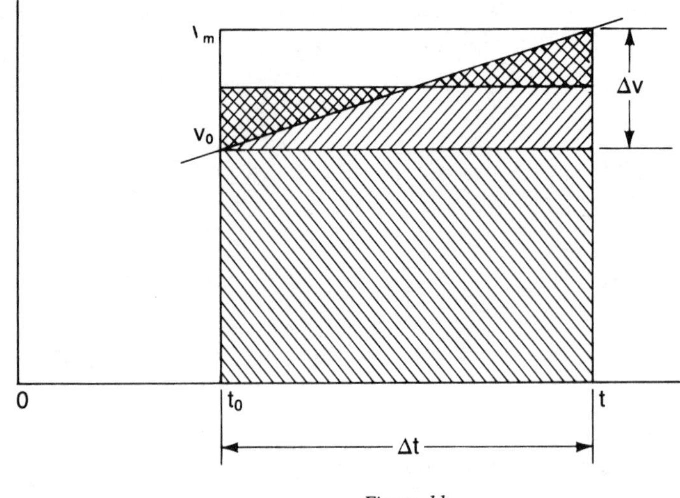

Figure 11

another rectangle of area $v_m(\Delta t) = \frac{1}{2}\, a(\Delta t)^2$. Since the area represents the distance traversed, we can write this analytically as follows:

$$\Delta x = v_0(\Delta t) + \tfrac{1}{2}a\,(\Delta t)^2$$

and if we assume that $x_0 = t_0 = 0$, it becomes

$$x = v_0 t + \tfrac{1}{2}at^2 \tag{6}$$

The physical interpretation is straightforward. The first term on the right of (6) represents the distance the object would have moved had its speed been constant, that is, had the object not accelerated. However, the object accelerated and covered an additional distance given by $\frac{1}{2}\, at^2$—that is, the distance it would have covered had it no initial speed at the beginning of the acceleration. When $v_0 = 0$, equation (6) reduces to (5a). A comparable interpretation can be given to the areas shown in Figure 11.

Let us summarize the primary results of the analysis. We began with equation (3), the definition of uniform acceleration and from it derived equation (6) which relates the distance traversed in free fall to the square of the time elapsed. The analysis shows that if one assumes that the speed of an object increases in proportion to the time elapsed, the distance traversed will be proportional to the square of the time elapsed (If A then B). Since we have two equations with three variables (the distance, the time and the speed), we may solve the two equations simultaneously to eliminate the time. We will have three equations that are considered to be the three fundamental equations of uniformly accelerated motion:

$v = v_0 + at$	(definition of uniform acceleration)	
$x = v_0 t + ½\, at^2$	(logical outcome of definition)	
$v^2 = v_0^2 + 2ax$	(simultaneous solution of the first two equations eliminating the time variable)	(7)
	$x_0 = t_0 = 0$	

It is an exercise for the reader to derive the third equation of (7).

The drawings we have used are turned upside-down relative to those that Galileo employed. This is not trivial. For example, Figure 9 is reproduced in 12. It is the same drawing, turned 180.° Galileo wanted to discuss free fall and to make the proposition that the descent of an object down an incline is a special case of free fall. The drawing in Figure 12 has nothing to do, except in a suggestive way, with the descent of objects down inclined planes. The line *Od* represents the *speed* of the object at any time. Such suggestiveness is part of the polemical aspect of Galileo's argument. It is his polemical style that made the treatise effective.

Following the demonstration that one can expect the distance traversed under constant acceleration to be proportional to the square of the time, there follows a corollary to demonstrate that the statement that the distance traversed is proportional to the square of the time is equivalent to the statement that in successive units of time, the distances traversed stand to each other as the odd numbers. This is proven in Figure 13 although it must be emphasized that Galileo did not use the

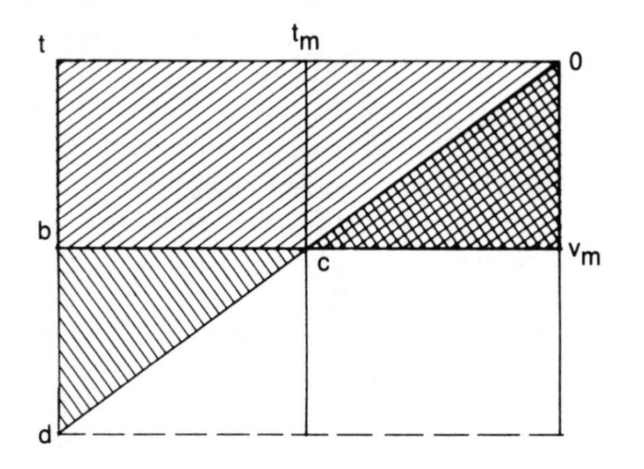

Figure 12

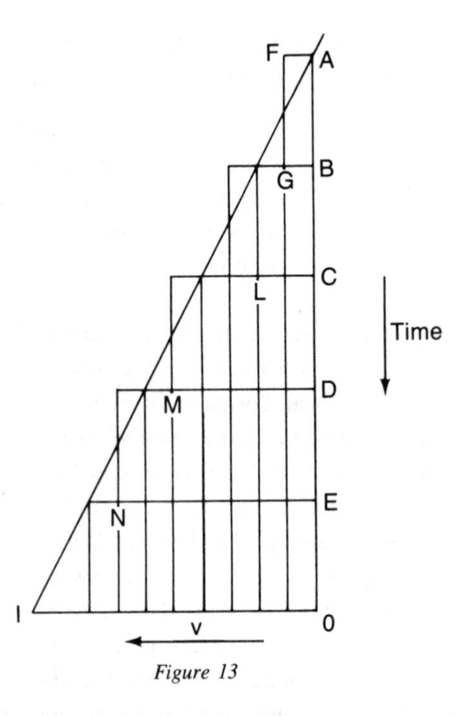

Figure 13

abstraction that the area under the curve could be used to represent distance. AI represents the speed of an object in *v-t* space. ABCDE is the time on the ordinate. Using the mean speed theorem ABFG represents the distance traversed in the time AB. In an equal time BC, the area is three times as great. In an additional time CD the area is five times as great, then seven times as great, etc. Thus in successive time intervals, the distances traversed stand to each other as the odd numbers. At the end of the first second, let us suppose that the acceleration ceases. The object will continue to move at the speed acquired during the first second so that during the second second, and each succeeding second thereafter, the distance traversed will be represented by the rectangle BKLC, twice ABGF. However, during that second the object continued to accelerate and covered an *additional* distance equal to ABFG. This argument can be made time interval after time interval, however small, as long as the intervals are equal.

At this point in the dialogue, Galileo has the Aristotelian, Simplicio, intrude with a challenge. "Indeed," Simplicio says, "I am very pleased with the mathematical demonstrations which have been provided. They are beautiful, they are clear. However, I am very doubtful that they have anything to do with any motion that one finds in nature. You seem to imply that there are many experiments which you have performed to demonstrate the truth of this. It seems to me that now is the time to tell us about them."

Salviati agrees. Experiment is the custom,

> . . . in those sciences where mathematical demonstrations are applied to natural phenomenon as is seen in the case of perspective, astronomy, mechanics, music and others where the principles once established by well chosen experiments, become the foundations of the entire superstructure.

This statement should be studied carefully, for in it, Galileo understood and pronounced the relationship between experiment and theory in any science. It is also significant that the call for experimental evidence in this dialogue is made by the Aristotelian. The evidence of the senses was an important concern for Aristotelian science. It was not so important to take the results of experiment as literally within the framework of Galilean or Newtonian science. The experiment that Salviati describes to test the new law of free fall is the measurement of the time required for objects to descend an inclined plane.

> A piece of wooden moulding or scantling about 12 cubits long, half a cubit wide, and three finger-breadths thick, was taken; on its edge was cut a channel a little more than one finger in breadth; having made this groove very straight, smooth, and polished, and having lined it with parchment, also as smooth and polished as possible, we rolled along it a hard, smooth and very round bronze ball. Having placed the board in a sloping position, by lifting one end some one or two cubits above the other, we rolled the ball as I was just saying, along the channel, noting in a manner presently to be described the time required to make the descent. We repeated this experiment more than once in order to measure the time with an accuracy such that the deviation between two observations never exceeded one-tenth of a pulse beat . . . .

Salviati says that, having done this measurement for one distance, they repeated it, many times, for other distances. (A cubit was a common unit of length). But clearly something is bizarre here. After all, what is a tenth of a pulse beat and how do you measure it? And is the pulse so regular that you would use it as a clock in an experiment which required total accuracy? Of course, in the absence of a clock, the pulse beat would have to serve, but that was not the case with Galileo. He informs us after the description that all of the measurements were made with an accurate water clock and that when the experiment was made many, many times, there was "no appreciable discrepancy in the results."

The reference to a tenth of a pulse beat is an example of the polemical nature of the argument. Galileo is intent on making his case to people who were not accustomed to working in the laboratory. Clocks were not common. The water clock measured time intervals by comparing the weight of water collected from a tall, usually conical vessel with a tiny hole in the bottom to allow water to drip out at a slow, regular rate. It could be calibrated in familiar units of hours, minutes and seconds by comparing the amount of water obtained to a regular period of the motion of the sun. (The clock is called a *clepsydra,* the Greek name for the device.) Some of the water clocks used by Galileo still exist and are on display at the Museum for the History of Science in Florence. Although Galileo had very fine water clocks, the claim that the time required for a ball allowed to descend an incline over a given distance did not vary by more than one tenth of a pulse beat on successive descents is hard to believe. If the heart is beating with great regularity at seventy beats per minute, the time between two pulse beats is somewhat less than a tenth of a second. A tenth of a pulse beat would be about a hundredth of a second. Galileo claimed accuracy of about one part in a hundred. Yet, today in beginning physics laboratory exercises, this experiment is repeated with smooth steel channels for the incline, steel ball bearings whose sphericity surpasses anything Galileo could have imagined, and electronic clocks which split the second into a thousand parts; even so, results to the degree of accuracy that Galileo claimed are difficult to obtain.

This is not to say that Galileo did not do the experiment. There is no question of that. Recently, data that Galileo collected were discovered among his papers in his own hand.* But what relationship does the performance of the experiment have to the claim that in free fall an object's acceleration is constant and the same for all objects? In the years immediately following Galileo's death when he was being canonized by romantic supporters as the first experimental scientist, it was believed that Galileo had demonstrated the law of free fall by experiment. For a time, in the early twentieth century when a reaction set in to the notion that measurement was the sole mark of doing science, several historians of science suggested that Galileo never did any experiments, that he invented results which he knew, from his intuitions, must be true.

Whereas there is no question that Galileo did the experiments, they do not give us the answer to the question: how do things accelerate when they fall? We measure neither speeds or accelerations, we measure distances and times. Any number of postulates might have been created to explain the results Galileo obtained. And if he has had a little fun with us by overstating the degree to which he reproduced his results, we can understand it because of his enthusiasm, his commitment and his passion.

The result of the experiment, Salviati assured his audience, was that for an incline which made a given angle with the horizontal, the distance traversed down the incline starting from rest was always proportional to the time elapsed, or in

* For a summary see Stillman Drake, *Galileo at Work: His Scientific Biography,* (Chicago, 1978) p. 85ff.

successive intervals of time, the distances traversed stand to each other as the odd numbers beginning with 1.

We have seen that the result can be accounted for by assuming that the speed of the object increases in proportion to the time, although such an assumption is not entailed. The next question is, what does motion down an incline have to do with free fall?

We can only summarize Galileo's argument. Interested readers may follow the details in Galileo's own words.* Using the premise that, regardless of the angle or length of an incline, the speed at the bottom is the same, as long as the vertical height of the inclines are equal, Galileo argued that the incline did nothing but dilute the effects of gravity.

PROJECTILE MOTION

This is the subject of the Fourth Day of the *Dialogue Concerning Two New Sciences*. We will not follow Galileo's treatment of the subject as closely as we have the problem of uniform accelerated motion.

The first theorem of the Fourth Day states that

> A projectile which is carried by a uniform horizontal motion compounded with a naturally accelerated vertical motion describes a path which is a semi-parabola.

This is easy to prove although, as with all other theorems, Galileo used only geometry. If Euclid was the authority for most subjects in geometry, Apollonius (third to second century B.C.) was the recognized master of conic sections. So the mathematics of the parabola was well known in geometrical terms.

We will first analyze the meaning of the above theorem. We will consider the motion of an object composed of two other motions: horizontal and vertical. By motion we mean "velocity" since we are speaking of motion in certain directions. So the meaning of "compounded" in the theorem (for which we used the term "composed") is to take the vector sum of two velocities. One of those velocities is in the horizontal direction and one in the vertical. The theorem states as a condition that the motion in the horizontal direction will be uniform. In each equal time interval, the distance traversed will be the same and the equation of such motion is

$$\vec{x} = \vec{v}t \qquad x_0 = t_0 = 0 \tag{8a}$$

* Galileo, *Two New Sciences,* "The Third Day," pp. 178–188.

We have plotted this in Figure 14 where the uniform motion is depicted across the top of the figure. In equal time intervals the object would traverse equal distances.

In the vertical direction, however, the motion is not uniform but uniformly accelerated. Of course an object cannot move in two directions at the same time. So the net velocity of the object is the vector sum of the horizontal and the vertical velocities. The vertical distance will be given by the equation:

$$\vec{y} = \tfrac{1}{2}\,\vec{a}t^2 \qquad y_0 = v_0 = t_0 = 0 \tag{8b}$$

We have chosen the symbol $\vec{y}$ to represent vertical distance, perpendicular to $\vec{x}$, the horizontal distance. We have plotted the vertical motion of the object along the ordinate of the Figure 14. Note that, as always for uniform acceleration, the distances traversed in successive seconds stand to each other as the odd numbers.

Yet as we have said, the object does not move in two directions at the same time. We are considering an object whose motion is *composed* of two components: a vertical uniform accelerated motion and a horizontal uniform motion. At any time the object must be identical with itself. Mathematically, this means that the time

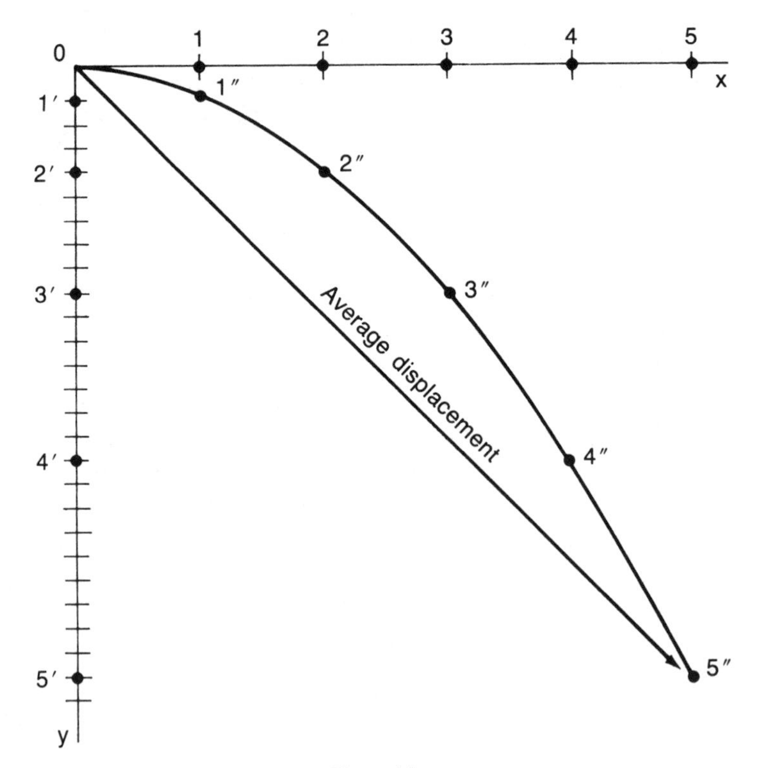

Figure 14

variable in equations (8a) and (8b) are the same. We can, therefore substitute for t in (8b) $x/v = t$ from (8a)

$$\vec{y} = \vec{a}x^2/2\ v^2 \tag{8c}$$

This *is* the equation of a parabola. We are determining the vertical distance as a function of the horizontal distance. So whereas equation (8b) is also the equation of a parabola, that equation relates the vertical distance to the time. Relating the vertical distance to the horizontal distance, equation (8c) is a statement of how the object will move in two dimensional space; the equation states that the trajectory will be a parabola. This is shown in Figure 14 in the same way as in Figure 8 of Chapter 1. Had the object not been accelerated in the vertical direction, it would have arrived at positions 1, 2, 3, 4, etc. On the other hand, had the object not been given a uniform horizontal motion, but simply accelerated, it would have arrived at positions $1'$, $2'$, $3'$, $4'$, etc. But neither is the case and instead of arriving at 1 or $1'$, the object arrives at $1''$; instead of arriving at 2 or $2'$ it arrives at $2''$, etc. We can see that the trajectory has the features of a parabola. The vectors drawn between points 0 and $1''$, $1''$ and $2''$ represent the *average* displacement during each successive time interval. To find the velocity of the object in any time interval, one need only divide that average displacement by the time. So the average velocity during each of the time intervals is the same as the depicted vectors and the average total velocity from the start would be the sum of those individual vectors.

GALILEAN TRANSFORMATION EQUATIONS

Suppose a person is sitting on a train and throwing a ball straight up in the air. We have already analyzed that motion and shown that it can be described in the equations in (7). On the other hand, to an observer on the embankment, the situation on the train cannot be so described. The observer on the embankment will agree with the observer on the train that certain events occurred. The two observers would agree that the ball left the hand of the thrower and would also agree on the occurrence of the event, "ball returns to hand of the thrower." They would not agree on the place where those events occurred nor on the trajectory of the ball. The description of the motion in the two frames of reference is shown in Figure 15. As we have indicated in the text the description of motion between two inertial frames of reference is given by the Galilean transformation equations. There are no bumps nor jossles, simply uniform motion in the same line. We have shown two such frames of reference in Figure 16. One of these frames of reference might be the embankment and the other the train. The observer on the train is always at the origin of his or her coordinate system, 0.′ In his or her frame of reference, $0'$ would need only one clock to time the flight of the ball, since the ball

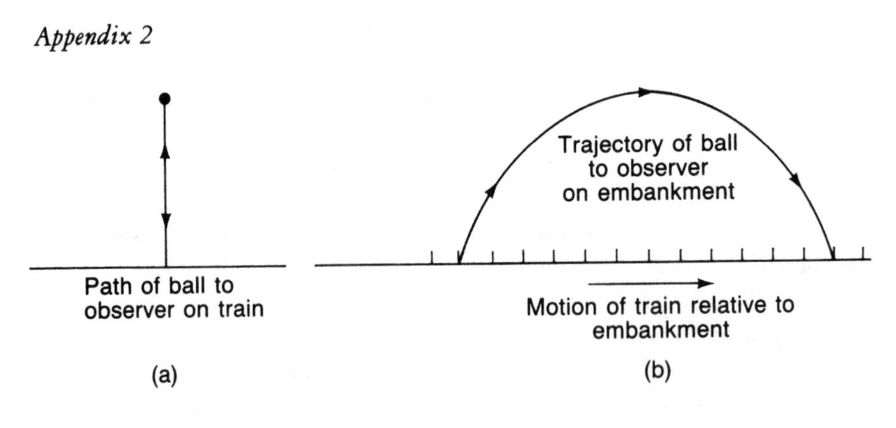

Figure 15

leaves his or her hand at $0'$ and returns to it at $0'$. But in the other frame of reference, the position of $0'$ changes. Let us assume that at time $t = 0$ the two observers, (with their x axes lined up along the direction of relative motion between the frames of reference, which, for the sake of argument, have virtually no separation between them) set their clocks to zero when the origins coincide. From that time, the following equations apply:

$$\begin{aligned} x' &= x - vt \\ y' &= y \\ z' &= z \\ t' &= t \end{aligned} \tag{9}$$

The origin of the primed frame of reference is a distance vt from the origin of the unprimed frame of reference at any time after they synchronized their clocks. If the flight of the ball is observed in the unprimed frame of reference, a series of

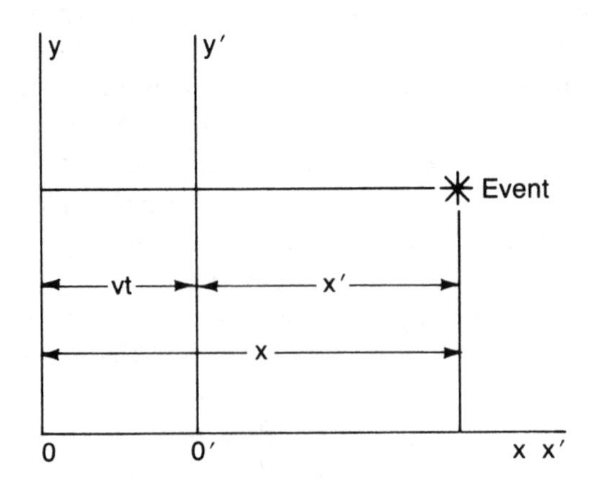

Figure 16

observers, each armed with clocks, is required. Everyone will agree that in only one frame of reference is it possible to make measurements of the time of flight of the ball with one clock. That is the frame of reference in which the events occur at the origin. In all other frames of reference, everyone agrees that more than one clock is required to measure the time interval of the events. From the point of view of the observers in the unprimed frame of reference, the ball has a vertical component of initial velocity as a result of being thrown in the air by the person on the train (at $0'$) and a horizontal component of velocity, constant and unchanging, resulting from the relative motion of the two frames of reference. In all other inertial frames of reference except the primed frame, the motion of the ball is a parabola. The exact shape of the parabola depends on the relative state of motion of the frames of reference. For frames moving very rapidly relative to each other, the horizontal displacement is very large. For frames moving slowly with respect to each other the horizontal displacement is very small. Although the description of the trajectory of the ball is different in different frames of reference, different observers will agree to disagree. They recognize that the disagreement arises from the relative motion of the frames of reference within which they describe the events. On the other hand, everyone agrees that only one observer made time measurements using one clock, that the ball left the hand of the person at $0'$ at a certain time and that it returned to the hand of $0'$ at a certain time. All of these agreements would be invariant. All observers also agree on the relative speed of the different frames of reference.

There is a special name given to the kinds of measurements of time that we have discussed. When measurements of the time between two events are made with one clock, it is said to be a *proper* time measurement. *Proper* is used in a technical sense and means "using one clock." All other measurements of the time between two events are said to be *nonproper*.

In classical mechanics, where the description of events in different frames of reference makes use of the Galilean transformation equations, we see that free fall is a special case of the general description of projectile motion. Free fall is the case in which there is no relative motion perpendicular to the direction of gravity between the observer and the projected object.

Appendix ·3·

Newtonian Mechanics

In this appendix I develop the basic outline of Newtonian mechanics with more mathematical description than in the text and show how, from the definitions and the premises, typical laws can be derived. The treatment is not historical. It does not represent how Newton developed his theory nor how it was elaborated on by those who came after him. The emphasis is on the relationship between the dynamical parts of the theory and the related descriptions of motion. At the same time, the treatment will help the reader to get a more complete sense of the power of the mechanical world view. In Appendices 4 and 5, the Newtonian analysis is applied to two cases, the kinetic theory of matter and the analysis of "ether drift" experiments.

PRIMITIVES

The following notions will be undefined: Mass, Length and Time. While descriptions associated with their meaning may be provided and recipes for measuring them and comparing them quantitatively will be specified, the concepts themselves are beyond discussion. Either you know what I am talking about or you don't. In context there is rarely a problem in communication just as there is none about the perception of the color of objects.

Length: Words that are associated with the concept of length are "extension," "distance," "span," or "stretch."

Time: The sense of time is a non-spatial continuum related to the following transitions: before and after; earlier and later; past, present and future; birth, growth and death.

Mass: The quantity of matter; associated with, but not the same as weight. It is that property of a body which gives it inertia.

It is virtually impossible to talk about these concepts. It is much easier to define how to measure a length, a time, or a mass. This gives rise to a view held by some scientists and philosophers that the meaning of the concepts is conferred by specifying how measurements are performed. We will discuss the measurement of mass, length and time in the section "Theory of Measurement."

DEFINITIONS

Velocity—Velocity is the ratio of the length traversed to the time required to traverse that length. Velocity is a vector quantity having both magnitude and direction. A recipe for determining velocity is provided in "Theory of Measurement," below. The symbol for velocity will usually be $\vec{v}$.

Speed—Speed is the magnitude of the velocity. The symbol for speed will usually be v.

Acceleration—Acceleration is the time rate of change of the velocity. It is a vector quantity. A recipe for determining the acceleration is provided in "Theory of Measurement," below. The symbol for acceleration will usually be $\vec{a}$.

Momentum—Momentum is the product of the mass of an object and its velocity. It is a vector quantity. The symbol $\vec{p}$ is usually used and thus $\vec{p} = m\vec{v}$, where m is the mass.

Density—The density is the ratio of the mass of an object to the volume occupied by the object. Volume is a measure of space and has dimensions of $(\text{length})^3$. The symbol for density is ρ, and $\rho = m/V$, where V is the volume.

Other quantities will be defined as the need arises.

THEORY OF MEASUREMENT

ABSOLUTE SPACE AND ABSOLUTE TIME

Physical events take place in absolute space and time. Absolute space is totally uniform. It is assumed that its properties are described by the postulates of Euclidean geometry (see Chapter 1). Contrary to personal perceptions and to experience with mechanical, electric, or even electronic clocks, absolute time is completely uniform and accurate. There is no way to measure position in absolute space or to know absolute time. The measures that are used for spatial and temporal coordinates are relative to arbitrary starting points. Measurements of time intervals are made using an arbitrarily chosen periodic process assumed to be uniform. Absolute motion would be the absolute time rate of change of position in absolute space. But since there is no way of making measurements of absolute time and space, absolute motion is not directly measurable. The measurements of time, space and motion that will be defined are, therefore, all relative.

LENGTH

There are many ways of measuring a length. Only one or two are specified here, but the premise is that different techniques lead to equivalent results.

"Proper length" is measured by comparing the ends of the object to the extension along a chosen standard length or a copy of that standard, when the standard and the object are at rest relative to each other. Other ways of making a proper length measurement involve the techniques of trigonometry as described in Appendix 1.

"Nonproper" lengths are measured when the object is in motion relative to you. One technique is to locate two observers whose position is virtually coincident with the ends of the object at the same time that it moves past them (Figure 1). The technique assumes that each observer has a clock synchronized to other clocks at rest with respect to him or her. Another way to measure the length of an object, in motion relative to an observer, is to use only one clock, and to time the intervals required for the ends of the object to go by. The length is determined by finding

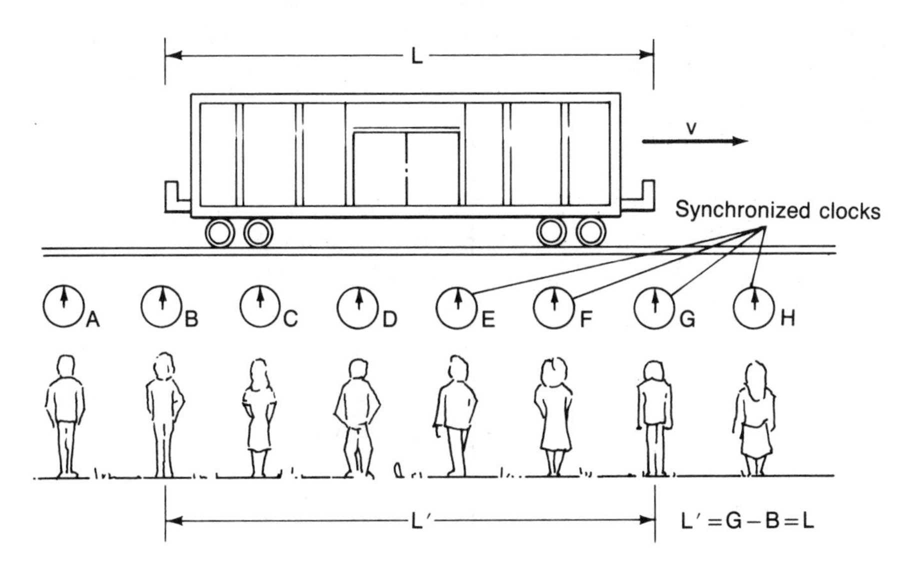

Figure 1

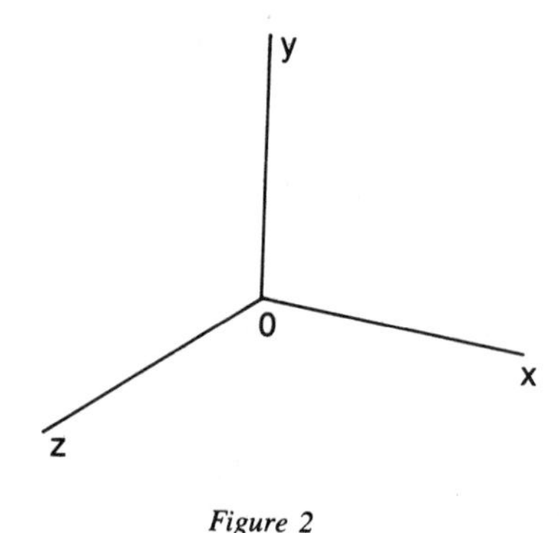

Figure 2

the product of the time and the speed with which the object moves relative to the observer.

In order to measure length and compare different lengths either at rest or in motion relative to an observer or a set of observers at rest relative to each other, it is useful to define a frame of reference or coordinate system, as in Figure 2. Three axes are erected in the three mutually perpendicular directions of space. By custom these three directions are labeled x, y and z. The three axes meet at 0, the origin. Lengths are measured by using a standard or a copy of a standard to mark distances from the origin. Such a coordinate system can be constructed at any point in space, in any orientation.

Such a coordinate system can be defined with other parameters, for example, a radius and two angles, (spherical coordinates), a radius, a height and an angle (cylindrical coordinates), etc. Relating different but equivalent coordinate system

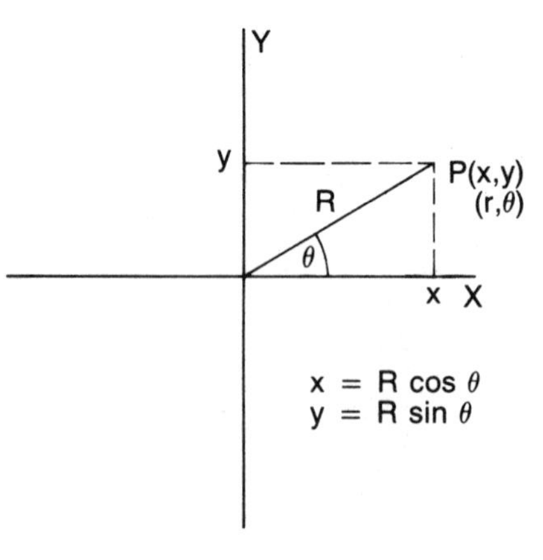

Figure 3

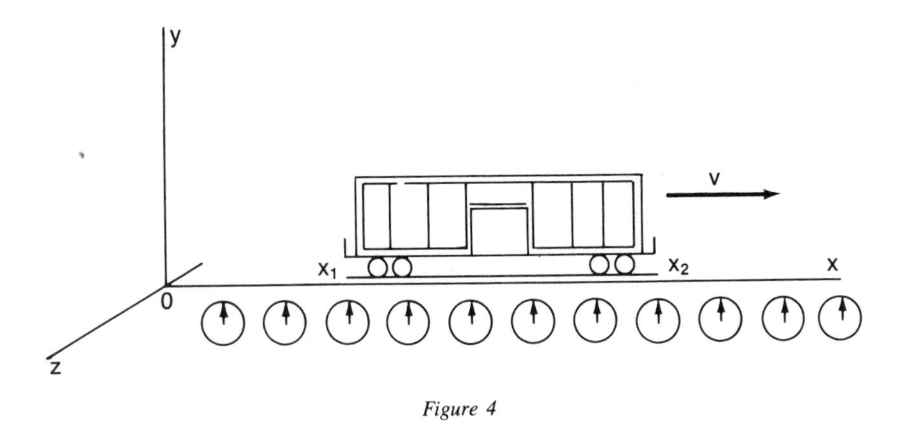

Figure 4

descriptions requires transformation equations. For most purposes we will use rectilinear coordinates as depicted in Figure 3.

To make a proper length measurement, a coordinate system is constructed so that one of the axes is parallel to the object. As depicted in Figure 4, the ends of the object coincide with two points, x_1 and x_2. And, within a frame of reference proper lengths can be determined by making use of the rules of trigonometry.

To make a nonproper measurement of length, suppose that the x coordinate in Figure 4 is lined up with observers, each carrying a clock synchronized with each other and running at the same rate. At a prearranged time, observers directly opposite the ends of the moving object identify their location as x_1 and x_2. The length being determined is then $x_2 - x_1$. The relative motion between the frame of reference of the object and the observers need not be uniform for this technqiue.

TIME

We have already spoken of time measurements with clocks. There are indirect ways of measuring time if other parameters are measured or calculated. For example, one method is related to the determination of a length by measuring the time for the ends of an object to pass a point when the object is in uniform motion. The length is calculated by taking the product of the time and the speed. If one knows the length of the object, the time required for the object to pass a point is the ratio of the length to the speed of the object relative to the observer.

Very often it is convenient to depict time as one axis of a coordinate system. Thus, in Figure 5, the vertical axis (ordinate) is used to measure time and the horizontal coordinate (abscissa) represents one spatial dimension. Curve A represents a point on an object at rest with respect to the spatial coordinate. The other

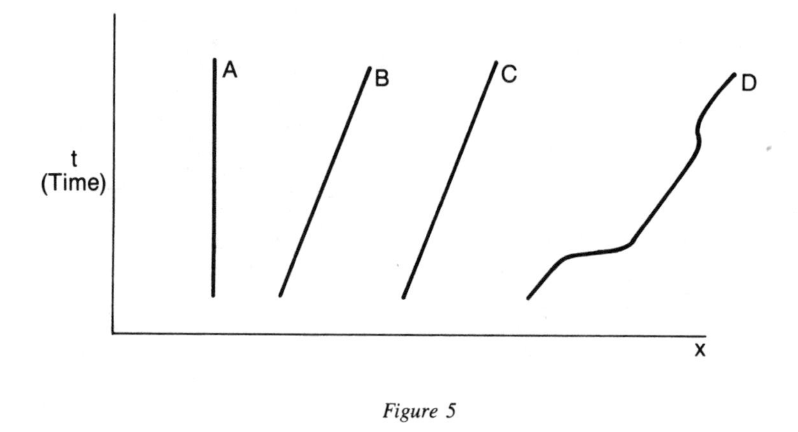

Figure 5

curves depict motion. For example, curves B and C represent the motion of the ends of a rod. Note that no curve in this space has a negative slope since that would represent time going backwards. It is also significant that no curve in the figure is horizontal.

MASS

Earlier, we referred to the mass as the quantity of matter. Here, we define the measurement of that quantity. Measurement of mass is defined by the interaction of any two bodies. The interaction can be due to a collision or to the temporary or permanent bonding of the bodies. The bodies are isolated from all other bodies in the universe. Two such bodies might be billiard balls colliding in a frictionless environment, or two iceskaters struggling against each other at the ends of a rigid, weightless pole. Whatever the physical situation, the *ratio* of the mass of two such bodies will always be the same:

$$m_1/m_2 = k_{12} \tag{1}$$

where k_{12} is a constant for those two bodies which can *always* be interpreted as the ratio of two other numbers: a_2/a_1. (The a's will represent the acceleration of the two bodies). If I pick a mass as a standard, equation (1) becomes $m_s/m_1 = k_{s1} = a_1/a_s$.

By making use of such a definition, I can consistently and independently define the magnitude of all masses, although the kinds of experiments I have described are, in practice, not possible.

The units of mass, length, and time have been discussed in some detail in Chapter 1. Let us turn now to the determination of the magnitudes of other physical quantities. The only kinds of direct measurements made in the physics here are measurements of mass, length and time. All other quantities are derived by calculation. This is true even when instruments are calibrated in terms of a derived quantity. For example, speedometers of cars are labeled in terms of speed, but they measure distances travelled per unit of time. A mercury thermometer is calibrated in degrees but it is the length of the column of mercury which is being measured. A spring balance is calibrated in terms of force, but the length of the extended spring is being measured.

In Chapter 2 and Appendix 2, the measurement of velocities, speeds, and accelerations were discussed. We note that if one makes a proper measurement of length in determining the speed of an object, a nonproper measurement of time must be made. If one makes a proper measurement of time, a nonproper measurement of length must be made. Note also that in speaking of uniform speed or velocity, one is forced to extrapolate between measured points.

THE VELOCITY ADDITION LAW AND THE CLASSICAL PRINCIPLE OF RELATIVITY

The classical velocity addition law is, formally, a direct consequence of the Galilean transformation equations, which have been discussed in the text and at the end of Appendix 2. In Figure 6a I have shown two coordinate systems moving uniformly with respect to each other. One frame of reference is designated by primes. I assume that the primed frame is moving to the right as viewed from the unprimed frame of reference. In Figure 6b, the problem has been simplified. We have eliminated one of the dimensions perpendicular to the direction of motion.

If we concentrate on the measurement of coordinate points in the frame, and if we assume that at time coordinate $t = t' = 0$, the origins of the two systems

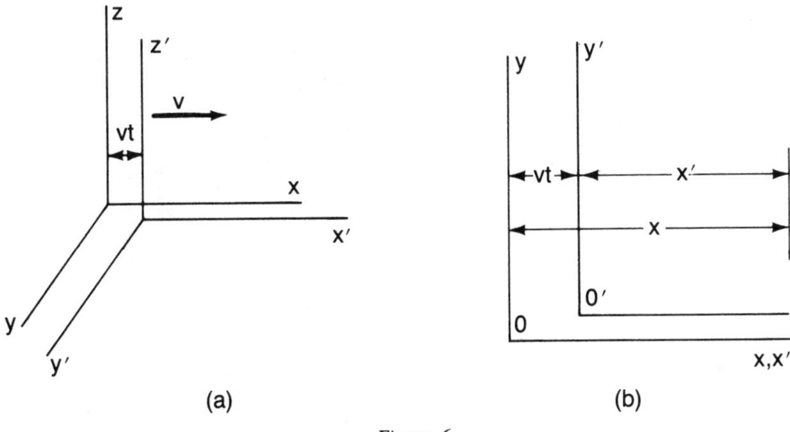

Figure 6

coincided and clocks in both frames were synchronized and set to zero, then at any other time

$$x' = x - vt$$

where v is the relative speed between the primed and unprimed frames of reference. Time coordinates and spatial coordinates perpendicular to the direction of motion have the same values in both frames of reference. These relationships between measurements of coordinates in the two frames of reference are summarized by the following set of equations:

$$\begin{aligned} x' &= x \pm vt \\ y' &= y \\ z' &= z \\ t' &= t \end{aligned} \qquad (2)$$

The plus sign is used in the first equation when the primed frame is moving in the negative x direction relative to the unprimed frame, and the negative sign is used when the primed frame is moving in the positive x direction.

Suppose that we wish to determine a length in the two frames of reference when the object is at rest in relation to the unprimed frame. The observers in the unprimed frame of reference make a proper time measurement. In general there are three ways for the observers in the primed frame of reference to proceed. Suppose that an observer at a particular place in the primed frame of reference made some observations of the time required for the ends of the rod to pass the point, x'_1, at which he or she is located. Furthermore, suppose that the object being measured is moving to the right relative to him or her. Then, at any time, t_a,

$$x'_{1a} = x_a - vt_a$$

and at time t_b,

$$x'_{1b} = x_b - vt_b$$

Since the observer has not moved, x'_{1a} and x'_{1b} are the same. If we take the difference between these two equations, $x_a - x_b = v(t_a - t_b)$ or

$$\Delta x = v(\Delta t)$$

In other words, the observer in the primed frame has used the product of the time elapsed for the object to pass him and the uniform speed between frames of reference to calculate the length of the rod.

The second technique is to locate observers simultaneously in the primed frame of reference who are positioned opposite the ends of the object as it moves past. For example, suppose that we agree that, at a particular time, those observers will indicate their x' coordinate. Those two observers would, then, be singled out from all other observers. At the instant chosen, t,

$$x_1' = x_1 - vt$$
$$x_2' = x_2 - vt$$

It follows that the length $x_2' - x_1' = x_2 - x_1$ or that $\Delta x' = \Delta x$.

In the third technique, no attempt is made to coordinate or synchronize measurements. At time t, the observer in the primed frame opposite the front end of the rod notes his or her coordinate position, and at time 2 a different observer in the primed frame notes his or her coordinate position opposite the rear end of the rod.

$$x_1 = x_1' + vt_1'$$
$$x_2 = x_2' + vt_2'$$

Taking the difference between the two yields:

$$x_2 - x_1 = x_2' - x_1' + v(t_2' - t_1') \quad \text{or} \qquad \Delta x = \Delta x' + v(\Delta t') \tag{2a}$$

The length can be determined by an observer in the primed frame of reference by subtracting the distance $v(\Delta t')$ from his measured value of the displacement of his frame, because the observer recognizes that his measurements were not made at the same time and the ends of the object moved *during the course of the measurement*.

Suppose that we wish to determine the speed of an object in both frames of reference. As the object traverses a certain distance, Δx, in the unprimed frame of reference in a certain time interval, it is measured as traversing a certain distance, $\Delta x'$, in the primed frame of reference in the same time interval, since $t = t'$. From equation (2a)

$$\Delta x/\Delta t = \Delta x'/\Delta t' + v(\Delta t')/\Delta t'$$

Or

$$\Delta x/\Delta t = \Delta x'/\Delta t + v \tag{2b}$$

Whether the sign of velocity (v) is positive or negative is determined by the relative direction of motion of the two frames of reference. If the velocity $\vec{W} = \Delta\vec{x}/\Delta t$ and $\vec{U} = \Delta\vec{x}'/\Delta t$ then (2b) becomes:

$$\vec{W} = \vec{U} + \vec{v} \tag{3}$$

This is the classical, or Galilean, velocity addition law, which states that the velocity measured in the two frames of reference in any direction will be the sum or difference of the velocity measured in one frame and the relative velocity between the frames. For example, if a plane is flying at 600 miles per hour relative to the ground, and a ball is thrown from the rear to the front of the plane at fifty miles per hour, an observer on the ground would measure the speed of the ball as 650 miles per hour. The formal relationship coincides with our intuitive sense of what the result will be.

When frames of reference are moving uniformly with respect to each other, they are said to be inertial frames of reference. It is important to note that, even if the observers in two inertial frames of reference disagree on the numbers arrived at, they would agree that if an object were moving uniformly as measured in one inertial frame of reference, it would be measured as moving uniformly in the other inertial frame.

The acceleration is defined as the time rate of change of the velocity of an object. Suppose that an object is accelerating in one of the frames of reference. In equation (3), $\vec{W}$ is no longer constant. Instead:

$$\Delta\vec{W}/\Delta t = \Delta\vec{U}/\Delta t' + \Delta\vec{v}/\Delta t'$$

but as $\vec{v}$ is the unchanging, uniform speed between the frames, the second term on the right of this equation is zero. If we use the symbol, $\vec{a}$, for the acceleration, the last equation becomes:

$$\vec{a}' = \vec{a} \tag{4}$$

In other words, whatever acceleration is measured in one frame of reference, all other inertial frames of reference will measure the same acceleration in that direction.

All of these results are derived directly from the Galilean transformation equations and summarize how observers in different frames of reference compare measurements. The result of equation (3) can be stated as follows: Transformation of velocities between inertial frames of reference is covariant. While observers might disagree about the numbers, they would agree on whether or not the velocity being measured is uniform or non-uniform. The result of equation (4) can be stated as follows: Transformation of accelerations is invariant between inertial frames of reference. In other words, not only will all inertial observers agree that the object is accelerating, they will also agree on the magnitude of the acceleration.

The Galilean transformation equations implicitly contain Newton's assumptions about the relationship between absolute and relative space and time. Although we can never know the absolute quantities, we can utilize the relative quantities to determine such things as uniform motion and acceleration and we know that the measurement of a relative acceleration is the same as the measurement of an absolute acceleration.

NEWTON'S LAWS OR AXIOMS

Newton's laws are actually postulates and are not amenable to test. Let me repeat that. Within the framework of Newtonian physics, Newton's laws cannot be tested. They are *assumed*. The historical roots of the postulates can be traced, but that is not the same as testing them. We measure mass, length and time. There is no way to measure directly quantities such as velocity, acceleration, or force. For convenience the postulate of universal gravitation has been included, although Newton stated it as a separate premise.

I. Objects in uniform motion will remain in uniform motion in the same direction in space unless acted upon by a net, external force.
II. The net force on an object is equal to the time rate of change of the momentum of the object.
III. When two objects interact, if object A exerts a force on object B, object B exerts an equal and opposite force on object A.
IV. Every object in the universe attracts every other object in proportion to the product of their masses and inversely proportional to the square of the distance between them.

Axiom I is sometimes referred to as the law of inertia. An object moving inertially is the origin, in effect, of an inertial frame of reference. How would an observer decide if he or she were in an inertial frame of reference? The only criterion is to test whether or not the law of inertia applied to objects within the frame. Another way of stating the law of inertia is to say that for an object at rest or moving inertially, the sum of the external forces on the body is zero:

$$\Sigma \vec{F} = 0.$$

A literal mathematical statement of axiom II as we have stated it would be as follows:

$$\sum \vec{F} = \frac{\Delta \vec{p}}{\Delta t} \tag{5}$$

where $\vec{p}$ is the momentum. But we have defined the momentum as the product of the mass of an object and its velocity. Whereas one can change the velocity of an object, that is, accelerate it, the mass is considered a propety of the body and not subject to change unless the nature of the body itself is changed by breaking up. Therefore, equation 5 can be rewritten as follows:

$$\sum \vec{F} = \frac{\Delta(m\vec{v})}{\Delta t} = \frac{m(\Delta \vec{v})}{\Delta t} = m\vec{a} \tag{5a}$$

Very often, in dealing with axiom II, or the second law as it is often called, we will drop the summation sign and write, $\vec{f} = m\vec{a}$; it is assumed that it is the net force referred to. Bear in mind that we measure mass, length and time. We do not *measure* directly the force exerted by bodies on each other.

There are ways of building objects for the purpose of indicating directly the amount of force being exerted on a body. One technique is to make use of Hooke's law. Robert Hooke was a contemporary and rival of Newton. Hooke showed that, when a net force is applied to the ends of a coil spring, the stretch of the spring is proportional to the magnitude of the force. Double the force and the change in dimensions is doubled. If the force is too great, the spring will stretch beyond its ability to be restored to its original state and will be permanently distorted. In such a case, the elastic limit of the spring is said to be exceeded. The relationship between the force and the extended length of a spring is depicted in Figure 7. The equation describing the physical situation is

$$\vec{F} = k(\Delta\vec{x}) \tag{6}$$

where $(\Delta\vec{x})$ represents the displacement of the spring from its equilibrium length. The constant k is called "the force constant" and is a number which is characteristic of any spring.

How might we calibrate such a spring? Consider an object, having mass m. Suppose, as in Figure 8, we attach a spring to the object, and by pulling on the opposite end of the object, accelerate it a constant amount. As long as the object is being accelerated by that amount, the spring is stretched to the same degree. Although it might be difficult to achieve in practice, such an experiment would

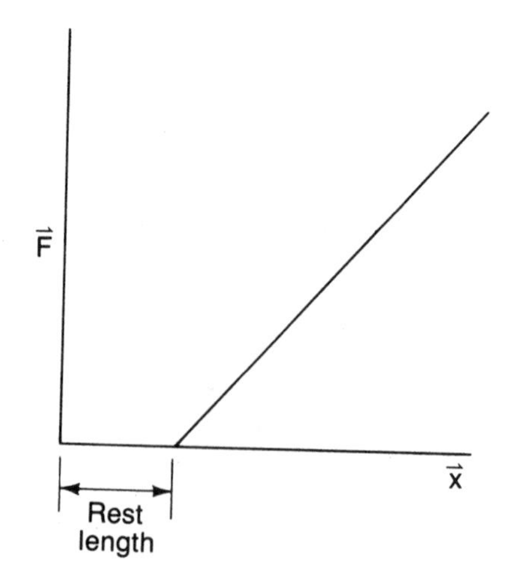

Figure 7

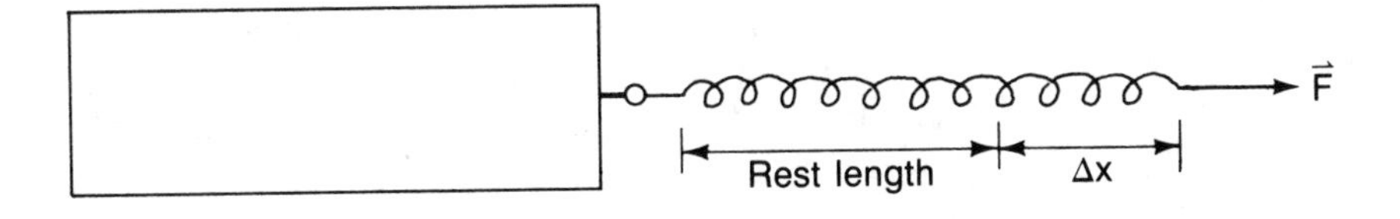

Figure 8

allow calibration of the spring by direct application of Newton's second law. In carrying out this procedure, the mass of the object is previously determined by making use of the operations specified in the theory of measurement. The acceleration of the object is determined by measuring its displacement over time. The stretch of the spring is measured by using any reasonable technique to measure length. Combining equations (5a) and (6)

$$\vec{F} = m\vec{a} = k(\Delta\vec{x})$$

Initially, to calibrate the spring we determine k.

$$k = \frac{m\vec{a}}{(\Delta\vec{x})}$$

The spring can now be calibrated directly in units for force.*

Axiom III or the third law, as it is usually referred to, is subtle and the most often misunderstood of Newton's axioms. It is important that the system in which it is being applied be clearly identified and the bodies be properly isolated.

For example, consider the system referred to in Chapter 2: a horse and a cart. We have depicted them in Figure 9a. To many people, the application of the third law in this situation is to say that regardless of the force exerted by the horse on the cart, the cart will exert an equal and opposite force on the horse and, therefore, how can a horse pull a cart? In applying Newton's laws, one must isolate the system and apply the forces involved to the objects. For this problem, it is sufficient to consider both objects as point masses. They are shown in Figure 9b. Let us examine the cart first.

The Force $\vec{F}_h$ is exerted by the horse on the cart. We will not consider the reaction to that force now, because it is not a force on the cart, but exerted *by* the

* The *unit* of force in the MKS system is as follows: When a one kilogram mass is accelerated by a force to one meter per second square, the force is said to equal 1 Newton. In the CGS system, a force of one dyne is the force exerted on a one gram object to accelerate it to one centimeter per second square. It is simple to show that one Newton contains 10^5 dynes. In the English system, the unit of mass used to be "the slug." It is now referred to as the pound-mass. Therefore, the pound force is defined as the force required to accelerate one pound-mass one foot per second square.

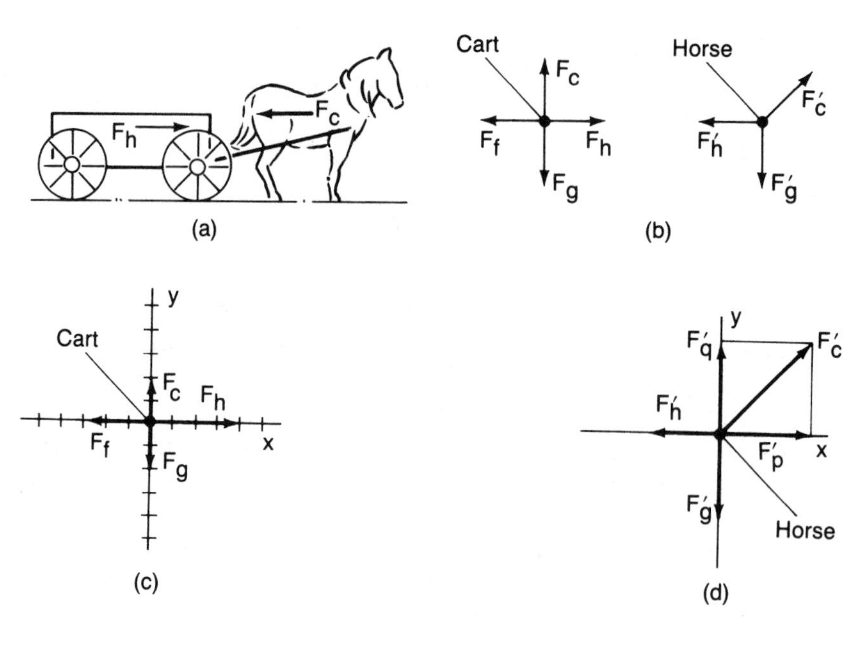

Figure 9

cart *on* the horse. In addition to $\vec{F}_h$, there are three more forces acting on the cart. The first is the frictional force, $\vec{F}_f$ in an opposite direction to the direction of that that the cart will move, the gravitational force of the earth on the cart, $\vec{F}_g$, perpendicular to the two other forces, and a force in reaction to the cart pressing on the surface of the earth, a force by the earth on the cart, $\vec{F}_e$ in the direction opposite to the gravitational force.

We have abstracted the problem in Figure 9c, where the forces are placed in a two dimensional Cartesian space. It is convenient to analyze the forces along each axis thus: As long as the cart does not leave the surface of the earth, $\vec{F}_e$ equals $\vec{F}_g$. In other words, the net force in the *y* direction, perpendicular to the surface of the earth, is zero. The only requirement to move the cart is that $\vec{F}_h$ be greater than $\vec{F}_f$. These forces are independent of each other. The horse need only pull hard enough to overcome the frictional forces. One can write that $\vec{F}_h - \vec{F}_f = m_c\vec{a}_c$; the subscript *c* refers to the cart. What about the horse? We have shown the forces on it in Figure 9b. As with the cart, these forces have been redrawn in a two dimensional Cartesian system in Figure 9d. $\vec{F}'_h$ is the reaction force to $\vec{F}_h$, that exerted by the cart on the horse.

As with the cart, there is the gravitational force upon the horse $\vec{F}'_g$. In reaction to the weight of the horse pressing on the surface of the earth, the earth exerts a force on the horse, $\vec{F}'_e$. In Figure 9b this force is not perpendicular to the surface of the earth. In 9d it corresponds to not being perpendicular to the x axis. When an animal or a person walks he or she does not push his or her foot straight down. He or she leans forward, lifts one foot forward and pushes back at an angle on the surface of the earth. The earth pushes back along the same line. It is this reactive force that is depicted in Figures 9b and 9d as force $\vec{F}'_e$. With the rules of analytical geometry, it can be resolved into two components. One along the x axis is $\vec{F}'_p$ and the other along the y axis is $\vec{F}'_q$. For the horse to move it must push on the surface of the earth with a force, $\vec{F}_e$, large enough so that the component of the reaction to that force $\vec{F}'_p$ is greater than $\vec{F}'_b$, the force exerted by the cart on the horse, although it is dangerous to anthropomorphize a problem, think of your natural inclination when you have difficulty pulling an object. In addition to straining your muscles, your tendency is to lean farther from the vertical. The effect is, in Figure 9d, to increase the magnitude of the component $\vec{F}'_p$.

We have gone through this analysis in detail because it is typical of the strategy that one employs in using Newton's laws. It is as though nature presents a puzzle and you must discipline yourself to solve it within the rules of Newton's axioms.

Our first step in this process was to abstract the objects to independent masses, sometimes called free body diagrams. This procedure is only allowed, strictly speaking, if the line of action of the forces passes through the center of mass of the body. If that condition is not met, then there is the possibility that the body will not only translate, but rotate as well. We will forgo that aspect of Newtonian analysis here and turn to the relationship between Newton's laws and the evidence provided by Kepler's laws for the motions of the planets, and Galileo's laws for motions near the surface of the earth. It is useful to prepare by analyzing the motion of objects moving in uniform circular motion.

UNIFORM CIRCULAR MOTION

I begin with a simple kinematical question: suppose an object were to move in a circle at a uniform speed. How, if at all, would it accelerate? We have shown such an object, a point mass, in Figure 10a. At any instant, the *velocity* of the object is tangent to the circle which represents its trajectory. Since velocity is a vector quantity which specifies both magnitude and direction, and the direction of the motion is constantly changing, the velocity of the object is also constantly changing. We therefore have an object whose speed is constant but whose velocity is constantly changing. The body is accelerating but the acceleration is in such a direction and of such a magnitude that its net effect is to alter the direction (but not the magnitude) of the object's velocity.

The analysis that follows is only an approximation. A rigorous analysis would require the use of vector calculus, a mathematical language which is not used in

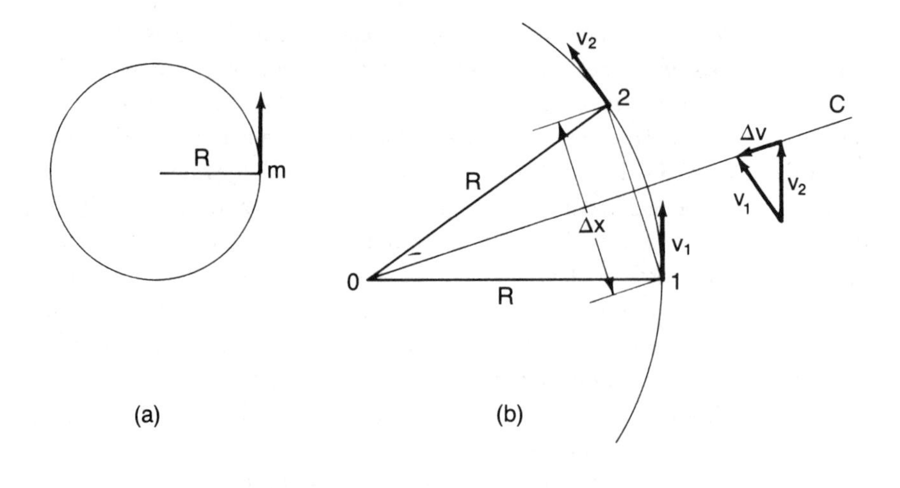

Figure 10

this book. We have enlarged a portion of Figure 10a in Figure 10b. The object is moving along the arc of the circle with radius $\vec{R}$. $\vec{v}_1$ is its velocity at one instant and $\vec{v}_2$ is its velocity at a later time. Between those two positions on the arc, we have drawn a chord and constructed a radius perpendicular to the chord that bisects the chord. That radius is extended beyond the arc and is labeled OC. (Using the midpoint between $\vec{v}_1$ and $\vec{v}_2$ averages out the errors which result from not using the mathematics of vector calculus). Vectors of equal length which are parallel to each other are equal. We move $\vec{v}_1$ and $\vec{v}_2$ so that they remain parallel to themselves.

Their tips fall on the perpendicular OC and the tail end of the vectors begin from the same point. The vector $\Delta\vec{v}$ is the difference between these vectors since $\vec{v}_1 + \Delta\vec{v} = \vec{v}_2$. It does not matter how long a chord I chose. Imagine it to be made shorter and shorter, corresponding to a smaller and smaller time difference between $\vec{v}_1$ and $\vec{v}_2$. As long as I allow the tips of $\vec{v}_1$ and $\vec{v}_2$ to fall on a line OC which bisects the chord between points 1 and 2, $\Delta\vec{v}$ will fall along the line OC. In other words, the change in velocity of an object moving with uniform speed in a circle must be directed along the radius, toward the center of the circle and so the acceleration must be directed toward the center of the circle at all times. That acceleration is a "centripetal" acceleration, meaning that it is "central seeking."

The next question is, how big must such an acceleration be? Examine Figure 10b again. The triangle $\vec{v}_1\vec{v}_2$ $(\Delta\vec{v})$ is similar to the triangle $\vec{R}_1\vec{R}_2(\Delta\vec{x})$. This is because, at any time, the velocity must be perpendicular to the radius (this is a condition of moving in a circle). Since corresponding parts of similar triangles are in proportion to each other:*

$$\frac{\Delta v}{v_1} = \frac{\Delta x}{R}$$

Imagine that point 2 moves closer and closer to point 1. If we divide both sides by that time inteval, we then have:

$$\frac{\Delta v}{(\Delta t)v_1} = \frac{\Delta x}{(\Delta t)R} \tag{7a}$$

But if we allow point 2 to get very, very close to point 1, then $(\Delta x)/(\Delta t)$ becomes equal to v_1, and $(\Delta v)/(\Delta t)$ is the acceleration. Equation (7a) is transformed into:

$$a_c = \frac{v^2}{R} \tag{7b}$$

The subscript c refers to the fact that this acceleration is centripetal. The interpretation of this analysis is quite straightforward. We asked what kind of acceleration is required if an object moves with uniform speed in a circle. The answer is that, given our definitions of speed, velocity, and acceleration, and the applicability of Euclidean geometry, the acceleration is directed toward the center of the circle and of magnitude v^2/R. There are many examples of objects moving uniformly in circles. For example, mechanical vehicles can be made to turn that way, ice skaters move that way, planets circle the sun in near circular paths, children sling objects around on the end of strings, etc. If Newtonian analysis is to be used, when an object moves along a circular trajectory there is a force exerted toward the center of the circle because according to the second law:

$$\vec{F} = m\vec{a}$$

But in this case $\vec{a} = \vec{a}_c$ of equation (7b) and so

$$\vec{F}_c = m\vec{a}_c = mv^2/\vec{R} \tag{7c}$$

where m is the mass of the object, and $\vec{F}_c$ is the centripetal force. As discussed in Chapter 2, the centripetal force has to be supplied by the manner in which the objects interact. If that is not possible, the object will not move in a circle.

In the *Principia,* Newton provided a proof for equation (7c) that followed one already published by Christian Huygens. But, in fact, Newton had already created

* Here and in other places, where only the magnitude of a vector quantity is being considered, vector notation is dispensed with.

a different and ingenious proof as a student. It can be found in his notebooks in the archives at Cambridge University.*

Suppose that a body moves within a circular barrier along AB as depicted in Figure 11a. When it reaches B, it collides with the perimeter and moves off along BC until it reaches C, etc. The trajectory is a square ABCD inscribed within the circle.

As the object moves at constant velocity along AB it has a momentum $m\vec{v}$. At B, it encounters the circular wall and exerts a certain force on that wall. According to Newton's third law, the wall exerts a force back on the object, thus changing its momentum. After the collision, the object is moving along the line BC, again with a velocity $\vec{v}$. We have shown this in detail in Figure 11b. The change in momentum $\Delta(m\vec{v})$ is as shown. The perpendicular MN bisects the angle between the momentum of the object before and after the collision, dividing it into two equal angles (θ). Examining the figure, it must be the case that before and after the collision, components of the momentum represented by MO and NO are the same. In other words, $m\vec{v} \sin \theta$ remains unaltered before and after the collision. On the other hand, if the horizontal component of momentum prior to the collision was $m\vec{v} \cos \theta$, after the collision it must be reversed and be $-m\vec{v} \cos \theta$.

* This information is from a mimeograph document received from Professor I. B. Cohen, November, 1961.

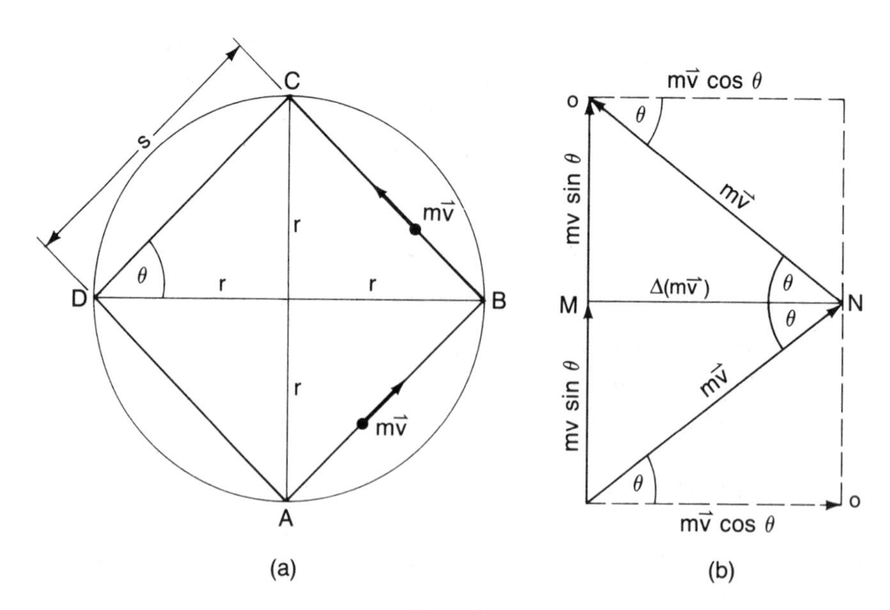

Figure 11

The total horizontal change in the momentum is:

$$m\vec{v} \cos \theta - (-m\vec{v} \cos \theta) = 2m\vec{v} \cos \theta \tag{8a}$$

Examining Figures 11a and 11b, we see that the $\cos \theta = r/s$ where r is the radius of the circle and s is the length of the chord.

$$\text{Change in momentum} = 2m\vec{v} \cdot r/s \tag{8b}$$

Just how large the force at the circular wall must be will depend on how fast the object is moving. The faster the object is moving, the greater the object's momentum and hence the greater will be the change in momentum if the trajectory is to be maintained along the path ABCD. At the same time, the faster the object is moving, the more collisions there will be with the wall during any unit of time. Examining Figures 11a and 11b, we see that the time between collisions is the length of the chord s, divided by the speed v. The number of collisions per unit time will be the inverse of that ratio or v/s. Referring back to equation (8b) we can now write

$$\begin{aligned}\text{total change in momentum per unit time} &= \frac{2mv^2}{s} \cdot \frac{v}{s} \\ &= 2mv^2 \cdot r/s^2\end{aligned} \tag{8c}$$

The left side of this equation, "The total change in momentum per unit time," is, by virtue of Newton's second law, nothing more than the force on the object. We have written equation (8c) in this way because, upon examining the figures, we see that according to the Pythagorean theorem, $s^2 = 2r^2$ and equation (8c) can be rewritten as follows:

$$\vec{F}_c = mv^2/\vec{r} \tag{8d}$$

We have arrived at the same result in (8d) as in (7c). Newton noted that equation (8d) was derived for a square trajectory, but that it could be generalized to any polygon, having as many sides as desired. The analysis of the events during collisions with the walls remains the same. When the number of sides is increased without limit and the size of any chord is reduced accordingly, the polygonal shape becomes a circle.

This analysis is different from the previous one in that it begins by examining the forces in collisions and is therefore a dynamical analysis. The earlier analysis was kinematical and raised the question of what the acceleration must be were motion to be uniform in a circular trajectory. The result is the same.

NEWTONIAN ANALYSIS AND GALILEO'S LAW OF FREE FALL

According to the law of universal gravitation, which we have labeled axiom IV, every body attracts every other body in the universe with a force given by:

$$\vec{F} = G\frac{mm'}{\vec{R}^2} \tag{9}$$

m and m' are the masses of the two bodies, $\vec{R}$ is the distance between the bodies, and G is a universal constant. $\vec{R}$ is a vector because the Force acts along the line between m and m'. The idea that a relationship pertains to all objects in the universe and that the same constant G applies is very bold. The value of G must be very, very small. This must be so because there is no sensible evidence that ordinary bodies are affected by such a gravitational attraction, regardless of how close they are to each other. The value of G was determined during the eighteenth century by using an extraordinarily sensitive instrument to measure the attractive force between two masses when the distance between them was known. The modern value is given as 6.67×10^{-11} newton-meter^2/kg^2. That is indeed a very small number.

If there exists such a gravitational force, it must result in the acceleration of bodies. According to the second law, the acceleration is proportional to the force. Consider two bodies, m_1 and m_2. According to (9), if they are a distance R apart,

$$\vec{F} = G\frac{m_1 m_2}{\vec{R}^2} \tag{10a}$$

But if we consider body 1 or body 2 alone, then

$$\vec{F} = m_1\vec{a}_1 = \mathrm{m}_2\vec{a}_2 = G\frac{m_1 m_2}{\vec{R}^2} \tag{10b}$$

This assumes that the mass referred to in the law of universal gravitation, the "gravitational mass," is the same as the mass of the second law, the "inertial mass."

Let us consider two objects which interact gravitationally when one is of a size comparable to the distance between the objects (Figure 12).

One object could be the earth and the second an object in free fall near the surface of the earth. We assume that the earth is a sphere of radius R, the total distance between the earth and the object (whose mass is m_1) is R', and the ratio R'/R is very close to 1. In other words, the object is very close to the surface of the earth. According to equation (10b)

$$\vec{F} = G\frac{m_1 M}{\vec{R}^2} = m_1\vec{a}_1 = M\vec{A}_1 \tag{10c}$$

M is the mass of the earth, $\vec{a}_1$ is the acceleration of the object in free fall, and A_1 is the acceleration of the earth. It is difficult to know what proper number to use for R in equation (10c). Each point in the earth contains a small but finite mass m.

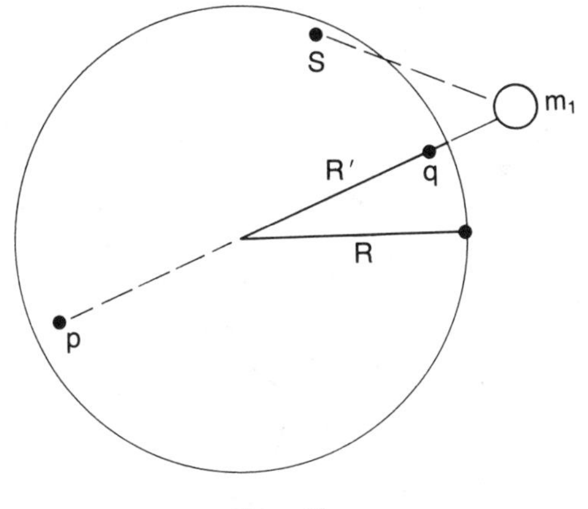

Figure 12

When those points are added, the sum of m's is M. Consider the points labeled q and p. The distance between q and m_1 is almost twice the distance between p and m_1. Or consider the distance between the point s and m_1. The direction of the gravitational force between s and m_1, which acts along the line between them, is virtually perpendicular to the direction of the force between m_1 and q or p. In order to determine the net force, must the force between individual point masses in the earth and m_1 be calculated individually and them summed vectorially? The answer is no. Newton showed, by using calculus, that one can treat a spherical object as though all of the mass were located at a point at the center of the sphere. We will not reproduce that argument, but indicate why it is reasonable.

In Figure 13, we have divided the earth into two hemispheres along the line between m_1 and the earth's center. Any two points which are chosen symmetrically relative to that diameter, for example points b and c, or points d and f, produce forces which, when added, lie along the diameter. Since this can be done for each point in the earth, the sum of forces between each point in the earth and the body will lie along the line passing through the center of the earth and the object m_1.

Now take a second diameter, perpendicular to the first. Choose pairs of points that are symmetrical with respect to this line, for example, the points h and i, so that the average distance will always be the distance between the center of the earth and m_1. Whereas this argument does not constitute a proof, it is possible to prove rigorously that for a spherical object, the mass acts as though it were concentrated at the center of the sphere. Consider m_1. According to equation (10c):

$$\vec{F} = G\frac{m_1 M}{\vec{R}'^2} = m_1\vec{a}$$

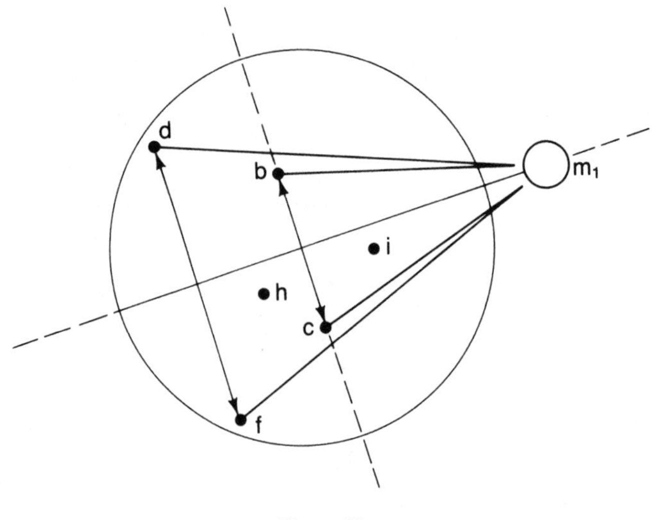

Figure 13

The first thing to notice is that, since m_1 appears in both terms, the acceleration is independent of value of m_1. In other words,

$$\vec{a} = GM/\vec{R}'^2 \tag{10d}$$

Earlier we pointed out that the distance R' was not very different from the radius of the earth. That is, R/R' is about 1. Think , for example, of an object in an airplane at a height of six miles. Since the average radius of the earth is about four thousand miles, the ratio of R/R' is $4000/4006 = .99850$. The ratio of $(R/R')^2$ is 0.99701. In other words, the difference in the acceleration of an object six miles high is about three parts in a thousand less than the acceleration of an object at the surface of the earth. For most objects which fall within a few hundred feet of the surface of the earth, the difference would be totally insignificant. It now becomes clear why the acceleration due to gravity near the surface of the earth is a constant, and seems to depend neither on the distance fallen nor the nature of the object in freefall. That constancy is only an appearance, an artifact of the geometry of the situation: The significant distance is not the distance to the surface of the earth, but rather to the center of the earth. Should our instruments be sensitive enough when we calculate the acceleration of a falling object, there would be a systematic, very small increase in the acceleration as the object plummets to the earth.

According to Newton, were the moon as close to the surface of the earth as most objects whose fall we observe, it would accelerate with the same value as all objects. Call this distance R_1. The moon is located a significant distance from the earth. In terms of the earth's radius, the moon is approximately sixty earth radii away. This distance is R_2. According to Newton's second law and the law of universal gravitation,

$$\vec{F}_1 = GmM/\vec{R}_1^2 = m\vec{a}_1; \; \vec{F}_2 = GmM/\vec{R}_2^2 = m\vec{a}_2 \qquad (10e)$$

A key analysis in Newton's development of the notion of universal gravitation and the orbiting of satellites like the moon was to compare the acceleration of the moon in "place" to what that acceleration would be were the moon near the earth's surface. That is,

$$a_1/a_2 = R_2^2/R_1^2 \qquad (10f)$$

Equation (10f) is derived directly from (10e). In other words, according to Newton's analysis, the moon is accelerating toward the center of the earth. In short, the moon is falling. Why it doesn't "fall down" is considered below.

According to Newton's third law, if the earth exerts a force on an object, the object exerts an equal and opposite force on the earth. The reaction force of free fall is given by

$$\vec{F} = G\frac{mM}{\vec{R}^2} = M\vec{A}$$

where A is the acceleration of the earth due to the gravitational force exerted on the earth by the falling body m. As in (10d)

$$A = G\vec{m}/R^2 \qquad (10g)$$

G is a very small number. Furthermore, the ratio of an ordinary mass falling to the surface of the earth (rocks, planes, ships, large buildings, etc.) to the square of the earth's radius will be insignificant. For a fifty ton object, the value of A in (10g) is of the order of magnitude of 10^{-10} ft/sec.[2] That would be an object whose mass is the equivalent of a dozen fully loaded cement trucks. In addition since objects are falling to the surface of the earth from all directions, on the average, one would expect that the net acceleration of the earth would be essentially zero.

How is the notion of the universal gravitation force related to the concept of "weight"? According to our analysis, all objects near the surface of the earth are accelerated at a constant rate, under the influence of the gravitational force. Once on the surface of the earth, the force does not cease. Rather, the *net* force becomes zero because, as depicted in Figure 14, the gravitational force is transmitted to the surface of the earth. The reaction to *that* force on the surface of the earth is a force exerted by the earth on the object. This force is sometimes called the "normal" force, the word normal meaning, "perpendicular" (to the surface of the earth). When something is lifted from the surface of the earth, the earth no longer

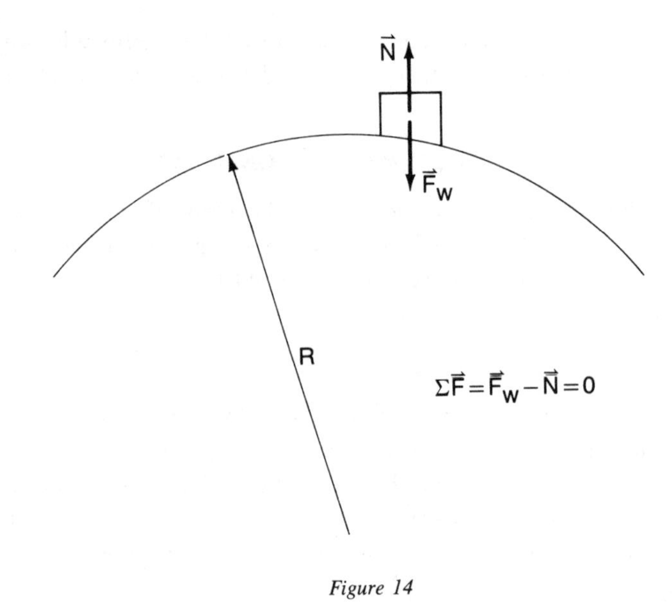

Figure 14

supplies a reaction force. The force is now supplied by whatever agent is lifting the object. It is the need to supply that force that we, when we are the lifter, sense as the "weight" of the object.

As near the surface of the earth the acceleration is a constant independent of any object, it is given a special symbol, g. The force resulting from the acceleration due to gravity is termed the weight and the equation $\vec{F} = m\vec{a}$ is transformed into $\vec{W} = m\vec{g}$ where $\vec{W}$ is the weight of the object and $g = GM/R^2$ as in equation 10d.

We see that Newton's analysis explains Galileo's law of freefall. Although we have not explained the concept of universal gravitation (and Newton explicitly refused to do so), the Galilean law of free fall and, by implication, projectile motions have been assimilated within the axioms of Newtonian mechanics.

NEWTON'S ANALYSIS AND KEPLER'S LAWS

In a series of theorems which appear in Book I and Book III of the *Principia,* Newton showed that one could derive Kepler's laws from Newton's laws and the law of universal gravitation. I do not follow him rigorously here, but only wish to show the logical relationship.

Newton first showed that the law of inertia, together with the hypothesis of a central seeking force, implies that a body will sweep out equal areas in equal time. In Figure 14a, a point mass is moving inertially along the line AD. In other words, in equal time intervals it covers equal distances. Take any point in space that is not

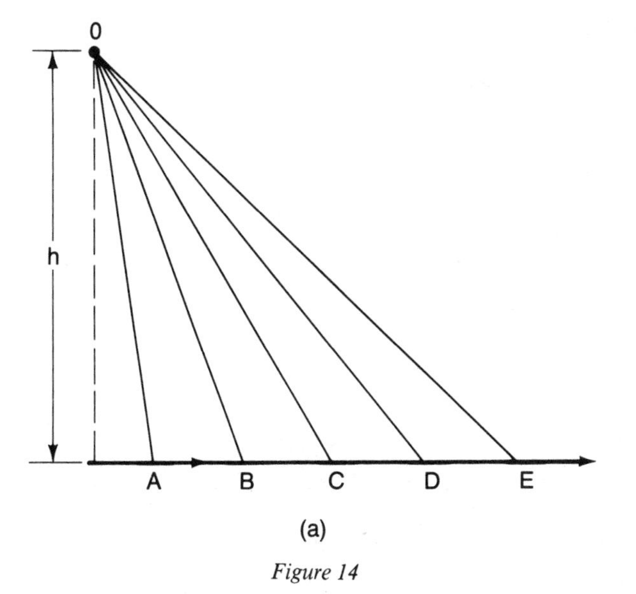

Figure 14

on the line, for example, the point O. The body will sweep out equal areas in equal times relative to that point.

By "sweeping out" we mean the following: When the object is at A imagine the line AO. As the object moves from A to B, point O remains fixed but the line remains attached to the body. (Of course the line will stretch.) By the time the body reaches B, an imaginary triangle ABO will have been generated or swept out. O can be any point in space. Triangles ABO, BCO, and CDO have the same base, since AB = BC = CD. But all of the triangles will also have the same perpendicular height, *h*. The area of a triangle (according to Euclidian geometry) is the product of half the base and the perpendicular height. Therefore a body moving inertially sweeps out equal areas in equal times, relative to any arbitrary point.

But now let us assume that when the object, moving inertially along AD, arrives at B, a blow is struck on the body along the line BO, toward O. The object will not continue to move toward C. Had it not been moving at all when the blow was struck, in the same time required for the object to move from B to C, it would have gone to B′. Not only does it *not* arrive at C in that time interval, it does not arrive at B′ either, but at the vector sum of BB′ and BC, as shown in Figure 14b. But the area of OBC′ is the same as the area of OBC, since they both have the same base, OB, and the same perpendicular height. Therefore, as long as the force exerted on the object is a central seeking force, the object will sweep out equal areas in equal times. Imagine now that when the object arrives at C′, another blow is struck on it toward O. The object now arrives at D′, still sweeping out equal areas in equal times. In this manner the object will follow a closed polygonal path

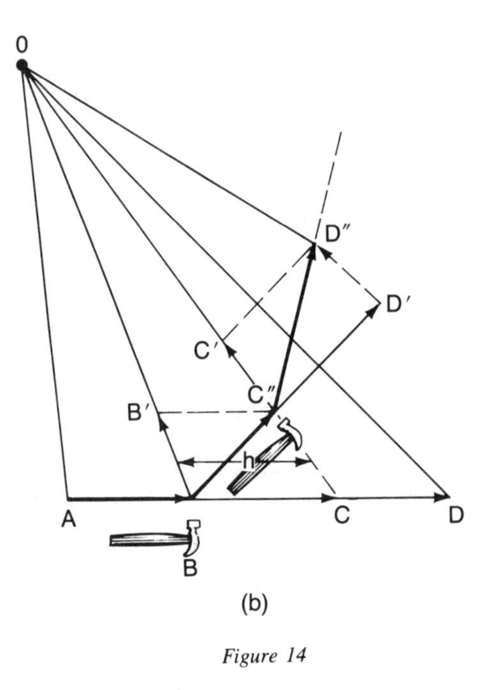

Figure 14

around O. The length of time between blows is now decreased. In the limit, as this time interval is reduced toward zero, the shape of the curve will be a smooth curve. In other words, we have shown that there is a direct logical connection between Kepler's second law (the law of equal areas) and the law of inertia. The fact that a planet sweeps out equal areas in equal times can be understood if we use the premise of inertial motion and a central seeking force.

If the shape of the orbit is an ellipse, not only must the force be a central seeking force, it must vary inversely as the square of the distance. Newton showed that under the influence of a central seeking force which varies inversely as the square of the distance, the trajectory is in the shape of one of the four conic sections: circle, ellipse, hyperbola or parabola. We have simplified the problem and show only that, assuming an ellipse, the force varies inversely as the square of the distance.*

Suppose that the orbit of the body is an ellipse such as that pictured in Figure 15. It represents the orbit of the moon around the earth with the point E at one focus of the ellipse the earth; or it could be the orbit of the earth around the sun, or a satellite of Jupiter around Jupiter, etc. The points A and P which intersect the

* The proof here is due to Colin MacLaurin (1698–1746), a disciple of Newton's. It is taken from a set of mimeographed notes received from Professor I. B. Cohen, Nov. 1961.

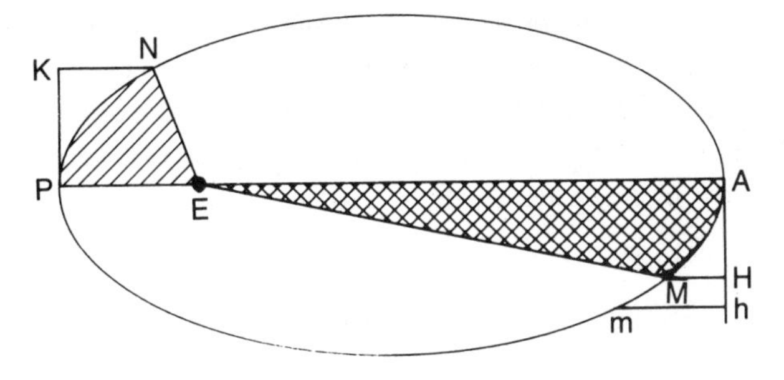

Figure 15

ellipse on its major axis are called the *perigee* and *apogee,* respectively; the points closest to and farthest from the point E.

Draw PK tangent to the ellipse (and perpendicular to the major axis) at perigee and draw AH tangent to the ellipse (and perpendicular to the major axis) at apogee. Consider the motion of the object on a small portion of its orbit near the points of apogee and perigee. These are the arcs PN and AM. (Although they are not of equal length, according to the law of equal areas in equal times, they represent arcs of equal time).

Let us suppose that we had total control over the motion of the orbiting object. Rather than allowing it to move in its normal orbit, we will hold it at P or A and allow it to fall toward E. (Recall that a central point of Newtonian analysis was to link the "fall of an apple with the fall of the moon.") The line KN is perpendicular to the tangent at P (the line PK). Similarly, the line HM is perpendicular to the tangent at A (the line AH). In the time that the orbiting object will have traveled along the arc PN or the arc AM, it will have fallen a distance given by KN or HM. Call that the time T. If PN and AM are very small portions of the orbit, the distances KN and HM are very small fractions of the total distance from the orbit to the central body. In other words, the acceleration over the distance KN or HM is constant.

But, according to the Galilean analysis, if the acceleration is constant, the distance traversed is given by:

$$x = \tfrac{1}{2}aT^2 \qquad (11a)$$

The symbols have the usual meaning. But if $\vec{F}$ is the force responsible for the acceleration, then according to Newton's second law, $a = F/m$ and we write that the distance traversed is given by:

$$x = FT^2/2m \qquad (11b)$$

Before continuing, it is convenient to transfer the arcs and distances from one side of the ellipse to the other. We transfer the arc PN with its associated tangent PK and perpendicular to the tangent KN, to the other side of the ellipse so that Am = PN, and Ah = PK, and KN = hm. According to equation (11b), if F_p is the force exerted by the central body at P, then:

$$\text{KN} = F_p T^2/2m = \text{hm} \tag{11c}$$

m is the mass of the orbiting object. Similarly, at A,

$$\text{HM} = F_a T^2/2m \tag{11d}$$

If we take the ratio of (11c) to (11d), we have

$$\text{KN/HM} = F_p/F_a = \text{hm/HM} \tag{11e}$$

It is clear why it was convenient to transfer the arc length PN to the other side of the ellipse at Am. There is another way to describe the ratio of hm/HM in (11e). It is, according to the law of free fall, the distance fallen and is proportional to the square of the time elapsed. If the object A had fallen a distance HM in the time T, and a distance hm in the time t, then

$$\text{hm/HM} = t^2/T^2 \tag{11f}$$

In developing the relationships in equations (11a)–(11f), we assume that the object falls from rest toward the central body. Let us perform another thought experiment. Rather than considering the fall of the object from P and A, let us assume that the object is in empty space and moving freely. In other words, we suppose that when the object is at A, we could "turn gravity off" and that the object would then move inertially with some velocity v along the line Ah. It follows that Ah = vt and AH = vT. In other words, in the time the object falls a distance hm, it moves tangentially distance Ah, etc. Taking the ratio of the inertial distances:

$$\text{Ah/AH} = \text{t}/T \tag{12a}$$

But in order to compare the inertial distances to the gravitational distances, let us square (12a)

$$(\text{Ah})^2/(\text{AH})^2 = t^2/T^2 \tag{12b}$$

Combining (11f) with (12b) yields:

$$\text{hm/HM} = (\text{Ah})^2/(\text{AH})^2 \tag{12c}$$

Let us now apply the law of equal areas to the problem. We have chosen very small arc lengths, PN and AM and we can consider the areas PEN and AEM to be approximately the areas of triangles PKE and AHE. These are right triangles because PK and AH are perpendicular to the major axis of the ellipse. Assuming we can approximate the pie-shaped areas by the right triangles, it follows that:

$$\tfrac{1}{2}\text{PK} \cdot \text{PE} = \tfrac{1}{2}\text{AH} \cdot \text{AE} \qquad (13a)$$

but we have constructed the problem so that PK = Ah and (13a) becomes,

$$\text{Ah} \cdot \text{PE} = \text{AH} \cdot \text{AE}.$$

In other words,

$$\text{Ah}/\text{AH} = \text{AE}/\text{PE} \qquad (13b)$$

We can square this last equation and combine it with (12c) to yield:

$$\text{hm}/\text{HM} = (\text{AE})^2/(\text{PE})^2$$

but since hm = KN by construction:

$$\text{KN}/\text{HM} = (\text{AE})^2/(\text{PE})^2 \qquad (13c)$$

As we saw in (11e), the ratio of KN/HM is the ratio of the gravitational force exerted on the body at P and A respectively. Therefore, we can write for (13c):

$$F_p/F_a = \text{AE}^2/\text{PE}^2 \qquad (13d)$$

This last equation states that the force on the object at points P and A varies inversely as the square of the distance between the object and the central body. The proof has been limited to the points immediately surrounding the apogee and perigee for an ellipse, but it can be generalized to all points on the ellipse and to other conic sections, as well. Taken with the earlier proofs, one can say that for the motion of one body relative to another body, should the trajectory be one of the conic sections, and should the body sweep out equal areas in equal times, these facts are manifestations of the following premises:

1. The law of inertia holds.
2. The force between the bodies is a central force.
3. The magnitude of the force varies inversely as the square of the distance between the bodies.

The fact that an object orbits a central body can be understood as the equilibrating result of combining universal gravitation with inertial motion. In Figure 16, we have depicted a portion of the object's trajectory. Point C, the object, is moving inertially along path CD. But in that same time interval, the object is falling so that it is carried back to the orbital trajectory at E. It then moves inertially along EF, returning to the orbital trajectory at G, etc. In the limit, as we consider smaller

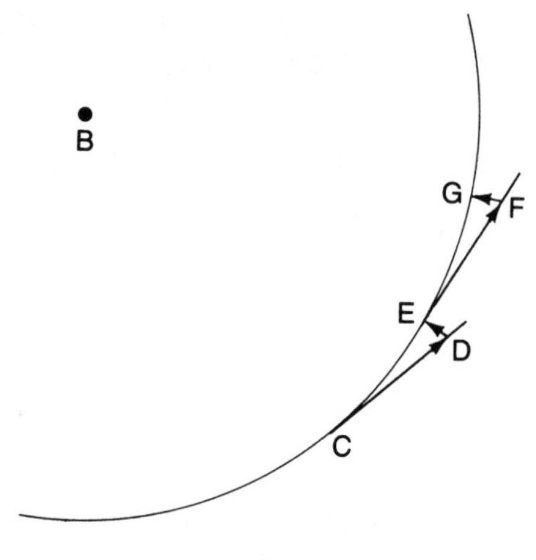

Figure 16

and smaller time intervals, the actual path of the object is along CEG. The object is always falling but its inertial motion prevents it from digressing from the orbital path, be that path a circle or an ellipse.

It remains to show the relationship between Kepler's third law and Newton's axioms. The third law states that the radius of the orbiting body and the time required to complete one revolution, the period, are related as

$$T^2/R^3 = k \tag{14}$$

where T is the period, R the radius and k a constant characteristic of the system.

We assume a central body of mass M around which another body orbits at a uniform speed along a circular trajectory. (This assumption is very nearly accurate for many of the planets orbiting the sun, the moon and many of the artificial satellites orbiting the earth, etc.). According to Newton's second law, the force on the object in orbit is given by $F = ma$, where m and a are the mass and acceleration of the body. But as the body is moving in a circular orbit, the acceleration is a centripetal acceleration:

$$F = ma = mv^2/R \tag{15a}$$

where v is the orbital speed and R is the radius of the orbit. The speed, however, can be determined, since the time T, the period, is the time required to travel once around the circle:

$$v = 2\pi R/T;\ v^2 = 4\pi^2 R^2/T^2$$

This can be placed in (15a) which becomes:

$$F = 4\pi^2 mR/T^2 \tag{15b}$$

In order to make explicit the role of Kepler's third law, we multiply the numerator and denominator of (15b) by R^2. This changes nothing, since the ratio of R^2/R^2 is simply 1:

$$F = 4\pi^2 mR^3/T^2R^2 \tag{15c}$$

As, according to Kepler's third law, R^3/T^2 is a constant, k, (15c) can now be written as

$$F = 4\pi^2 mk/R^2 \tag{15d}$$

In other words, the force is inversely proportional to the square of the distance between the bodies.

The term $4\pi^2 k$ is a constant in (15d). As k is a constant for the system, Newton assumed that it had something to do with one object that is common to any central body-satellite system: the central body. In fact, the attribute of the central body which governs, according to Newton, is the mass of the central body.

$$4\pi^2 k = GM$$

where M is the mass of the central body and G is a constant of proportionality; (15d) becomes

$$F = G\frac{Mm}{R^2} \tag{15e}$$

which we recognize is the law of universal gravitation.

The inventiveness and boldness of the derivation in transforming Kepler's third law into a statement hypothesizing universal gravitation should not be overlooked. Newton did not discover the law. He invented it. Although one can now calculate, as Newton did, the trajectory of the moon by calculating the acceleration that is expected at the distance of the moon from the earth, and consequently how far the moon would fall in each unit of time, this does not change the fact that the theory is an invention: a creation of a man who saw how planetary motion can be understood by the hypothesis of a universal principle.

The power of Newton's accomplishments cannot be overestimated. He was able to demonstrate that the motion of objects near the surface and on the surface of the earth, and in the heavens, could be understood by the application of a very limited number of premises and postulates. There was no need to invoke special cases, special powers, special actions. The same hypotheses which explained free fall and inertial motion, explained as well the trajectories of the planets. It is this stunning synthesis that was the technical and philosophical basis for what has been termed the mechanical world view. We turn now to showing how Newton's laws imply certain conservation laws and that the whole Newtonian system is covariant under a Galilean transformation.

Appendix 3

CONSERVATION OF MOMENTUM

Momentum is conserved in Newtonian physics. That conservation is a direct outcome of the application of the second and third laws to physical situations. But the concept of that conservation law hinges on being able to specify *the system* to which it is applied, and to speak of that system in isolation from the rest of the universe.

First we consider a body in isolation without external forces on it. The body is either at rest or moving uniformly along a straight line. Technically, according to the second law, the time rate of change of the momentum is zero and the momentum of the body remains constant over time.

Let us complicate the problem a little. Suppose we consider two bodies having mass m_1 and m_2. The bodies are held together by a string around the perimeter of both bodies with a spring between them. As shown in Figure 17, when the string is burned through, the bodies will move apart. Consider carefully the role of the spring. At the instant the rope is burned through, m_1 is exerting a certain force on the spring and in reaction the spring is exerting a certain force on m_1. All the spring does, ideally, is transmit the force being exerted on it by m_1 to m_2. But, according to the third law, m_2 is exerting an equal and opposite force on the spring. This is what is meant when we say that the spring is under tension. Once the rope burns through, each body is free to accelerate under the action of the force being exerted on it, and the compression of the spring is reduced. Thus the force is reduced, the acceleration is decreased and finally, at the point that each of the masses loses contact with the spring, the force on each body and on the reaction forces on the spring are reduced to zero, the bodies move off inertially and the spring drops.

Suppose that in the beginning the whole system is at rest. The total momentum of the system is zero. Until the bodies separate from the spring, the force exerted on each of them is equal and opposite, as shown in Figure 18. In fact, in Figure 19, we have made use of Hooke's law and plotted the force as a function of the time in which the bodies are in contact through the spring. Note that the forces on each body are the negatives of each other (Newton's third law) and that the areas under

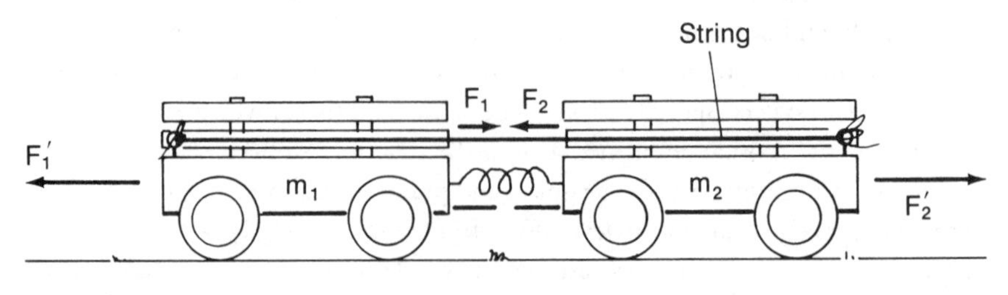

Figure 17

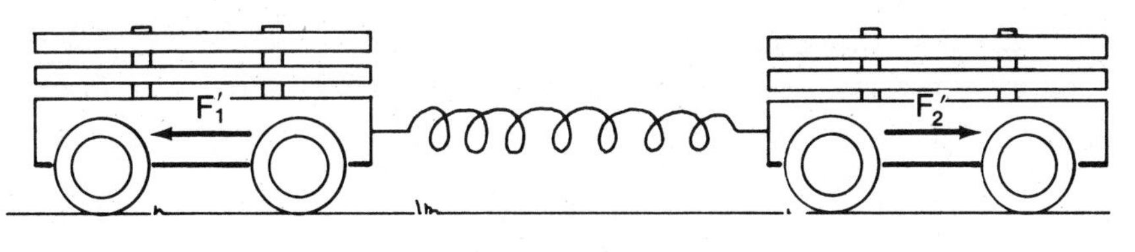

Figure 18

the curves are mirror images of each other. For convenience, we define a new abstract quantity called the *impulse.* The impulse is the product of the force and the time over which the force acts on a body. In Figure 19 the impulse is represented as the area under the triangle because the force varies in proportion to the time. Should the force be constant, for example, when a horse pulls with a constant force on a cart, the area under the curve of a Force-Time diagram is a rectangle. A physical interpretation of Figure 19 is that as long as the two bodies are in contact with each other through a spring, whatever the external force exerted on one body, an equal and opposite force is exerted on the other body.

According to Newton's third law: $\vec{F}_1 = -\vec{F}_2$ and as this is true during the entire period that the bodies interact, once the rope is broken,

$$F_1(\Delta t) = -F_2(\Delta t) \qquad (16a)$$

But, according to Newton's second law, the force is equal to the product of the mass and the acceleration of each body. If one considers the entire time over which the bodies interact, the average acceleration on the bodies is given by:

$$a = \Delta v/\Delta t \qquad (16b)$$

Combining (16a) and (16b):

$$m_1a_1(\Delta t) = -m_2a_2(\Delta t);\ m_1(\Delta v_1)(\Delta t)/(\Delta t) = -m_2(\Delta v_2)(\Delta t)/\Delta t$$

$$m_1\Delta v_1 = -m_2\Delta v_2 \qquad (16c)$$

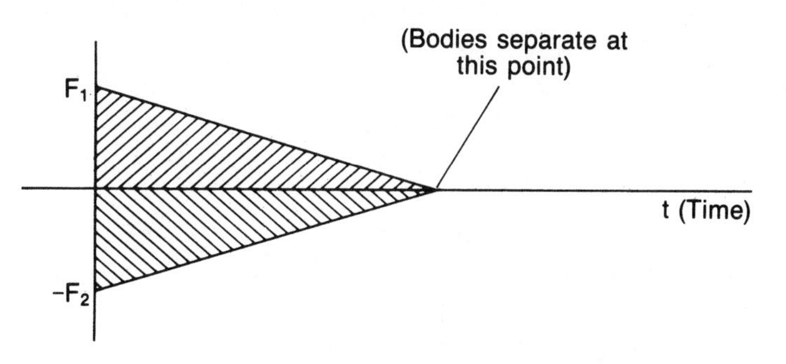

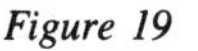

Figure 19

If we write out the terms in (16c), recognizing that Δv is the change in the velocity of the object during the time the bodies are free to interact, that is, that Δv is given by $v_f - v_i$ where the subscript f refers to final and the subscript i refers to initial, equation (16c) becomes:

$$m_1 v_{1f} - m_1 v_{1i} = -m_2 v_{2f} + m_2 v_{2i} \tag{16d}$$

Let us rearrange the terms of this equation so that the initial and final momenta are collected on the same side of the equation

$$m_1 v_{1f} + m_2 v_{2f} = m_1 v_{1i} + m_2 v_{2i} \tag{16e}$$

Whatever the total momentum of the system is before the interaction (we assume that the spring had no mass at all and hence no momentum) it is also the momentum of the system after the interaction. In this particular case, before the explosion (the breaking of the string) the momentum of the system was zero. After the explosion, the momentum of the system is still zero. The momentum carried off by one of the masses in one direction must be the negative of the momentum carried off by the second mass. The bodies move along the same line and their speeds are inversely proportional to their masses. This is directly derivable from (16c):

$$m_1 / m_2 = \Delta v_2 / \Delta v_1 \tag{16f}$$

Suppose that the masses had been equal. According to this analysis, the bodies would have moved with equal and opposite velocities. Suppose that one of the masses were twice as large as the other. Then they would move from each other so the speed of the more massive is oppositely directed and half the value of the speed of the lighter. In any event, whatever the total momentum of the system before the interaction, the momentum after the interaction is unchanged. We summarize this result by saying that, in the absence of net external forces, the momentum of the system is conserved.

Suppose that, rather than being at rest before the breaking of the string, the system of two masses were moving at a speed V in one direction. The equations (16a) through (16f) predict that after the masses move off from each other, the total momentum of the system is the same:

$$m_1 v_{1f} + m_1 v_{2f} = (m_1 + m_2)V.$$

No matter how the velocity of the first object changes, that change would be compensated for by the change in velocity of the second body. The momentum of the system adds up to the same as it was before the breaking of the string.

In endowing the entire system with a speed V we have in effect described the same events from two different frames of reference. In one frame the bound masses are at rest. In another frame, which is moving at a uniform speed $-V$ along the x axis relative to the first frame, the bound system is moving along the positive x' axis with a velocity V.

Although observers in each frame of reference would assign a different momentum to the bound system before the breaking of the string, and observers in each frame of reference would assign different momenta to each part of the system after the interaction, both sets of observers would agree that the momentum of the system had been conserved. You can derive this result by applying the Galilean velocity addition law directly to equation (16c) or (16d). In other words, the law of conservation of momentum is preserved under a Galilean transformation. When quantities are related to each other in this way, we say that they are covariant. The law of the conservation of momentum is covariant under a Galilean transformation.

Although we have described the conservation of momentum for a special case, it can be expanded easily. Rather than considering an initial situation in which a single body breaks up into two (or more) bodies, think of two objects moving toward each other, colliding, and rebounding. For example, in Figure 19a two bodies of equal mass move toward each other along a line passing through their centers. If they have the same mass and the same speed, the momentum of the system is zero before the collision. During the collision, Newton's second and third laws apply. The elastic forces supplied by a spring in the last example are now being supplied by the internal structure of the bodies themselves. It might be that the bodies interact with such great forces that one or the other becomes permanently deformed. That does not alter the fact that the second and third laws can be employed during the interaction. The application of these laws leads to the same analytic formulation of the problem as represented in equations (16a)–(16f). The momentum after the collision remains the same as it was before the collision. Note that we specified two objects of the same mass moving toward each other with the same speed. Examine once again equation (16e). There are still two parameters to be decided, and given the one equation, there is no way to predict in the absence of any other information, what the final velocity of each object is going to be. All that we can say is that whatever their final velocities, the sum of those velocities adds up to the sum of their velocities before the collision. In this case, that sum is zero. If we further generalize the problem, there is no need for the masses and initial velocities

Figure 19a

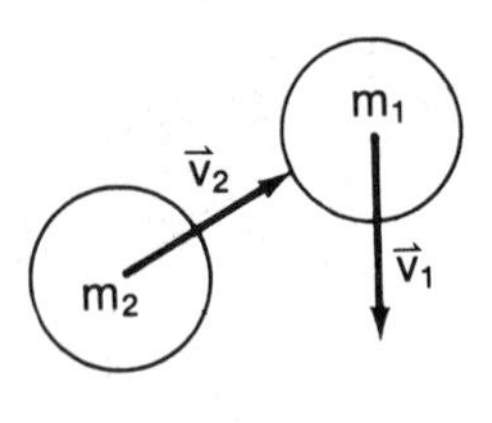

Figure 20

to be the same. The only requirement is that whatever the initial momentum of the system, after the collision the magnitude of that total momentum be unchanged.

Momentum is a vector quantity. Suppose two objects undergo a collision which is not along a line connecting the center of their masses, as in Figure 20. In general, they will not reverse themselves after the collision but will move off at an angle. Nevertheless, according to Newton's laws, the momentum is conserved. In order to analyze this problem, it is useful to employ a Cartesian coordinate system. The momentum can be resolved into components and the law of conservation of momentum is a statement that the algebraic sum of the components along any one axis are preserved before and after a collision.

The problem can be further generalized by considering any number of objects to be included within the system. The law of conservation of momentum remains, in principle, derivable. The formal, analytical part of the problem is no different for a system containing only two objects. The only requirement is, as before, that the system be isolated.

A statement of the law in its most general form is that if one considers a system in an inertial frame of reference in which the momentum is conserved, the momentum is conserved for that system in all other inertial frames of reference, as well. Of course, if there *are* external forces, the net change in the momentum per unit time for the system will be equal to the magnitude of that external force. We have already seen that all inertial observers will agree, not only on whether or not such a force is being applied, but about the magnitude of the force. Newton's laws are *invariant* under a Galilean transformation.

THE CONCEPT OF ENERGY AND ITS CONSERVATION

During Newton's lifetime there was considerable discussion and debate about the importance of the role of the quantity that we have defined as the momentum $m\vec{v}$ and another constructive quantity known at the time as *vis viva* or "living force." It is defined as the product of the mass and the square of the speed, mv^2. Using the speed rather than the velocity, makes this a scalar rather than a vector quantity. According to Christian Huygens, *vis viva* was conserved in all interactions which are perfectly elastic between bodies, that is, in all collisions in which the bodies do

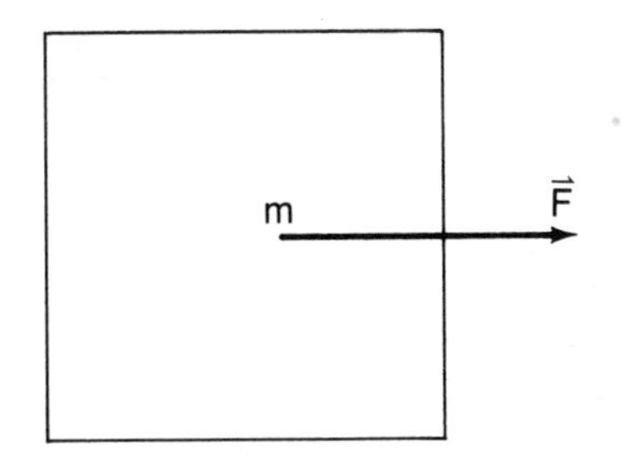

Figure 21

not undergo permanent deformations or other losses. If one sums up the *vis viva* before the reaction, according to this principle, that sum is preserved after the collision.

It might be said that such a quantity cannot be of much practical use since there is no such thing as a perfectly elastic collision. Furthermore, because of friction and other dissipative interactions, there are large losses in any real system. Yet one can derive the *vis viva* directly from Newton's laws. It is a very important derivation because it can be generalized, expanded and philosophized about until the content becomes quite general and a powerful tool for physical analysis.

To begin with, we consider a very simple system as depicted in Figure 21, an object of mass m to which a net constant force is being applied. We could make a plot of the force versus the time, as we did for the conservation of momentum. However, if the force is constant, the acceleration will be constant and that implies a definite relationship between the force and the distance traversed. We plot, therefore, in Figure 22, the force over the distance that the object travels under the action of that force. That quantity we refer to as the *work*. *Work* is used in a technical sense. It is the product of the force and the distance over which the force is being applied. Of course, the object is accelerating but at a constant rate since the force is constant. According to our analysis of cases of constant acceleration, the distance traversed under the action of a constant acceleration is given by

$$v_f^2 - v_o^2 = 2a(\Delta x) \tag{17a}$$

where v_f and v_o are the final and initial speeds of the object, Δx is the distance traversed and a is the magnitude of the acceleration.

Let us now develop the analysis in a formal way:

$$F = ma \tag{17b}$$

For the case of a constant force (17b) can be combined with (17a),

$$F = \frac{v_f^2 - v_o^2}{2(\Delta x)} \cdot m$$

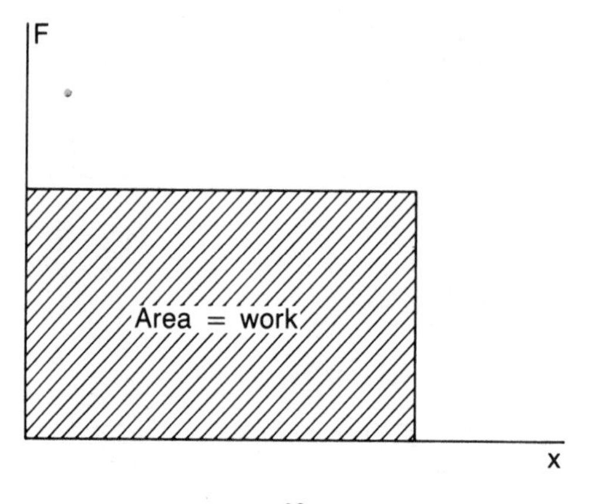

Figure 22

which can be arranged as follows:

$$F(\Delta x) = \tfrac{1}{2}mv_f^2 - \tfrac{1}{2}mv_o^2 \tag{17c}$$

The left side of this equation is the quantity we have defined as the work. The right side contains two quantities that are related to the *vis viva;* one term is half the final *vis viva,* one term is half the initial *vis viva* of the object. This quantity, half the *vis-viva,* we define as the *kinetic energy.*

According to (17c) the work done on an object is equal to the change in the kinetic energy. Should the work be zero whatever kinetic energy the body has would be unchanged. As with the result of the application of any logical system, this is simply a restatement of the postulates. Consider Newton's first law. According to that law whatever state of motion an object has, in the absence of external forces, that motion will persist. That means that the product of the mass of the object and the square of the speed are constant.

Equation (17c) should be compared with (16a). According to (16a) the result of a force acting through time is to change the momentum of the object in a precise way. According to (17c) the result of the force acting over a distance is to change the kinetic energy in a precise way.

The concept of kinetic energy would not be very useful if we had to restrict ourselves to constant forces. Although the precise treatment is beyond the range of mathematical languages in this book, the concept can be generalized to any force whatsoever. In Figure 23, a varying force acting on an object over some distance has been plotted. Rather than defining the work as the product of the force and the distance, a concept which no longer makes any sense since the force is constantly changing, we define the work as the area under the curve.

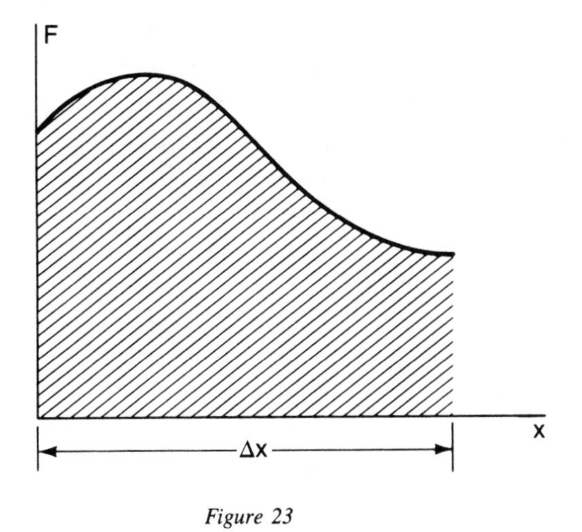

Figure 23

Figure 24 shows a way to evaluate such an area. The area has been broken up into a series of smaller areas composed of rectangles and shapes that are almost triangles. There will be some error if the area under the curve is thought of as the sum of the rectangles. The error would be reduced if the area under the curve is taken to be the sum of the rectangles and the triangles. On the other hand, if the distance labeled Δx were made very, very small, and the number of rectangles was increased proportionately, the area of the offending "triangles" would be reduced. One can imagine reducing size Δx and thereby increasing the number of rectangles that would be required. At first glance it would seem that carrying the

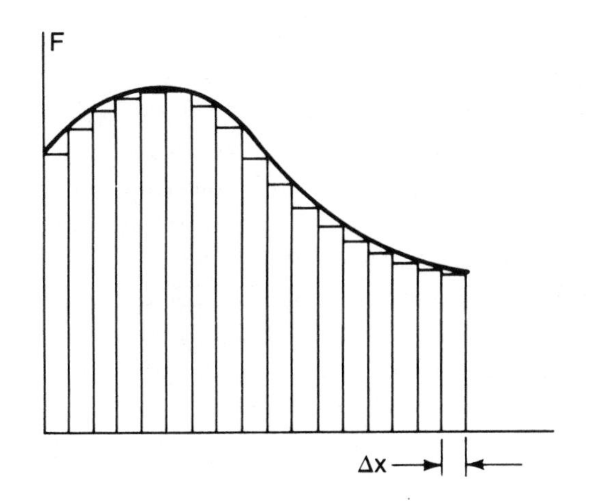

Figure 24

process to its logical limit, the addition of what would be approaching an infinite number of rectangles of infinitely small width (the area of the triangles would approach zero) is an impossible task. But this is one of the things that the mathematical language known as the integral calculus allows us to do.

Let us return to the physical situation described when we considered the conservation of momentum: the collision of two masses in isolation. Viewed over all, since the net force on the system is zero (that is, what we mean by an isolated system) whatever the kinetic energy of the system it is conserved. Before the collision, the two masses are moving toward each other along the same line with speeds v_1 and v_2. The total kinetic energy before the collision will be:

$$½m_1v_{1i}^2 + ½m_2v_{2i}^2 \tag{18a}$$

As the system was not perturbed, no external forces were introduced and, by Newton's third law, forces exerted by one of the objects on the other was at all times equal and opposite. For the same length of time, whatever change is effected in the kinetic energy of one of the particles will be effected in the other particle as well. (Kinetic energy is not a vector quantity, and the *direction* of motion is not a factor.) The net effect is that the kinetic energy after the collision is the same as before the collision.

$$½m_1v_{1i}^2 + ½m_2v_{2i}^2 = ½m_1v_{1f}^2 + ½m_2v_{2f}^2 \tag{18b}$$

If we combine (18b) with (16d), we find that we have two independent equations describing the same situation. Once we specify the masses of the two objects and the initial speeds of the two objects, the equations can be solved simultaneously to determine the final speeds. Suppose, for example, that the two objects have equal mass and are moving toward each other with equal speed. The two equations predict that, after the collision, the speed of the objects will be reversed but the magnitudes will be unchanged. Or suppose that one of the objects is at rest and the second begins with a speed v. If the masses of the two objects are equal, after the collision, the first object will come to rest and the second will move off with a speed v. If the masses are not equal, a different distribution of energy and momentum will occur, but the problem is always solvable when restricted to two bodies if there are no losses and no external forces.

The concept of kinetic energy as it stands is not very useful. In fact, one can think of many situations in which there seems, *prima facie,* to be no external intervention, but at the same time, the kinetic energy in those situations increases. For example, recall the objects at rest with a spring between them. After the binding was released, the kinetic energy which had been zero is no longer zero. Or as in the case of an object in free fall, or descending an inclined plane. Clearly, as time passes, the kinetic energy of the body increased. There was no intervention from outside. Cases like these appear to limit the usefulness of the concept of kinetic energy.

Let us examine the gravitational case with some care. Recall that it does not matter which path the object takes, as long as it moves from one height near the surface of the earth to another, the speed of the object will change by the same amount. With an inclined plane, for example, the steepness of the plane is irrelevant in determining the final speed.

The concept of work can be applied here, as there is a force acting on any object near or on the surface of the earth. All objects exert forces on all other objects. Work, however, has a special meaning in physics. It is the product of the force on an object and the distance the object moves *in the direction of the force.*

In Figure 24a, the situation in the gravitational case is shown. The force on the object is constant since that force is given by the mass of the object and the acceleration due to gravity. In falling through the distance h, an amount of work has been done *on* the object *by* the gravitational force given by:

$$W = Fh = mgh \tag{19a}$$

As a result of the force, the object has accelerated and the kinetic energy of the object has increased:

$$\Delta(\text{K.E.}) = \tfrac{1}{2}m(\Delta v)^2 \tag{19b}$$

If our earlier analysis has any general validity at all, then the work done by the gravitational force is the cause of this increase in the kinetic energy. In other words,

$$mgh = \tfrac{1}{2}m(\Delta v)^2 = \tfrac{1}{2}mv_f^2 - \tfrac{1}{2}mv_i^2$$

but since the mass of the object appears in each term of this equation, it can be cancelled and the equation rewritten as

$$2gh = v_f^2 - v_i^2 \tag{19c}$$

It is not surprising that this result is the same as one of the kinematical equations derived in Appendix 2 for the relationship between speed and distance traveled under the influence of constant acceleration. It is the relationship we used to define the work on an object—equations (17a)–(17c). Just as in the case of applying any other force to a body, one can say that in the gravitational case, the work done on the body is equal to the change in the kinetic energy of the body. We already know that the speed of the object does not depend on the path that it follows. In Figure 24b, the same object which was falling in Figure 24a, is descending an inclined plane through the same height, h. Recall that a key element in the Galilean analysis of free fall is the assumption that the speed of an object at the bottom of an incline depends on the vertical distance. That being the case, the

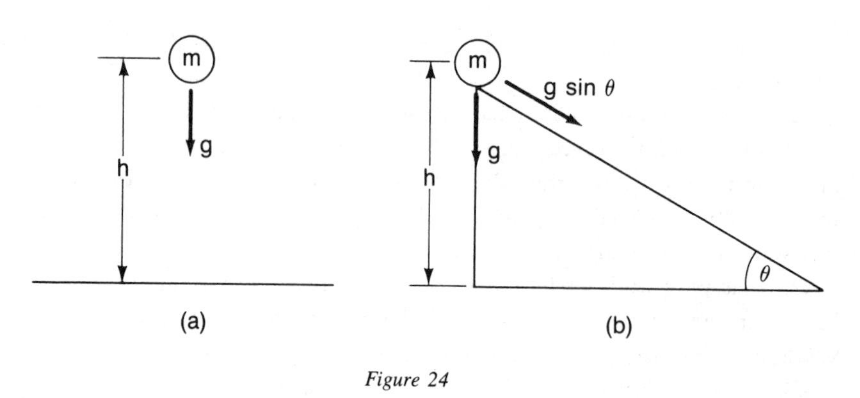

Figure 24

equations derived in (19a)–(19c) for a freely falling body apply as well to the inclined plane. Although the object has traveled a much longer distance than it did in falling vertically, the work on the object is the same as before: The product of the gravitational force and the distance the object moved *in the direction of that force*. Rather than a straight incline, suppose that we build a device like the one in Figure 25. Once at the top, the object rolls down the various inclines, accelerating as it goes, and will roll at a constant speed along the horizontal portions. The speed at the bottom will be the same as the speed acquired in falling the distance h in Figure 24a, or in descending the incline in Figure 24b, since the object will still have descended a total distance h. The work done is equal to the work along individual sections. In descending the incline from a to b the object would have "fallen" a distance 1_1. Later it would fall a distance 1_2 and still later a distance 1_3. In going from a to b one can write, as we did in equations (19a)–(19c)

$$2g1_1 = v_b^2 - v_a^2 \tag{20a}$$

Similarly for the other two portions of the path in which the object descended an incline

$$2g1_2 = v_d^2 - v_c^2 \tag{20b}$$

As work is not a vector quantity, the total work will be the sum of that done in the individual parts:

$$mgh = mg1_1 + mg1_2 + mg1_3 \tag{20c}$$

This is the case since $h = 1_1 + 1_2 + 1_3$. Furthermore, it is the case that $v_b = v_c$ and that $v_e = v_d$ since the path at those points is horizontal and the object is moving inertially. The net effect of following the course described in Figure 25 is, as before:

$$v_f^2 = v_i^2 = 2gh.$$

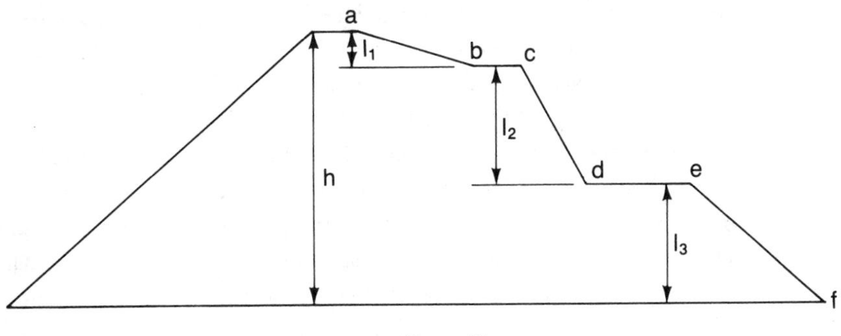

Figure 25

Thus far we have talked about work being done on an object by the force of gravity. This has helped to make understandable the increase in the kinetic energy of the system composed of the object and the inclined plane even though the system appears to be isolated. The earth itself has been left out of the system, and should be included. The increase in the kinetic energy of the object as it moves down the incline is the result of the interaction of the earth and the object.

The analysis can be taken one step further. Suppose that we consider how the object got to the top of the incline in the beginning. Someone or something lifted it and carried it to the top. In order to lift the object, as shown in Figure 26, a force equal to the weight of the object had to be exerted in a direction opposite to the direction of the force of gravity. Since force of gravity is essentially constant for a great distance above the surface of the earth, the amount of work required to lift the object from the bottom of the incline to the top is mgh, exactly the amount of work the gravitational force will do in the descent of the body from the top of the incline to the bottom. In effect, lifting the object from the bottom of the incline to the top makes it possible for the force gravity to do work on the object—it gives the object the *potential* for acquiring kinetic energy. We define the *gravitational potential energy* in the vicinity of the earth as

$$\text{P.E} = mgh \qquad (20d)$$

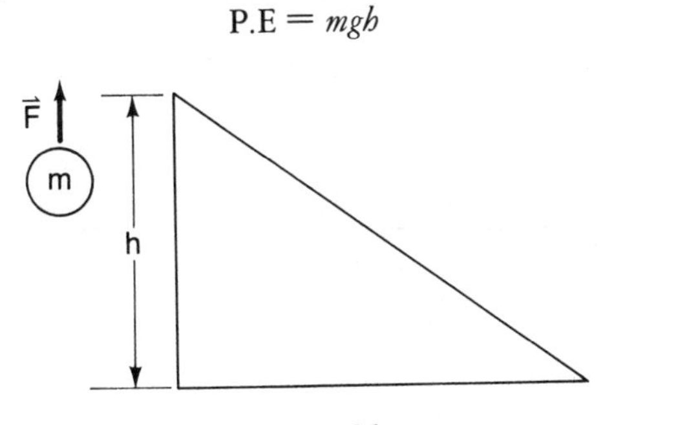

Figure 26

Assuming that there are no losses nor external interferences, the sum of the kinetic and potential energy for the system is a constant. The work done *against* the force of gravity in lifting an object is equal to the work done *by* the force of gravity in increasing the kinetic energy of the object as it falls.

We have termed the work done against the force of gravity *gravitational potential energy*. We have termed the work done by the force of gravity to be *kinetic energy*. It is possible now to make a first, limited statement of the principle of the conservation of energy: In an isolated system in which no forces other than the gravitational force operate, the amount of energy in the system is constant.

$$E = T + K = mgh + \tfrac{1}{2}mv^2 \qquad (20e)$$

E is the total energy, T is the potential energy and K is kinetic energy.

As an example of the operation of this principle, consider the motion of a simple pendulum as depicted in Figure 27. Initially, the total energy of the system is zero and the pendulum is at rest in its equilibrium position. When the pendulum bob is drawn to point a, it is actually lifted a height, H. The total energy in the system is now given by $E = mgH$. After the pendulum bob is released, if there is no outside intervention and if one assumes that there are no frictional or other dissipative losses, the total energy will always remain mgH.

$$\begin{aligned} mgH &= mgh + \tfrac{1}{2}mv^2 \\ gH &= gh + v^2/2 \end{aligned} \qquad (20f)$$

When the pendulum is at b, $h = 0$ and all of the energy is kinetic. The speed can be found directly from (20f). When the pendulum bob is at an intermediate position, say c, which is a height h_1, the energy is divided between potential and kinetic. If the height of the bob is known, the speed can be found. When the bob is at position d, all of the energy is again potential and the bob has momentarily

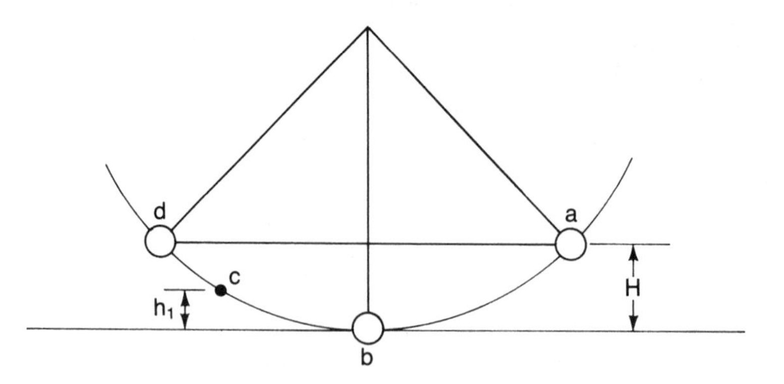

Figure 27

come to rest. Using the concept of gravitational potential energy, the analysis of the motion of the pendulum can be considerably simplified relative to the analysis in which Newton's second law is applied directly. Note that we are only interested in the *relative* height of the bob—the height above equilibrium position. Suppose that suddenly the cord breaks. The bob would begin to accelerate and fall. The bob has a certain amount of potential energy which will not be available unless the bob becomes free to fall. In other words, the point chosen to represent the zero of energy is arbitrary and depends on the nature of the system being analyzed. Relative to the floor, the energy of a painting on a wall of a penthouse is the product of the weight of the painting and its distance to the floor. But relative to the earth, the potential energy is much, much greater. Short of the collapse of the entire building, most of this potential energy would not manifest itself in work on the object.

Another system in which the analysis is identical to the analysis of the motion of the pendulum bob is shown in Figure 28. Two inclined planes are separated by a horizontal floor. A ball is released from point a, a height H above the floor. The total energy in the system is mgH. When the ball is at position b, all of the energy is kinetic and the ball rolls across the floor with a speed given by $v^2 = 2gH$. This is derivable from (20f). When the ball begins to climb the other incline, it slows down. The energy, which was all kinetic, is now divided between kinetic and potential until finally, when the ball is at position d, the energy is again all potential energy. It is important to know that using the concept of energy gives us an alternative language with which to describe what was heretofore spoken of in terms of motion from net forces. It is simpler because the quantities are scalar rather than vector.

The fact that a potential energy can be defined for the gravitational force depends, ultimately, on the fact that the work done by the force of gravity is the same as the work done against the force of gravity in moving an object between two points a and b, near the surface of the earth. As in the case of the inclined plane or the pendulum, the amount of work is independent of the path followed. Such a force is termed a *conservative force*. The potential energy can only be defined for conservative forces. Near the surface of the earth, the gravitational force is given by mg. As we have seen, that is simply a special case of the law of universal

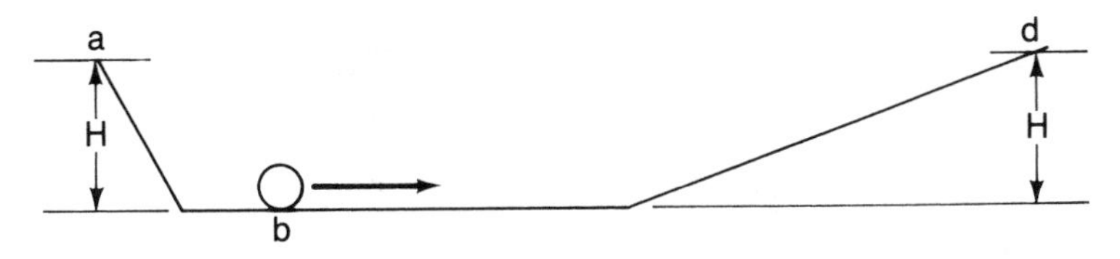

Figure 28

Gravitation. The force between any two objects in the universe is, according to Newton's postulate of universal gravitation,

$$F = G\frac{mM}{R^2} \tag{21a}$$

That a potential energy function can be defined near the surface of the earth depends ultimately on the fact that the force (21a) is itself a conservative force for which a potential energy function can be defined. Consider two masses, *a* and *b*, as shown in Figure 29. Initially, the masses are a distance R apart. If forces are exerted on the masses to separate them an additional distance ΔR (where ΔR is very small), the amount of work done against the force of gravity is given by:

$$\Delta W = G\frac{mM}{\Delta R} \tag{21b}$$

We will not prove that relationship here, since the precise derivation requires the integral calculus. Once the objects *a* and *b* have been separated an additional distance ΔR, and they are released and begin to accelerate toward each other, the work done by the gravitational force to return them to R apart, is equal to the work done in separating them in the first place. The amount of work done against the force of gravity between any two points is independent of the path followed in going between those points and hence, the force is conservative.

Before writing the equation for the conservation of energy for the system, a decision must be made concerning the position to be used as the arbitrary zero of potential energy. Recall that with the pendulum, the point chosen was the lowest in the swing of the bob. Examine again equation (21a). There are two obvious choices that might be made for the zero of potential energy. One is when the separation between the objects approaches zero. But in that case, the force would be very, very large. The other is when the distance between the objects is very, very large; in fact, when that distance approaches infinity. The force between the objects is then zero. Any slight change in the distance from that point means that the objects would begin to accelerate toward each other. By defining the potential energy at an infinite distance to be zero the strange result is that, at any other separation, the potential energy is negative. This is nothing more than an artifact

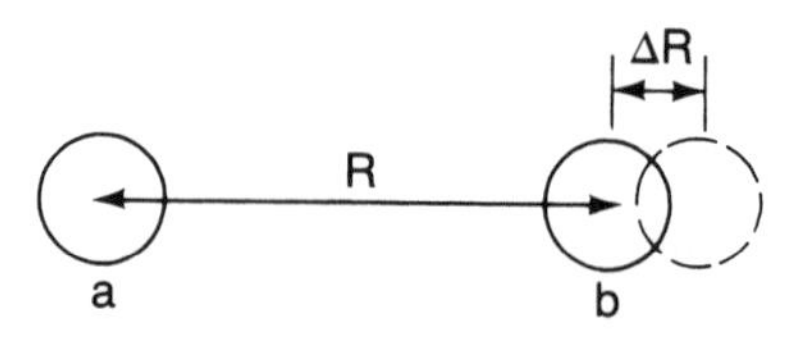

Figure 29

of the mathematics used to describe the situation. As in (20e), the total energy is given by:

$$E = T + K$$

T is the potential energy and K is the kinetic energy. The total work done against the gravitational force in separating two particles initially in contact is given by GmM/R. When the particles are an infinite distance apart, the potential energy is designated T_∞. At any other point the potential energy is less because the work required to separate the particles is less than that required to separate the particles an infinite distance:

$$T_\infty - T_r = -GmM/R \tag{21c}$$

T_∞ is zero, and

$$T_r = -GmM/R \tag{21d}$$

Putting this into equation (20e) gives, for the general gravitational case:

$$E = K - GmM/R \tag{21e}$$

This seems very abstract and of little use, but let us see how it might be used. Suppose an object is resting on the surface of the earth. Its gravitational potential energy is $-GmM/R_e$, where R_e is the distance from the surface of the earth to its center. If I try to hurl the object with a certain amount of energy, how fast would I have to throw it so that it would never return to the earth but would escape? Clearly, the only time the object is not attracted to the earth is when it is an infinite distance away. At that point, the potential energy is zero and if I gave it enough kinetic energy to reach that point, it would never return. Therefore, at the surface of the earth, if the *total* energy is zero, and there were no losses due to friction, air resistance, etc., the total energy always remains zero. In equation (21e) the kinetic energy, K is given by $\frac{1}{2}mv^2$ where v is the initial velocity I must give the object. E, the total energy, is zero. Therefore, the equation becomes:

$$0 = \tfrac{1}{2}mv^2 - GmM/R_e \qquad v^2 = 2GM/R_e \tag{21f}$$

M is the mass of the earth and R_e is the radius. The speed thus specified is just enough to take the object an infinite distance away and is known as the "escape speed." A rocket launched with that speed will not return.

In order to show the usefulness of the concept of potential energy let us examine another kind of conservative force: the force associated with Hooke's law. According to that law, the force required to change the length of a spring from its equilibrium position depends only on the distance the spring is extended or compressed. The force on an object attached to the spring is a function only of position. It meets the requirements we have specified for a conservative force.

As we have seen, the force is given by:

$$F = k(\Delta x)$$

Δx is the distance the spring has been compressed or extended from its equilibrium position. If we call the equilibrium point *zero,* the equation becomes:

$$F = kx \qquad (22a)$$

as shown in Figure 30.

The amount of work done in extending or compressing the spring from its equilibrium position is the product of the force and the distance. But since the force also depends on the distance, the force is not constant. There is an easy way to determine the work. In Figure 31, we have made a graph of the force required to extend a spring from its equilibrium position as a function of the distance.

In a manner directly analogous to our analysis of the distance traversed in the case of uniform acceleration, we can make use of the concept of the area under the curve. In force-distance space, the area under the curve represents the work done in extending the spring. But the area under the curve is $\frac{1}{2}Fx$ or

$$W = \tfrac{1}{2}Fx = \tfrac{1}{2}kx^2 \qquad (22b)$$

Consider an object moving at a constant speed v. Suppose it encounters a spring bumper. The object compresses the spring bumper a certain distance x, slows down and stops. The bumper continues to exert forces on the object, accelerating it in a direction opposite to its initial speed. When the bumper reaches its equilibrium position, the object is again moving at a speed v. There has been an exchange of energy first from the kinetic energy of the object to the potential energy of the compressed spring. That energy is equal to the work done by the moving object (equation 22b) in compressing the spring. That work, in turn, is equal to the work done by the spring on the object in accelerating it in the opposite direction. The

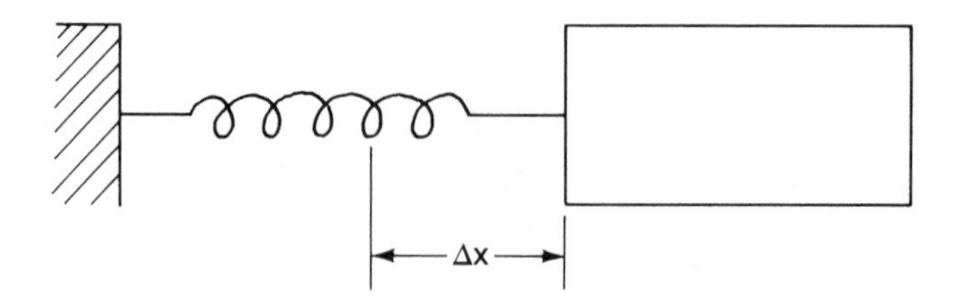

Figure 30

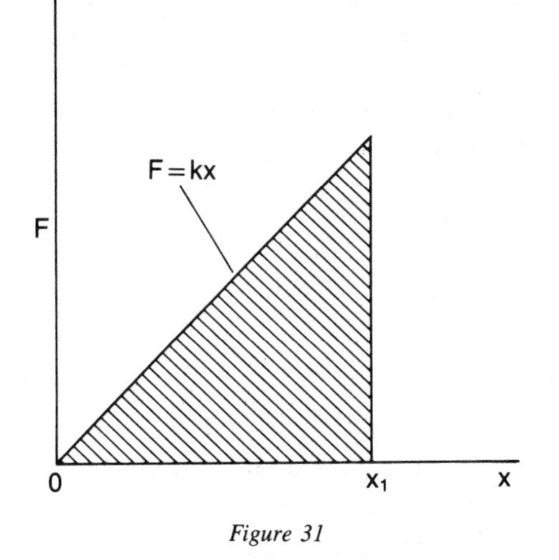

Figure 31

total energy of the system is equal to

$$E = \tfrac{1}{2}kX^2 \tag{22c}$$

X is the maximum compression of the spring. At that instant, the object is not moving and hence contains no kinetic energy. At any other time that the object and the spring bumper are in contact, that total energy is divided between the kinetic energy of the object and the potential energy of the spring.

$$E = \tfrac{1}{2}kX^2 = \tfrac{1}{2}kx^2 + \tfrac{1}{2}mv^2 \tag{22d}$$

We can now understand our earlier example of the two objects that are tied together with a spring compressed between them. The total energy of the system before the breaking of the string is the potential energy residing in the spring. After separation, the total kinetic energy of the two particles carried that energy away. We can say, therefore, that assuming that there were no losses, the total energy of the system remains constant.

It was not until the middle of the nineteenth century that the concept of the conservation of energy was fully developed. The idea was in the air. There were a number of claimants to the discovery of the principle.* The key to the elucidation of the law was the recognition that heat is not a substance, but a form of mechanical energy: the manifestation of the random motion of the particles (atoms) presumed to compose all matter. (Appendix 4 details the analysis of the relationship between Newtonian mechanics and the mechanical concept of heat.)

* T. S. Kuhn, "Energy Conservation as an Example of Simultaneous Discovery."

If we consider the kinds of dissipative heat losses like friction and air resistance to be forms of energy, it is possible to announce a general law of nature. In any isolated system, the total energy remains constant. Energy is conserved. At any time, the energy of the system is given by the kinetic energy of motion, the various forms of potential energy that might be found in the system, and the heat being generated as a result of the interaction of the various parts. In symbols this becomes:

$$E = T + K + Q \tag{22e}$$

Q is the heat being generated or absorbed. We may extend this analysis if we include electromagnetic energy. Ultimately for any system, equation (22e) states that, whereas energy may be interconverted from one form to another, there are never gains or losses. The amount of energy in an isolated system is constant. This notion can also be expanded. For example, consider the burning of a piece of wood. The energy that one obtains when the wood is burned can be considered the transformation of chemical potential energy, which was stored in the wood while the tree was growing. That chemical energy was converted from the energy of the sun into growth of the tree. Ultimately, when the wood is placed in a suitable environment, that energy is converted to heat. When all the processes are accounted for, there are no net gains or losses. Ideally, all of the energy will be accounted for.

During the last half of the nineteenth century, the concept of energy, which is nothing more than the ability to do work (that is, work as defined in physics), became established and generalized. It also became the dominant concept in the analysis of physical systems. Before the electron theory of Lorentz, the most serious challenge to the mechanical world view, in which Newton's laws are the primitives, was a view sometimes known as "energetics" or "energy monism." According to the leading exponent of this view, Wilhelm Ostwald, the only observables in the world are changes and transformations in energy.

We do not perceive matter directly. Analysis shows that energetic interactions are responsible for the reaction of our nervous systems to external stimulation. That stimulation is the interplay of various kinds of energy. The energeticists made energy the bedrock of their understanding of the world. The only reality was energy. The most radical energeticists denied the reality of matter, as such, altogether. It was nothing more than a manifestation of the ultimate substrate: energy. The hypotheses about such entities as atoms were not acceptable. In fact, energeticism is a form of empiricism in which all knowledge is based on the perception of changes in energy. Energeticism was never a major challenge to the mechanical world view, but the existence of the movement points up the importance of the concept of energy in the development of natural philosophy during the period after Newton.*

THE CONSERVATION OF ENERGY IN INERTIAL FRAMES OF REFERENCE

As with the principle of the conservation of momentum, if energy is conserved in one frame of reference it will be conserved in all inertial frames of reference.

Let us reexamine one of our earlier examples of two billiard balls which collided along the line of their centers and rebounded. We saw that as long as the system is isolated, the kinetic energy before the collision is equal to that after the collision. Now imagine the situation from another perspective. If we assume that the initial speeds of the balls were equal and uniform in the room frame of reference, from the perspective of an observer on one ball, the other ball will have twice the speed as measured in the room, before and after the collision. The two observers will disagree about the spatial and temporal description of the events, but they will agree that the kinetic energy before and after the collision will be conserved. All that has been done, informally in this description, is to apply the Galilean transformation equations in the form of the Galilean velocity addition law to the events.

* For a discussion of the relationship of the development of energeticism to other developments in nineteenth century physics, see, C. C. Gillispie, *The Edge of Objectivity,* pp. 500–502. It should be pointed out that two principles formed the basis of the energeticist perspective. The first was the conservation of energy. The second was the so-called second law of thermodynamics which gives the basis for understanding how and when energy transformations occur spontaneously. According to that law in any transformation process, a quantity defined as "the entropy" will never spontaneously decrease. At best, its value will remain constant. The entropy is defined in thermodynamics as the ratio of the heat involved in the exchange to the temperature at which the exchange is being made.

In kinetic theory and statistical mechanics, the entropy is a measure of the disorder of the ensemble. The thermodynamic concept of entropy is quite abstract and difficult to understand in physical terms. The equivalent concept in statistical mechanics can be understood as follows: The most probable state for a system is one in which the disorder is maximized. The equivalent statement in thermodynamics is: It is impossible to build a heat engine which takes energy from a high temperature reservoir and does work without delivering some heat to a low temperature reservoir. For example, consider the operation of an automobile engine. The high temperature source is the exploded gasoline-air mixture. The work done is in making the engine run, the low temperature reservoir is the atmosphere. If there is no low temperature reservoir, the engine immediately stops running. Try it: Jam a potato into the exhaust pipe of an automobile that has the engine running.

Appendix ·4·

The Kinetic Theory of Matter and the Mechanical World View

THE KINETIC THEORY OF MATTER and the theories which succeeded it, statistical mechanics and quantum statistical mechanics, are attempts to understand the gross behavior of matter with assumptions about its nature. As measuring processes and technological skills become more sophisticated, each of these theories is found wanting. Thus statistical mechanics replaced kinetic theory largely because the assumptions made in kinetic theory are judged to be extraordinarily artificial, and the number of phenomena to which the kinetic theory can be applied is extremely limited. This process has continued from the time these theories appeared in the middle of the nineteenth century to the present. The important element of the relationship between evidence and belief for our purposes is to explain the gross properties of matter in terms of assumptions that make use of hypothetical entities: atoms or molecules. Our predilection for particulate assumptions is very strong. Indeed, most scientists who work in fields in which theories are developed about the fundamental nature of matter and in which experimental evidence is gathered, almost invariably reject the notion that ultimate constituents of matter are hypothetical. At each level of explanation, as the number of fundamental particles grows embarrassingly large, a new synthesis emerges in which the classes of particles previously thought to be fundamental are no longer seen that way. Rather they become manifestations of the conglomeration of still more fundamental particles.

For example, at the beginning of the nineteenth century, when the chemical atomic theory made its first appearance, atoms were thought to be unanalyzable billiard balls. Copper was the way it was because copper atoms were the way they were. Copper sulfate, which chemical analysis showed to contain copper, sulfur, oxygen, and nothing else, was the way it was because of the manner in which the three elements arranged themselves under the right physical conditions. That is, they formed a molecule which acted like an atom. By the end of the nineteenth century, the notion that the atom was unanalyzable was no longer acceptable. At the very least, it was thought to contain two kinds of electric charge: positive and an equal quantity of negative. By the second decade of this century, the arrangement was seen as the positive charge residing in an extremely dense core surrounded by tiny particles of negative charge. By the third decade of this century, the central core was believed to be composed of two kinds of particles: protons, each carrying an equal amount of positive charge, and neutrons, having essentially the same mass as protons, but electrically neutral. This dense core was surrounded by a number of negative charges, electrons, each having negative electric charge quantitatively equal to the charge on the proton. It was also believed that the neutron could decay into a proton and an electron, thus accounting for the emergence of electrons, or "beta" rays, from the dense nucleus. But since the beta particles from the same species emerged with any amount of kinetic energy up to a maximum, a new particle was proposed: the neutrino, whose job was to carry away the remainder of the energy. As no one had other physical evidence for such a particle, it must be very, very hard to detect.

This process continued through the 1960s. By 1962 there were over a hundred fundamental particles, all of them hypothesized on the basis of the physical evidence of detection. The word "detection" is somewhat misleading. The experiments are extremely elaborate. The significance of the behavior of the detectors is enmeshed in an elaborate theoretical structure containing intricate inferential chains from "seeing" a track, such as a trail of condensed moisture, supposedly the result of the passage of a charge particle, to saying "proton." Very often, detection involves the simultaneous triggering of two or more detectors with precise spatial orientations to each other. This results in the "blip" on an oscilloscope face, or in recording an "event" by a computer. The entire process, while essentially simple, is abstract and removed from sensory responses usually associated with "evidence." By the middle of the 1960s, some particles were not detected, but emerged from the analysis which showed that, under certain conditions, there is an "energy resonance" of the interaction of other fundamental particles.

It is significant that a new synthesis emerged in which many fundamental particles were considered to be manifestations of still more fundamental particles.

In some quarters, the hope was expressed that it would be possible to understand the nature of matter by using two such fundamental entities. This was not possible. But it is hypothesized that there are only two *kinds* of fundamental particles: quarks and leptons. (The term quark comes from a sentence in James Joyce's *Finnegan's Wake:* "Three quarks for Mr. Marks.") Initially, three kinds of quarks were hypothesized. But, of course, each quark required an anti-quark. Their properties were given whimsical names: "charm," "color," "strangeness." Now, within ten years, the number of quarks has grown to six (and for each quark, there is a corresponding lepton) and each of the six has three aspects. So, as counted in some circles, we are already up to twenty-four of the "really" fundamental particles. By some people's count, the number is forty-two. It has been suggested that perhaps this newly hypothesized class of fundamental particles are themselves the conglomerate product of still more fundamental particles.*

I would be remiss if I did not point out that the kind of relationship between evidence and belief in this region of natural philosophy, known as "high energy physics," is the epitome of the relationship between evidence and belief in any field of science. In the past, there has been a kind of condescension, at least informal, by physicists and other scientists toward other fields because these other fields are "soft". Soft means that there is not an ironclad relationship between evidence and belief. For example, a favorite target is clinical and experimental psychology. But, in fact, the relationship between evidence and belief in all of these fields is about the same and always has been. I consider Sigmund Freud's struggle to try to understand his evidence in terms of a theory of the development of personality to be an exemplar of good science.† (This should not be confused with the technology which emerged in which one attempted to manipulate the behavior of neurotic individuals by using Freudian techniques. Freud himself participated in this.)

KINETIC THEORY: THE EVIDENCE TO BE EVALUATED

Let us return to the kinetic theory as illustrative, not only of the typical relationship between evidence and belief, but typical also, of the kind of reduction that was hoped for within the mechanical world view.

* An excellent summary of the relationship between theory and belief in this area of natural philosophy is unwittingly supplied by R. W. Wilson, "The Next Generation of High Energy Accelerators," *Scientific American* (Jan.), 1980. See also H. Harari, "The Structure of Quarks and Leptons," *Scientific American* (April), 1983.

† See, Phillip Rieff, *Freud: The Mind of a Moralist* (1st ed., New York, 1959; 3rd ed., Chicago, 1979).

The evidence to be used in our example is the simple behavior of gasses which had been known since the time of Newton and Robert Boyle. According to Boyle's law for a trapped sample of gas, the product of the pressure of the gas and its volume are a constant as long as the temperature remains the same.* It is often argued that Boyle's law is empirical. All that one need do is measure the volume, the temperature and the pressure, and see that the product of the pressure and volume remain constant at a given temperature regardless of how the volume or the pressure is altered. But the measurements are no simple task, and the measurement process itself contains a myriad of inferential chains.

Let us take the measurement of pressure as an example. Pressure is defined as the Force per unit area:

$$P = F/A \tag{1a}$$

In Figure 1a we have shown a surface on which a force equal to F is being exerted. Assuming that the force is distributed uniformly over the surface, the pressure is given by F/A where A is the surface area. On the other hand, even if the force is *not* distributed uniformly, *assuming* that it is and knowing the pressure, allows one to calculate the average force.

A liquid or a gas in a container is presumed to exert an equal force in all directions. (Consider the shape of the container when it is allowed to take on any shape whatsoever: for example, a balloon.). Therefore, one can calculate the force at any point if one has a measure of the force per unit area. In Figure 1b, we have shown a cube filled with water. The weight of the water in the volume is given by the product of the density (the mass per unit volume), the volume and the acceleration due to gravity (Recall that weight is a force.). That is,

$$W = \rho V g = mg \tag{1b}$$

ρ is the density, V the volume, and m the mass.

But consider any point in the water, for example, the point A. We observe that it is at rest. Therefore, the forces on it from all directions are equal. So too, the force per unit area is equal in all directions; that is, the pressure is equal in all directions. Furthermore, since the force in a downward direction is equal and opposite to the force in the upward direction, the force per unit area is a function only of the depth of the water.

* Robert Boyle did not discover Boyle's law for the simple reason that he wasn't looking for it. Boyle was intent on another purpose and the fact that one could manipulate his data to reveal Boyle's Law was pointed out to him by someone who, presumably, was on the lookout for such a relationship. See, J. B. Conant, "Robert Boyle's Experiments in Pneumatics," in J. B. Conant (ed), *The Harvard Case Studies in Experimental Science* (Cambridge, 1948), Case #1.

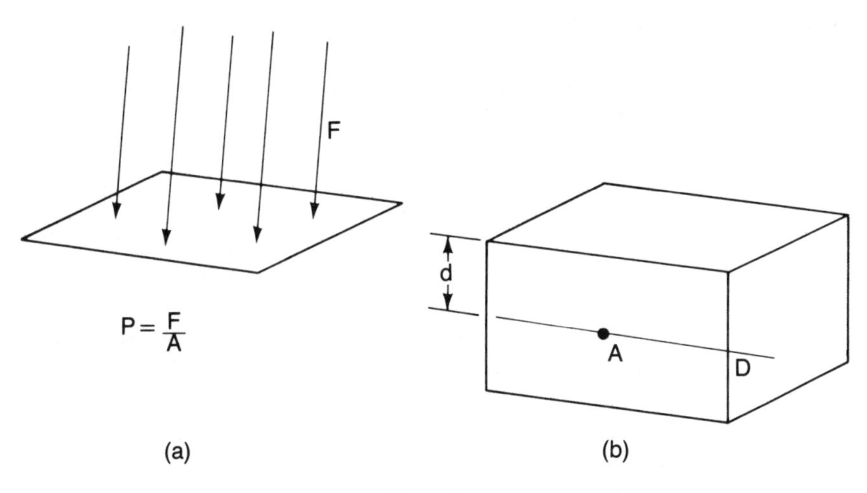

Figure 1

The point A is on the line CD. Consider the projection of the surface at depth *d*. At any point on this surface, the Pressure is given by:

$$P = F/A = \rho Sdg/S = \rho dg \tag{1c}$$

The weight of the water to the depth *d*, is given by the density of the water, the volume of that part of the water that is the product of its cross-sectional area S and the depth d. We see the immediate use of the concept of pressure. In a fluid, the force per unit area, the pressure is directly proportional to the depth. The pressure is the same in all directions. This comes directly from Newton's laws.

Given all of this theorizing, how does one measure a pressure? One must measure a force and a cross-sectional area, but we have seen that we can measure only masses, lengths and times. One way a pressure is measured is by attaching a tube of known cross-sectional area to the sample. The other end of the tube is connected to a "pressure gauge." The gauge contains a flexible diaphragm or some sort of spring arrangement, attached to a needle. When a force is exerted against the diaphragm, the needle is deflected. Applying a known force to the diaphragm allows one to immediately calibrate the deflection of the needle in terms of pressure. But, note that in order to do that, one must make use of the concept of pressure since the force will be applied through the small tube. We assume that the force per unit area at the outlet of the tube will be uniformly applied across the area of the diaphragm. We see that the measurement of pressure is not the simple, empirical task so often claimed. It is sophisticated, complex, and shot through with theoretical inference.

In addition to the relationship between pressure and volume at a fixed temperature in a gas, two more relationships can be expressed: one relates the volume and temperature at a fixed pressure and the other relates the pressure and the temperature at a fixed volume. All of these relationships can be combined into "the ideal gas law."

$$PV = nRT \tag{1d}$$

R is a universal constant known as the ideal gas constant and n is a constant characteristic of the particular sample of gas being used. Given such a sample, according to (1d), if the variables P, V and T are related as PV/T, that number will never change, regardless of what happens to the pressure alone, the temperature alone or the volume alone. In other words:

$$P_1V_1/T_1 = P_2V_2/T_2$$

where the subscript 1 refers to one particular time and the subscript 2 refers to any later time.

At a given temperature, the above equation becomes Boyle's law, PV = a constant.

There are several important things to learn about this law. First, it is called the *ideal* gas law because it does not apply, strictly speaking, to any real sample of gas. The law is extrapolated from real behavior. One way to characterize a gas is to note to what extent it conforms to ideal gas behavior. A gas like nitrogen dioxide (largely responsible for that red haze over major populations centers) does not conform very well at all. A gas like helium conforms very closely. The gas law is an abstraction which applies to a gas that has no real existence: an ideal gas. Second, note that thus far we have not said anything about the nature of the gas.

All the laws of thermodynamics have that character. They are macro laws which make no assumptions about the nature of the material to which they are being applied. We are going to attempt to derive the macro laws of thermodynamics, including laws like Boyle's law, by hypothesizing about the nature of the matter, its composition, and how the parts interact.

THE KINETIC THEORY

In our first attempt, we concentrate on ideal gases. The premises are as follows:

I. An ideal gas is composed of atoms, particles which are not further analyzable. They are to be considered billiard balls. All of the properties of the gas are the result of the mechanical interaction of the particles with each other or with the walls of the container.

II. Any finite volume of gas contains a very large number of particles. No matter how small a subvolume is considered, the number of particles in that subvolume is virtually uncountable.

III. The particles, which are in constant motion, have dimensions that are small compared to the distance between adjacent particles. The ratio of the volume of the particles to the volume of the container is essentially zero.

IV. The particles exert no forces on each other except when they collide. Therefore, between collisions with each other and with the walls of the container, the particles move inertially—at uniform speed in a straight line.

V. Collisions of the molecules with each other and with the walls of the container (assumed to be smooth) are elastic. There are no deformations, no loss of energy, nothing bizarre happens as a result of such collisions. When colliding, the particles obey Newton's laws of motion.

VI. In the absence of external forces, the particles are distributed randomly. One small region of the entire volume is indistinguishable from any other region.

VII. All directions of particle speed are equally represented.

VIII. The particles can have any speed from zero up to the maximum, the speed of light.

Some of the implications of the premises should be made explicit before seeing what they entail for the behavior of the gas. Premises I, II, VI, VII and VIII taken together can be given the following physical interpretation: No matter where one looks in the volume containing the particles, one part looks like the other. Should we put a very small dipper in the gas and remove a sample, that sample will be indistinguishable from any other sample we choose to extract. Furthermore, suppose we have placed a small arrow in each of the particles representing the direction of motion of the particle. Should we now extend that arrow so that the head pierces an imaginary sphere with which we have surrounded the entire volume, the points on the sphere would be equally dense over the entire surface of the sphere.

Suppose that we erect a three dimensional Cartesian coordinate system in the gas. These assumptions allows the conclusion that on the average, a third of the particles will be moving in the x direction, a third in the y direction and a third in the z direction. Obviously, this will never be true. But it is true that, if I resolve the actual velocities of the particles, the components of those velocities will be equal in all directions. Thus, to assume that at any time a third will be moving parallel to each axis is acceptable.

You or I might object to such an assumption on the grounds that, because of collisions, a particle moving in any particular direction would be knocked asunder and move off in another direction. But for every particle thus deflected, another will be deflected in a converse way. This also means that if we consider a certain number of particles heading for a wall of the container, the number striking that wall over a certain period of time will be the same because, for every particle which, because of collision with other particles, is deflected from its trajectory, another particle will be deflected from its original trajectory and will now move toward the wall. Despite this chaos, the picture of the gas that we have drawn is mostly empty space. Prediction is possible because we are dealing with such large populations.

For example, whereas from moment to moment I cannot predict who among the population will be unemployed, will suffer a heart attack or die, I can make predictions for the entire population about these variables. In the same way, although I cannot predict the behavior of individual particles on the gas, I can make very precise statements about the average behavior. Indeed, the mathematics for dealing with large populations was perfected first for the kinetic theory and its successors.

THE DERIVATION OF THE GAS LAW

Given the model that we have built for the gas, our understanding of the concept of gas pressure is as follows: As the particles move randomly in any sample of gas, they strike the walls of a container at a constant rate. The gas pressure is a manifestation of the forces exerted by the gas particles on the walls. Since we are dealing with such large numbers of collisions, we will have to employ an averaging technique to quantitatively calculate the force and the force per unit area, which is what we mean by pressure.

Consider the container shown in Figure 2. It is a cube of length 1 on all sides. Let us consider one particle moving in the x direction. We assume that it has a certain speed, v_{x1}. That being the case, the time required for the particle to traverse the length of the box is given by:

$$t = 1/v_{x1} \tag{2a}$$

We have labeled the wall B. As the particle moves back and forth in the x direction between B and the facing wall, the time between collisions of the particle with B is the distance traversed, divided by the speed of the particle:

$$t = 2l/v_{x1} \tag{2b}$$

Let us examine the interaction of a particle with the wall of the container. We have shown the particle approaching the wall in Figure 3. The particle has a mass m and a speed v_{x1}. Therefore, its momentum is mv_{x1}. Since the collision is elastic and the walls are essentially unmoveable, the particle will rebound from the wall

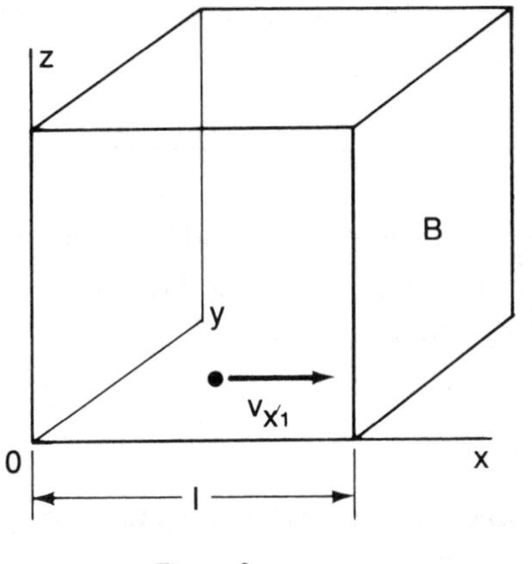

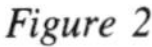

Figure 2

with the same magnitude of momentum with which it approached, but in the opposite direction. The force exerted on the wall by the particle during the collision is equal to the time rate of change of the momentum. For each collision by this particle, the change in momentum will be given:

$$\Delta p = mv_{x1} - (-mv_{x1}) = 2mv_{x1}$$

The time required for the particle to hit the wall is given by (2b), the number of collisions per unit time due to this one particle will be the *inverse* of (2b).

The average force exerted by this particle on this wall over time is going to be given by:

$$\begin{aligned} \overline{F}_{x1} &= 2mv_{xl}/2l/v_{x1} \\ &= mv_{x1}^2/l \end{aligned} \qquad (3a)$$

The bar above the symbol for force is to be read as the average force. In other words, $\overline{F}_{x1}$ is to be read, "The average force resulting from the particle moving with speed v_{x1}." The average pressure exerted by this one particle on the wall, B, is

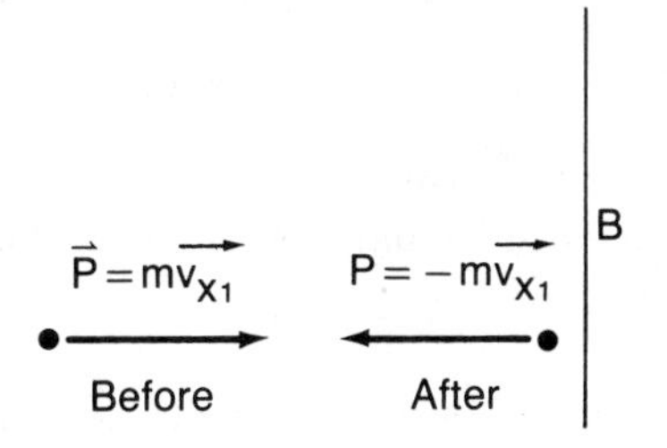

Figure 3

given by the average force per unit area. If the total area of the wall B is given by A, then

$$P_1 = \frac{\overline{F}_{x1}}{A} = \frac{mv_{x1}^2}{Al} = \frac{mv_{x1}^2}{V} \tag{3b}$$

V is the volume of the container. We could have done the same calculation for any other particle in the box moving in the x direction. For example, for a particle moving with a speed v_{x2}, equation (3b) is:

$$P_2 = \frac{mv_{x2}^2}{V} \tag{3c}$$

Since the pressures from each of the particles are additive, the total pressure P, on the wall of the container is:

$$\begin{aligned} P &= P_1 + P_2 + P_3 + \ldots \\ &= \frac{m}{V}(v_{x1}^2 + v_{x2}^2 + v_{x3}^2 + \ldots) \end{aligned} \tag{4a}$$

We can express this result in terms of the average speed of all the particles in the x direction. N is the total number of particles in the container, and the average of the square of the speed in the x direction is given by:

$$\overline{v_x^2} = (v_{x1}^2 + v_{x2}^2 + v_{x3}^2 + \ldots)/N \tag{4b}$$

N is the total number of particles in the container.

By combining equations (4a) and (4b), we obtain

$$P = Nm\overline{v_x^2} \tag{4c}$$

I have expressed the pressure on the wall labeled B in terms of the interaction of the particles which compose the gas and the wall. But, as we have already seen, the pressure is the same in all parts of the container. I could have picked any other direction and any other wall with which to perform the analysis. The magnitude of the average of the squares of the speeds in any other direction are found in the same way. We can express the average of the square of the speed in any direction as

$$\overline{v^2} = \overline{v_x^2} + \overline{v_y^2} + \overline{v_z^2} \tag{4d}$$

But since the Cartesian coordinate system is arbitrary, and no direction is inherently different from any other direction, we know that:

$$\overline{v_x^2} = \overline{v_y^2} + \overline{v_z^2} \tag{4e}$$

Replacing the last two terms in (4d) with their value in the x direction gives:

$$\overline{v^2} = 3\overline{v_x^2} \tag{4f}$$

This can be substituted into equation (4c) as follows:

$$PV = \frac{1}{3}Nm\overline{v^2} \qquad (5a)$$

We can see that we are arriving at a relationship that is an expression of the ideal gas law. Let us now manipulate (5a) to make the connection more explicit. Begin by multiplying and dividing the right side by 2:

$$PV = \frac{2}{3}N \cdot \frac{1}{2}m\overline{v^2} \qquad (5b)$$

The last term in (5b) is the average kinetic energy of a particle in the gas. This energy is sometimes referred to as the internal energy of the gas and given the symbol, U. Heating the gas, for example, will increase its internal energy. This can now be interpreted as increasing the average kinetic energy of the particles which make up the gas. Of course, heating the gas in this way will cause a change in the temperature of the gas. If we assume that the absolute temperature of the gas is a measure of the internal energy, then:

$$PV = \frac{2}{3}U \qquad (5c)$$

U is given by $\frac{1}{2}Nm\overline{v^2}$, the total internal energy of the gas. If we compare equation (5c) with (1d), we can make a precise statement about the relationship of the internal energy to the temperature:

$$\frac{2}{3}U = nRT \qquad (5d)$$

Let us reexamine the process. We began by proposing what the relationship is between pressure temperature and volume for an ideal gas, the ideal gas law. We recognize this law as the extrapolation of measurements of these variables.

We then made a series of assumptions about the nature of the gas, including the notion that the gas is composed of point masses. We applied Newton's laws to those particles by using an averaging technique and derived the same ideal gas law by identifying the total average energy of the ensemble of particles with the temperature of the gas.

We could now extend this analysis to other properties of the gas, derived from observation. These were originally expressed in laws, similar to the ideal gas law which makes no assumptions about the nature of the gas, but relates such quantities as volume, temperature, and pressure to each other in different ways under different conditions. To the extent that the relationships are derivable within the kinetic theory, the laws have been reinterpreted, as was the ideal gas law, in terms of Newtonian mechanics. When such derivations fell wide of the mark, physicists reexamined the assumptions and modified the analysis, making it more sophisticated and a better predictor. Thus, rather than basing the result on the average of the square of the speed of the particles, they realized that they could derive a relationship expressing how the speeds of the particles are distributed at different temperatures. This analysis became the successor to the kinetic theory, statistical mechanics. Later, as more subtle measurements became possible the assumptions behind the derivation were questioned because of the failure of the theory to comprehend the data.

A development like the kinetic theory can be interpreted as the reduction of the subject of thermodynamics to Newtonian mechanics. As we can see, it requires a number of assumptions about the nature of matter. It is this kind of reduction that was termed "the mechanical world view." It is not claimed that this was the explicit program of nineteenth century physicists. The notion of the mechanical world view is an abstraction that historians of science like me have applied to try to make sense out of the development of physical theory after Newton in the context of nineteenth century values. The subsequent development of "the electromagnetic world view" in Chapter 2 is also a creation of historians. Such categorizations are useful in organizing our understanding of late nineteenth and early twentieth century physics. Always bear in mind, though, that, like the laws of physics with which we have been dealing, these concepts are abstractions.

Appendix ·5·

Ether Drift Experiments: The Search for the Absolute Frame of Reference

IN CHAPTER 2, WE SAW that nineteenth century physicists hoped to determine the nature of the interaction of ether and matter and to measure the speed of the earth through the luminferous ether by performing certain kinds of optical experiments. In this appendix we elaborate on a number of these experiments. They have been chosen because they are typical. Many, many experiments of these types were performed.

The physical description of the apparati will be kept to a minimum. People who are not familiar with the day-to-day workings of physicists often have difficulty relating experimental design to theoretical questions. Therefore, virtually nothing is left to the imagination about how formal mathematical manipulation relates to experimental design and physical arrangements. Such manipulations provide the link between the question and the answer, presumably given by nature.

THE BRADLEY ABERRATION EXPERIMENT—c. 1727

James Bradley was a contemporary of Newton. The following experiment is one that he undertook to discover and measure stellar parallax, the apparent shift in the position of stars as a result of being viewed from slightly different perspectives as the earth moves around the sun. The phenomenon of parallax is illustrated in Figure 1. Whereas Bradley found that the apparent positions of the stars changed, it was not in the manner to be expected from parallax. For example, in Figure 1, we have shown a star located on a line perpendicular to the earth, sun plane

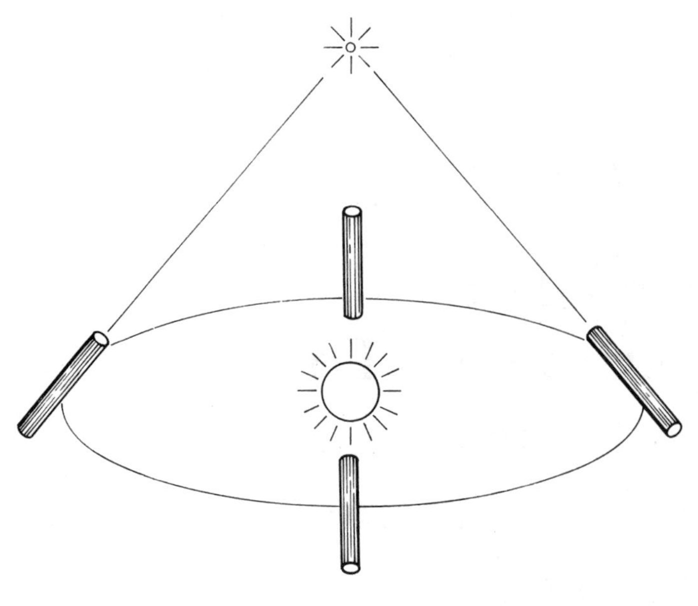

Figure 1

directly over the sun. The result of parallax is to see a change in the orientation of the star relative to other stars. This is a function of the position of the telescope on the earth. But the shift in the apparent position of the stars discovered by Bradley is not a function of position of the earth, but of the direction of motion of the earth (Figure 2). The shift was ninety degrees out of phase with the shift that Bradley sought. It puzzled him for a long time and if his description of how he came to properly interpret the results can be believed, it was a sudden experience and the solution appeared in a flash. It is one of those fascinating byways that we will not be able to pursue here.* The result of Bradley's work was an experimental determination of the speed of light. Later, it was reinterpreted to provide information about the interaction of the earth and the ether.

Suppose that we are examining a star that is directly overhead. *Experimentally,* Bradley found that over time, in order to keep the star in the center of the field of view, the telescope had to be tilted a small amount in the direction of the motion of the earth. This is shown in Figure 3.

We follow a light wave, represented by one ray (see Chapter 2) as it proceeds from the top to the bottom of the telescope. At time t_0 (Figure 3) the ray enters the telescope at e, when the telescope is at a. In the time that the telescope moves from

* See A. B. Stewart, "The Discovery of Stellar Aberration," *Scientific American,* (March), 1964, *210*:100.

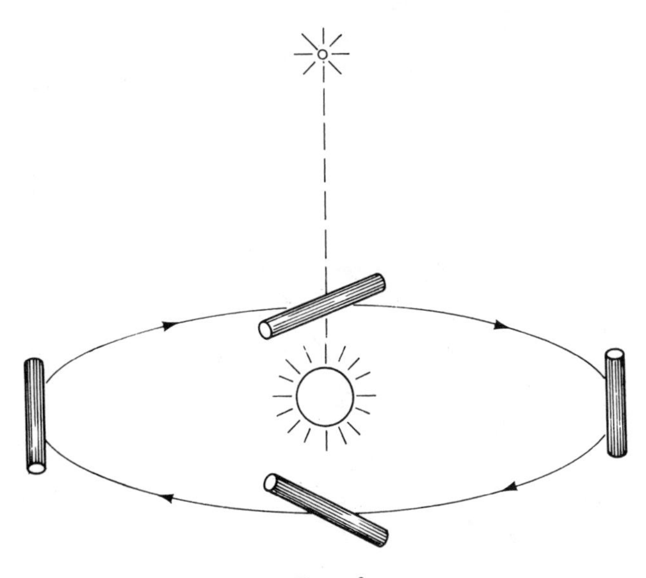

Figure 2

a to b, the ray of light arrives at the bottom of the telescope at b. The speed of light is given by c (We assume that the speed of light in empty space is the same as the speed of light in air; see Chapter 2), and the distance that the light has traveled from the top to the bottom of the telescope is given by:

$$eb = c(\Delta t) \tag{1a}$$

where $\Delta t = t_1 - t_2$. In other words, the light has continued on its vertical pathway. In that same time the telescope, firmly attached to the surface of the earth, has moved from a to b. Had the telescope been vertical and the light entered e, the center of the tube, the light would have no longer been at the center when it arrived at b. In order to keep it in the center the telescope was tipped *very* slightly in the direction of motion. (The degree to which the telescope has to be tipped has been exaggerated in Figure 3 for illustration). This is because as the light moved from e to b, the back wall of the telescope would move toward the line eb. Tipping the telescope in the direction of motion allows the back wall to fall away in proportion to the distance the telescope is traveling toward the line eb. This phenomenon is known as aberration and the angle in Figure 3 is known as the angle of aberration.

The distance the telescope has moved, ab, is given by:

$$ab = v(\Delta t) \tag{1b}$$

v is the orbital speed of the earth. Examining Figure 3, we see that

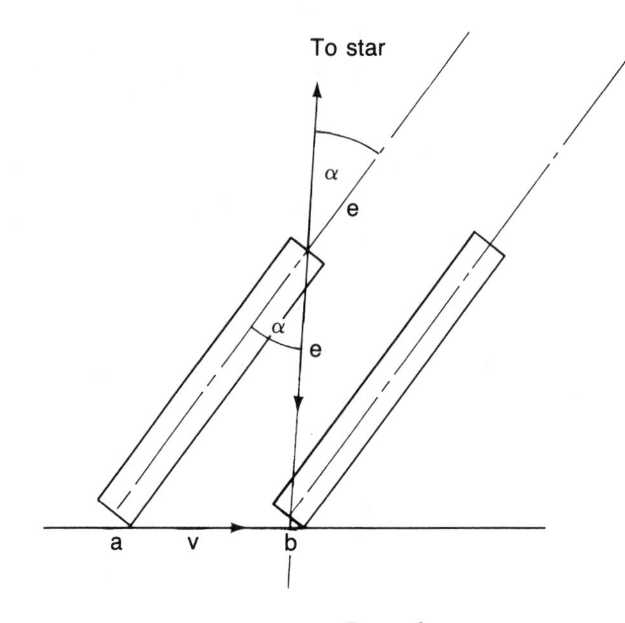

Figure 3

$$tan\ \alpha = \frac{v(\Delta t)}{c(\Delta t)} = \frac{v}{c} \qquad (1c)$$

But, as explained in Appendix 1, for very small angles the tangent is approximately equal to the angle itself when the angle is expressed in radian measure. Therefore (1c) becomes:

$$\alpha = v/c \qquad (1d)$$

Assuming that we know the orbital speed of the earth (as it takes one year for the earth to travel in approximately a circular path whose radius is 9.3×10^7 miles, it is not difficult to calculate the orbital speed of the earth), and that we measure α, we use equation (1d) to calculate the speed of light. The value thus calculated by Bradley was consistent with other experimental determinations of the speed of light that had been made.

For our purposes we have assumed that, as the light moves in a vertical direction and the earth moves in a horizontal direction, the two motions remain independent of each other. In other words, the Bradley experiment implies that the motion of the earth and the motion of the light are independent of each other. But if the motion of the light is the propagation of a disturbance in a medium, the motion of the earth does not affect the medium. The medium remains absolutely fixed.

THE MAXWELL-ROEMER EXPERIMENT

This is an experiment that has never been carried out. It is included because the conception is magnificent and it beautifully illustrates the concepts employed in attempting to determine the absolute velocity of the earth and the manner in which ether and matter interact. The idea of using the experiment to measure the motion of the earth through the ether was suggested in the nineteenth century by the British physicist Clerk Maxwell. It is based, however, on a technique for measuring the speed of light which was first devised by the Danish natural philosopher Olaf Roemer. The results of that experiment were published in 1676. Roemer was working in France at the time as a paid member of the French Academy of Sciences.

Before Roemer, several serious attempts to measure the speed of light had been made. Galileo had proposed that two men with cloaked lanterns be stationed some distance from each other. One of the men would uncloak his lantern. Upon seeing the light from that lantern, the second participant in the experiment would uncloak his lantern. The time interval between the first man's uncloaking of his own lantern and his sighting of the light from the second man's lantern would be twice the time required for light to traverse the distance between the two lanterns. Galileo had even attempted the experiment but had found the results inconclusive. It was generally recognized that if the speed of light were not infinite but very very fast, the distances required to detect the effect would have to be very, very large.

During the same period, Descartes, convinced that the speed of light was infinite, proposed that one measure the observed time for the start of a lunar eclipse and compare it to the predicted time. Any significant discrepancy could be the result of the time required for light to traverse the distance between the earth and the moon.* Although correct in principle, the method was unworkable because the expected time for an eclipse of the moon was not known to the accuracy required.

For approximately eight years before Roemer's work, astronomers at the Royal Academy had observed a curious variation in the period of the first of Jupiter's moon's, Io. This time was rather easy to establish since one could use the start or the end of an eclipse of Io as a reference point as it moved behind or emerged from behind the planet, viewed from the earth. There was general agreement that the period of Io was somewhat longer than forty-two hours, but there were discrepancies noted of as long as twenty-five minutes. Roemer suggested that the variations

* In fact, the discrepancy would be due to the delay in the cessation of light coming to us from the moon.

in the measured period of the moon were because light has a finite velocity and requires a certain time to travel between Io and the earth. In fact, he proposed using the discrepancy to measure the speed of light.

Io is eclipsed by Jupiter every time it enters the area marked as Jupiter's shadow (Figure 4). If τ is the period of the moon, the time required for the moon to move once around Jupiter, then the time required for Io to reenter Jupiter's shadow on two successive occasions is τ. Observing this from the earth, when the earth is a distance L from Jupiter, a certain time interval is necessary for the information that Io has entered Jupiter's shadow to arrive at the earth. If L is the distance from the earth to Jupiter, then that time interval is given by L/c where, as usual, c is the speed of light.

The period of Jupiter around the sun is very slow compared to the period of the earth. It requires twelve years for Jupiter to make one circuit around the sun. Therefore, in the course of the forty-two hours required for Io to go once around Jupiter, the planet is virtually at rest. The earth, on the other hand, changes its

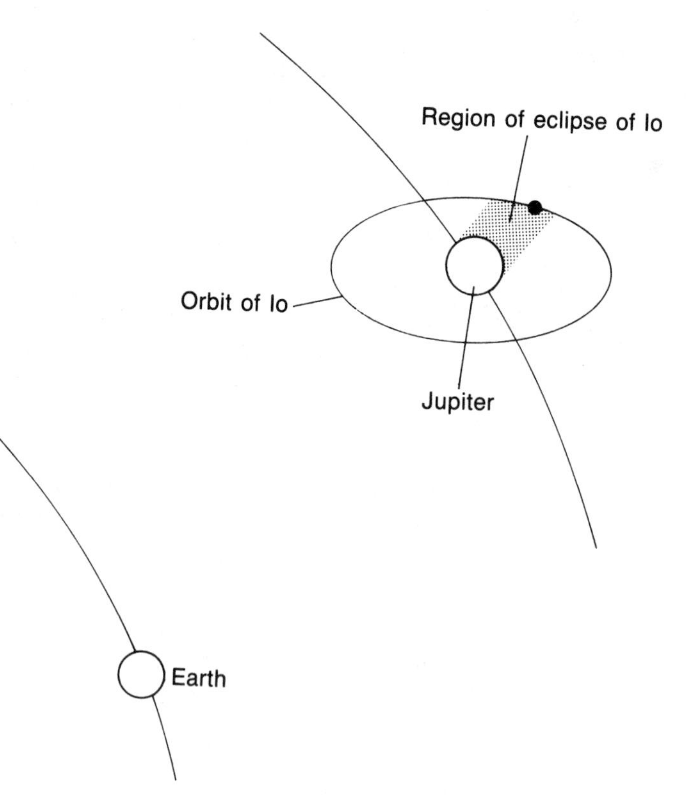

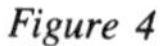

Figure 4

position relative to Jupiter by a significant amount during that period. In Figure 5, when the earth is at position 1, the time required for the information that Io has entered Jupiter's shadow requires a time L_1/c. When the moon again enters the shadow, the earth is at position 2 and the time required for the information to reach the earth is L_2/c. The period is longer than had the earth been at rest. Let us work out the details.

Suppose the earth is at position 1 at time t_1. The information that Io has entered the planet's shadow arrives at a time given by:

$$T_1 = t_1 + L_1/c \tag{2a}$$

The next time Io enters Jupiter's shadow, the earth is in position 2 at time t_2. The information of this event arrives at the earth at a time given by:

$$T_2 = t_2 + L_2/c \tag{2b}$$

But we see from the figure that:

$$L_2 = L_1 + \Delta 1 \tag{2c}$$

$\Delta 1$ is the distance that the earth has moved in the time required for Io to go

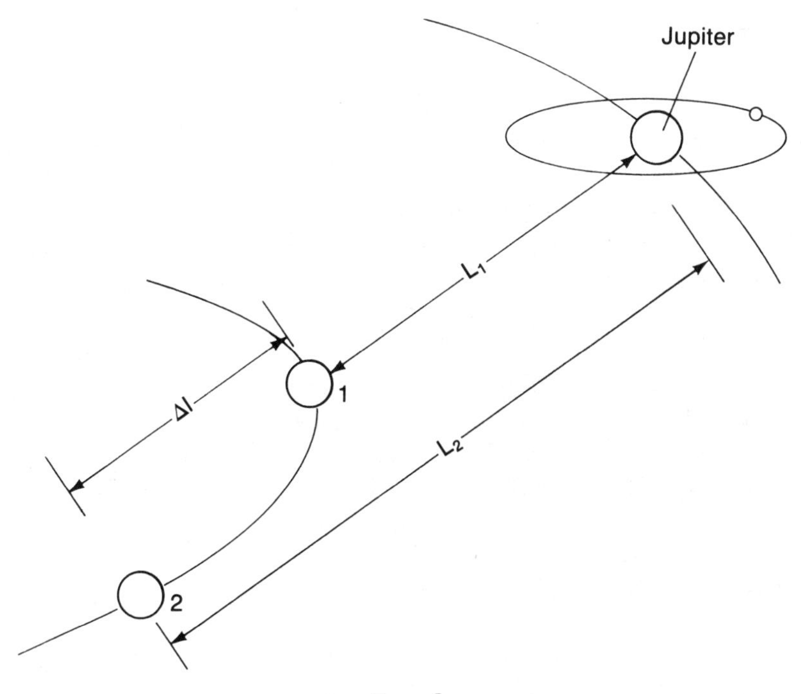

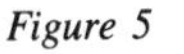
Figure 5

once around its orbit. Combining (2c) and (2b) and subtracting (2a) from the result gives the following for the measured period for the moon:

$$\Delta T = T_2 - T_1 = (t_2 + L_1/c + \Delta 1/c) - (t_1 + L_1/c)$$
$$= t_2 - t_1 + \Delta 1/c \qquad (2d)$$

But as $t_2 - t_1$ is the period that would be measured when the earth is at rest—the "true" period—(2d) can be rewritten as

$$\Delta T = \tau + \Delta 1/c \qquad (2e)$$

We must examine equation (2e) very carefully. What is observed is ΔT. Since the earth *is* in motion, we cannot measure τ and there is no way to measure $\Delta 1$, the change in distance between the earth and the moon during the time of one period. We can see from Figure 5 that $\Delta 1$ will be different at different times of the year. It will be the projection of portions of the earth's orbit onto the line between the earth and Jupiter. As the earth moves toward Jupiter the observed period will be shorter; as it moves away from Jupiter, the observed period will be longer.

We are in the unenviable position of having one equation with three unknowns because we also do not know the velocity of light. But there is something that can be done about it. Suppose that, rather than measuring one period, we measured the time required for a number of periods. Let that time be T_n and let the number of periods be n. Equation (2e) now becomes:

$$T_n = n\tau + \Delta 1_n/c \qquad (2f)$$

where τ is as before, the true period of Io and $\Delta 1_n$ is the total distance the earth has moved in the time T_n. In (2e) the notion of the observed period, ΔT has been replaced by the total elapsed time for n periods of the planet to be counted.

Let us now pick a time period to measure so that $\Delta 1_n$ is zero. This requires approximately one year. Equation (2f) becomes:

$$T_n = n\tau \qquad (2g)$$

and, having measured the time interval and counted the number of periods, the magnitude of the period of Io as seen from Jupiter could be calculated.

There are several techniques for insuring that $\Delta 1_n$ is zero. For example, at the time of closest approach between the earth and Jupiter, Jupiter crosses the meridian at midnight. In other words, Jupiter is directly opposite the sun relative to us. The planet is said to be *in opposition*. This is shown in Figure 6.

Having determined the true period of Io, we can return to equation (2f) and use it to determine the velocity of light.

Rather than counting the number of periods between successive oppositions of the planet, let us now count the number between one opposition and the time

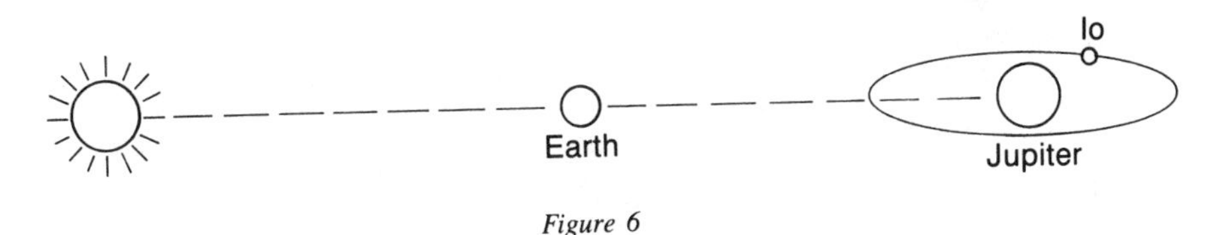

Figure 6

when Jupiter and the sun are rising at the same time. In other words, let us count the number of periods for Io in the time that it takes the earth to go halfway around its orbit. Examine once again equation (2f). We are going to measure T_n and count the number of periods n. We have already determined τ. The only variables left are $\Delta 1_n$ and c. But in this case $\Delta 1_n$ is nothing more than the diameter of the earth's orbit, and the discrepancy between T_n and $n\tau$ will be the time required for light to traverse the earth's orbit.

Roemer reported that the time required for light to traverse the earth's orbit was twenty-two minutes. Contemporaries of Roemer took this figure, and using the then accepted values for the size of the earth's orbit, calculated the speed of light to be between 130 and 140 thousand miles per second. (Using the modern value for the earth's orbit, Roemer's data leads to a value of 139,000 miles per second, consistent with Bradley's later determination. The modern figure for the time required for light to traverse the earth's orbit is seventeen minutes.).

When Roemer did this experiment, it caused a sensation. Roemer's colleague at the Academy of Sciences, Cassini, saw it as "one of the most beautiful problems in physics," but was not convinced by the evidence. Robert Hooke, the curator of the Royal Society, considered to be one of the outstanding "virtuosi" of the time, thought that the experiment was inconclusive. Others renounced it as the "ingenious and deductive hypothesis of (the wave theory). . . . "* Nevertheless, acceptance came. Both Newton and Huygens threw the weight of their authority behind the technique.†

Roemer's conception is wonderful, and clever. It is one of those jewels of an idea.

No less clever was Maxwell's proposal to use the Roemer determination of the speed of light to measure the absolute motion of the earth through the ether. Recall that twelve years are required for Jupiter to make one complete turn of its orbit. Suppose that observers on earth constantly monitor the speed of light with

* C. B. Boyer, "Early Estimates of the Velocity of Light," *Isis,* 1941, *33*:24–40.

† For an account of the acceptance of Roemer's work, see I. B. Cohen, *Roemer and the First Determination of the Velocity of Light* (New York, 1942). Cf. N. E. Dorsey, "Farenheit and Roemer," *Wash. Acad. Sci,* 1946, *36*:361–72. I. B. Cohen, "Roemer and Farenheit," *Isis,* 1948, *39*:56–58.

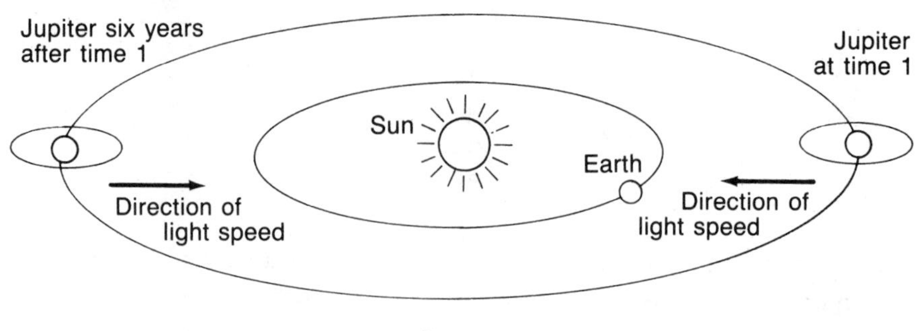

Figure 7

the Roemer technique. As shown in Figure 7, if the speed of light is determined at one point in the orbit, six years later, the direction traveled by the light will be directly opposite. Any serious discrepancy can be accounted for by assuming that the earth is moving relative to the medium through which the light is moving.

As I said at the start of this discussion, the experiment has never been done. The reason is, very simply, that it has been beyond our technological prowess. Suppose that we were looking for a discrepancy as large as one second. There are a total of 189 million seconds in six years. We would be looking for a discrepancy of one part in 189 million, a rather small number requiring fantastically accurate clocks. But, in fact, it is more than likely that the discrepancy we are looking for cannot be any greater than one part in a million, if that large. This would require a clock whose accuracy was greater than one part in 10^{15}. Such accuracy in clocks is probably no longer beyond our means but by now, given what has happened to our attitudes toward ether drift experiments, the experiment is not worth trying.

THE FIZEAU EXPERIMENT

In Chapter 2 we pointed out that in 1810, Arago performed the simple experiment of focussing a telescope on a star. He found that over time, the focus of the telescope did not change. If the motion of matter does not affect the ether, one can expect the focus to change since a change in the relative speed of light through matter (for example, through the glass of the lenses of the telescope) should change the index of refraction and hence the focus point.

This negative result was explained in 1818 by Fresnel by assuming that ether is partially dragged by matter. If v is the relative speed between matter and ether, and n is the index of refraction, the speed of the ether inside matter is given by:

$$v(1 - 1/n^2) = fv \qquad (3a)$$

That no motion of the ether is observed outside of matter is accounted for by assuming that there is a discontinuity at the boundary between ether and matter so that inside the speed is as in (3a), and outside the speed is zero. The quantity $f = 1 - 1/n^2$ is often referred to as the *Fresnel dragging coefficient.*

This coefficient was put to a direct test in 1851 by the French physicist A. Fizeau, who had devoted a good part of his career to working out terrestrial techniques for measuring the speed of light. Fizeau compared the speed of light in transparent matter at rest to the speed of light in the same matter when that matter was in motion.

Several things made the attempt very, very difficult, and the most serious is that the speed of light is so much faster than any other speed familiar to us. In performing the experiment, Fizeau was searching for very small changes in a very large number and he took the only route possible. He built an apparatus that was very sensitive to changes in the speed of light. The detection of the changes were manifested by shifts in the pattern of fringes from the interference of two light beams.

Figure 8 is a schematic representation of the experiment. Three full-silvered mirrors, P_1, P_2, and P_3 are arranged with respect to a half-silvered mirror at M. A half-silvered mirror is what the name suggests: a piece of glass is lightly silvered so that the intensity of light reflected from the surface is equal to the intensity of light

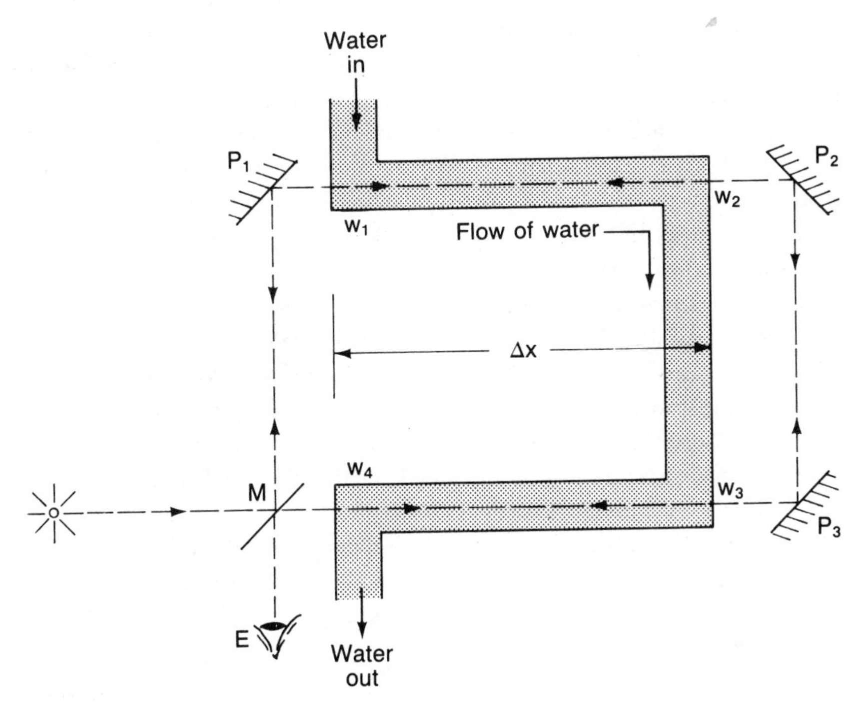

Figure 8

that passes through the mirror to the other side. Thus, if a beam of light is incident on the surface, half the light is reflected and half is transmitted.

A beam of light originates at 0. It strikes M so that the angle of incidence is 45°. As the angle of incidence must equal the angle of reflection, the beam is split into two parts, one is transmitted by M and the other reflects at 45° to M. Given the arrangement of the full-silvered mirrors, two paths for light are established: one is clockwise from M to P_1, P_2, P_3 and then back to M, and the second from M to P_3, P_2, P_1 and then back to M. Once the beams arrive at M, they are again partly reflected and partly transmitted. For the beam that traveled in the clockwise sense, $MP_1P_2P_3M$, the reflected beam from M arrives at the eye of the observer. For the beam that traveled in the counterclockwise sense, $MP_3P_2P_1M$, the transmitted beam arrives at the eye of the observer. These two beams of light interfere with each other, and the observer sees a series of dark and light bands.

As shown in Figure 8, the mirrors are arranged so that for part of the time, the light is moving through water in the pipes W_1W_2 and W_3W_4. (The water piping would have to be transparent where the light entered and left the water system). As shown, Fizeau arranged for the water to be set in motion by a pumping system which could quickly get the water up to a speed of seven meters per second.* Seven meters per second is such a small fraction of the speed of light that you may well wonder that any change could be observed. The fact that Fizeau had such expectations speaks for the incredible sensitivity of interference techniques. It is to be noted that the clockwise beam is traveling in the same direction as the water, and that the counter clockwise beam is traveling in a direction opposite to the direction of the water. It is also to be noted that along the parts of the path between M and P_1 and between P_2 and P_3 there is no water. We can ignore those two parts of the light path as the clockwise and counterclockwise beams will be unaffected in those parts of their trajectory.

With the water at rest, an observer at E will find that the time for a ray of light to traverse $MP_1P_2P_3ME$ is the same as the time for a ray of light to traverse $MP_3P_2P_1ME$. This manifests itself in a particular inference pattern. Once the water is moving, the speed of light in the directions MP_3 and P_1P_2 will no longer be the same clockwise and counterclockwise and this manifests itself as a fringe shift. It was that *shift* that Fizeau looked for as evidence that the speed of light was being affected by the motion of the water.

Let us suppose that before turning on the water, light takes a time Δt to traverse the length Δx in the arms P_1P_2 or MP_3. The total time for a ray of light to traverse

* The pumping system was a very simple arrangement. A large tank of water was located four floors above the experimental apparatus. Using a valve much like the valve in a flush toilet, the tank could be rapidly emptied.

both of these lengths is $2\Delta t$, in either the clockwise or counterclockwise direction. When the water begins to flow, three possibilities are presented:

1. The motion of the water does not affect the velocity of the light. This we will refer to as "the no drag hypothesis."
2. The motion of the water is communicated *in toto* to the light. This will be referred to as "the total drag hypothesis."
3. The motion of the water is only partially communicated to the velocity light. This will be referred to as "the partial drag hypothesis."

In order to see how the relationship between the physical apparatus and the conclusion is reached, let us work out the mathematical implications of the three possibilities.

THE NO DRAG HYPOTHESIS

This hypothesis assumes, as in the Bradley experiment, that the ether is absolutely fixed, not only in free space, but in matter as well. Since there would be no affect on the speed of light, we would observe no shift in the fringe pattern at E.

THE TOTAL DRAG HYPOTHESIS

All of the velocity of the water is communicated to the propagating light. If c' is the speed of light in water, in one direction the speed is $c' + v$ (the light is moving in the same direction as the water), and in the other direction the speed of light is $c' - v$ (The light is moving in a direction opposite to the direction of water.). Let ΔT_1 be the time required for the clockwise ray to traverse the distances MP_3 and P_1P_2. Call these distances $2\Delta x$. Then

$$\Delta T_1 = \frac{2\Delta \mathrm{x}}{\mathrm{c}' + \mathrm{v}} \tag{3b}$$

For the counterclockwise beam, let the time required to traverse be MP_3 and P_1P_2 be designated by T_2. Then

$$\Delta T_2 = \frac{2\Delta x}{c' - v} \tag{3c}$$

The time difference for the beams moving in opposite senses will be given by:

$$\Delta T_2 - \Delta T_1 = \frac{2\Delta x}{c' - v} - \frac{2\Delta x}{c' + v}$$

$$= \frac{2\Delta x(c' + v) - 2\Delta x(c' - v)}{c'^2 - v^2} = \frac{4v(\Delta x)}{c'^2 - v^2} \tag{3d}$$

In still water, the total time required for the beam to traverse $MP_1P_2P_3ME$ is the same as the time required for the beam to traverse $MP_3P_2P_1ME$. Call this time Δt. It is composed of two time intervals. The first, Δt_1 is the time interval required to traverse $MP_3 + P_1P_2 = 2\Delta x$. The second, Δt_2, is the time interval required to traverse $P_1M + P_2P_3 + ME$. We can compare the times required before and after the water begins to move by remembering that Δt_1 is nothing more than $2\Delta x/c'$. We can now substitute directly for Δx in equation (3d). Since Δt_2 is the same before and after the water begins moving, the time *difference* in equation (3d) is due solely to the alteration in Δt_1.

$$\Delta T_2 - \Delta T_1 = 2\Delta t_1\left(\frac{c'v}{c'^2 - v^2}\right)$$

In order to focus on the moving water more clearly, divide both the top and the bottom terms of this last equation by c'^2. The equation then takes the following form:

$$\Delta T_2 - \Delta T_1 = 2\Delta t_1\left(\frac{v/c'}{1 - v^2/c'^2}\right) \tag{3e}$$

but, as speed of the water will never be anything but the smallest fraction of the speed of light in water, the ratio v/c' will be very small. It follows that the ratio of v^2/c'^2 will be even smaller. Therefore we can approximate the denominator in (3e) by 1 and the equation becomes:

$$\Delta T_2 - \Delta T_1 = 2\Delta t_1(v/c') \tag{3f}$$

Let us examine this equation carefully. First, it says that once the water begins moving there is a time difference for the light to traverse the apparatus clockwise compared to the time required for the counterclockwise path. This difference is proportional to the ratio of v/c'. (Hence the experiment is called a "first-order" experiment because the time difference is proportional to v/c' raised to the first power.)

The equation can be manipulated further to advantage. Before the water is turned on, the time required to traverse the arms parallel to the flow of water is designated as Δt_1. After the water is turned on, the time in the direction opposite to the flow of the water is ΔT_2 and the time required in the direction of the water is ΔT_1. The fractional change in the time required to move around the apparatus is given by:

$$F = \frac{\Delta T_2 - \Delta T_1}{\Delta t_1} \tag{3g}$$

This is the left side of equation (3e) if we divide it by Δt_1. Equation (3f) can therefore be written as follows:

$$F = 2v/c' \qquad (3h)$$

Let us recall what this means physically.

Before turning the water on, a certain fringe pattern is established in the apparatus as the light waves moving clockwise and counterclockwise, are brought together and allowed to interfere at E. When the water is turned on, a certain fractional change in the time for the light beams to run the course will occur, as given in (3h) and that time difference manifests itself as a shift in the pattern of fringes. One can predict how much of a shift will take place in terms of the known parameters of the apparatus, the expected speed of the water, and the known speed of light in water.

THE PARTIAL DRAG HYPOTHESIS

If the motion of the water only partially drags the ether (and hence the light), only a fraction, f, of the water's motion will be impressed on the moving light rays. Equation (3h) becomes:

$$F = fv/c' \qquad (3j)$$

This is derived from the beginning by substituting fv for v in equations (3b) and (3c). Everything else proceeds as before. The value of f will be between zero and one. Equation (3j) becomes the general equation for the effects of moving water on the fringe pattern. When $f = 1$, (3j) reduces to (3h).

When Fizeau did the experiment, he found a fringe shift. The magnitude of the fringe shift corresponds to a time difference that indicates that the ether is only partially dragged along by the moving water. Since Fizeau calculated the time difference by measuring the amount of fringe shift, he was in effect measuring the left side of equation (3j). He knew the value of v and of c'. He therefore solved equation (3j) for f. The value of f corresponded quite closely with his prediction. That is:

$$f = 1 - 1/n^2$$

This experiment was done in the middle of the nineteenth century. As it is a terrestrial experiment in which the velocity of light through moving media are compared, it is direct and powerful experimental evidence in support of the partial drag hypothesis. At first it appears to contradict the result of the Bradley experiment, which suggests that there is no drag whatsoever. Toward the end of the nineteenth century, H. A. Lorentz and others were able to work out a reconciliation which applied the partial drag hypothesis to the Bradley experiment.

But in the meantime, an experimental modification of the Bradley experiment by the British astronomer George Airy confirmed the partial drag hypothesis in a rather dramatic way.

THE AIRY EXPERIMENT

Airy's experiment differed from Bradley's in that he used a telescope whose tube was filled with water. The index of refraction of water is 1.33. Therefore it requires a longer time for the light to descend the telescope tube in Figure 3. The index of refraction is the ratio of the speed of light in vacuum to the speed of light in another medium. In this case, the light is moving in water at three-fourths the speed it moves in air. Therefore, in the time required for the light to descend the telescope tube filled with water, the earth moves farther than it did when the telescope was filled with air.

There is another related effect to be considered. When light moves from one medium to another, the change in velocity also changes the direction of the light as it enters the second medium. It is this property which allows glass to be shaped so as to focus light rays from different directions on a point—in other words, to be a lens. The transition between media is described by Snell's law, or the law of refraction. According to this law, if the angle of incidence, θ, of the light is defined as the angle between the approaching ray and the perpendicular to the surface boundary between two media, and the angle of refraction, θ', is defined as the angle between the ray after it has entered the second medium and the perpendicular to the boundary between the two media,

$$\frac{\sin\theta}{\sin\theta'} = n = c/c' \tag{4a}$$

This relationship is illustrated in Figure 9.

Not only is there a change in the speed of light as it moves from one medium to

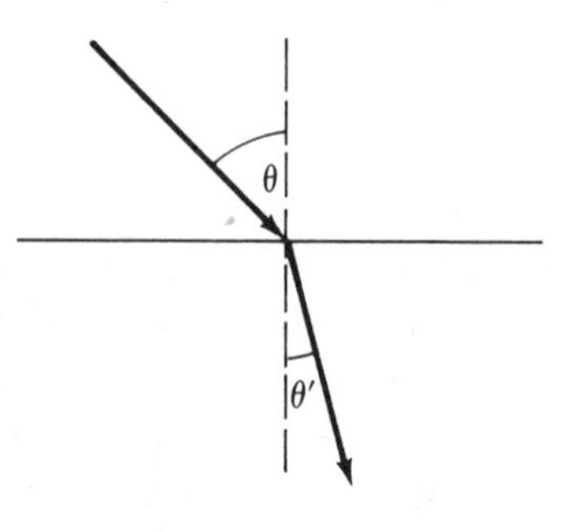

Figure 9

another, but that change manifests itself as a change of direction of the light. The amount of the change of direction depends, as we see from (4a) on the angle of incidence. Should the angle be zero (light ray perpendicular), no change in direction will occur. Should the index of refraction be a number greater than one, the sine of the angle of incidence will always be greater than the sine of the angle of refraction. This means that the angle of incidence will be greater than the angle of refraction. In other words, the light will be bent *toward* the perpendicular to the boundary between the two media.

A schematic of the Airy experiment is shown in Figure 10. Experimentally, *Airy found that, although the telescope was filled with water, the angle of aberration was precisely the same as it was when Bradley performed the experiment with an empty telescope*. At first glance this is surprising. Had the telescope been empty, the path of the light down the tube would have carried it to D, as in the Bradley experiment. But because of the water, the light in the tube is refracted toward the perpendicular. In other words, the light arrives at some point, say C. But in fact, since the aberration angle is the same as in the Bradley experiment (We must tip the telescope the same amount as before in the direction of motion to keep the star centered.) and the velocity of light in water is less than in air, the light will have arrived at E.

Recall that in the Bradley experiment the light traveled down the telescope tube a distance $AD = c\Delta t$. During that time the earth had moved a distance $BD = v\Delta t$. The velocity of light in water, c' is less than c. The path of light in the telescope tube, assuming that the motion of the light is independent of the motion of the earth, is AC. In traveling from A to C the light requires a time ΔT. In other words, $AC = c'\Delta T$. But since $c' = c/n$, where n is the index of refraction we have:

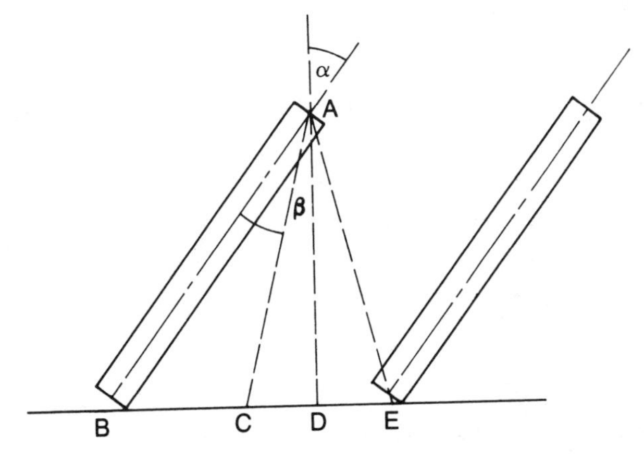

Figure 10

$$AC = c'\Delta T = c\Delta T/n \tag{4b}$$

The ratio of the distance traveled by the light with the telescope filled with water, to the distance traveled with nothing but air in the telescope is given by:

$$AC/AD = \frac{(c/n)\Delta T}{c(\Delta t)}$$

It is important to remember that we are dealing with very, very small angles. The aberration angle in the Bradley experiment was approximately twenty seconds of arc. This means that in practice, AC is approximately equal to AD. Therefore, the left side of this last equation is approximately equal to one:

$$\frac{AC}{AD} \simeq 1 = \frac{(c/n)\Delta T}{c(\Delta t)}; \ \Delta T \simeq n(\Delta t) \tag{4c}$$

Since the light arrives at point E, in the time, ΔT, the earth travels a distance $BE = v\Delta T$. Making use of (4c) we can write

$$BE = vn\Delta t \tag{4d}$$

Examine Figure 10 once again. For this case, Snell's law can be written as:

$$\frac{\sin \alpha}{\sin \beta} = n$$

Since α and β are *very* small angles

$$\sin \alpha \simeq \frac{BD}{AC}; \ \sin \beta \simeq \frac{BC}{AC}$$

Therefore,

$$\frac{\frac{BD}{AC}}{\frac{BC}{AC}} = n = BD/BC \tag{4e}$$

In the Bradley experiment we saw that $BD = v\Delta t$. We substitute this into (4e) and write:

$$BC = \frac{v}{n}(\Delta t) \tag{4f}$$

By combining (4c), (4d) and (4f) we can calculate the velocity of the ether (and the light) in the direction BE if the light is to arrive at E rather than C. In keeping with earlier terminology, let us call this the drag velocity, v'.

Referring again to Figure 10, in order to compute v' in terms which have already been defined and derived, we express the distance CE in terms of known parameters. The time required for the light to move from C to E is precisely the time required for the light to move, in the perpendicular direction from A to C. According to (4b) that time is ΔT. Therefore:

$$v'(\Delta T) = \mathrm{CE} = vn\Delta t = \mathrm{BE} - \mathrm{BC} \tag{4g}$$

Substituting for BE and BC by making use of (4d) and (4f) we have

$$v'n\Delta t = vn\Delta t - (v/n)\Delta t$$

or

$$v'n = vn - v/n$$

Dividing by n, the index of refraction, and collecting terms,

$$v' = v(1 - 1/n^2) \tag{4h}$$

Equation (4h) says that the velocity of drag is, indeed, proportional to the velocity of the earth, v. As n is greater than 1 (the index of refraction for water is 1.33) the term $1 - 1/n^2$ is a number less than 1. So the Airy experiment suggests that the ether is partially dragged along by the earth since the light has somehow acquired a speed in the direction of the earth's orbital velocity in order to arrive at E rather than at C. We also see that the amount of drag is dictated by the Fresnel dragging coefficient, the same factor which Fizeau found in his experiment on the speed of light in moving water. Fresnel had predicted this on the basis of Arago's fixed focus telescope experiment (Chapter 2).

It is important to point out that the experiments by Arago, Fizeau, and Airy are representative of a large number of ether drift experiments performed during the nineteenth century. All of them suggested that the ether was partially dragged by matter. All of these experiments were first order experiments and as discussed in the text of Chapter 2, the theory of electrons that H. A. Lorentz perfected in the early 1890s accounted for the mechanical, thermal and electromagnetic behavior of matter and radiation, and for the various first order ether drift experiments, also. Lorentz's theory corresponded to a principle which held that to the first order in v/c (where v is the speed of an object relative to the fixed ether and c is the velocity of light), the laws of physics have the same form in all inertial frames of reference. However, one expects some detectable effects of motion relative to the ether, if an experiment sensitive enough to reveal second order or higher effects, can be carried out. (By second order effects, I mean effects which depend on v^2/c^2).

Such experiments had begun to be performed. There are a host of them, but the most famous was the first one by the American, A. A. Michelson, in 1881 and in 1887, by Michelson with E. W. Morley. Since that time, the experiment has been repeated under a variety of conditions. The sensitivity of the apparatus is rather remarkable. When Michelson constructed his first version of it (the instrument is commonly referred to as an *interferometer*) he was working in a basement laboratory in Berlin. But because the vibrations from people walking on the sidewalk outside the building perturbed the proper working of the instrument, Michelson was forced to move the apparatus to the quiet of a country setting near Potsdam.*

The results of this experiment in and of themselves are not so astounding. As we will see, there was no measurable drag of the ether by the earth. What was devastating about that result is that it was a contradiction to the mass of evidence accumulated from first order ether drift experiments. This suggests that the matter drags ether, as in the Fresnel dragging coefficient. As Lorentz's theory was tailored around that assumption, the Michelson-Morley experiment and the other second order ether drift experiments were expected to show some effects of the earth moving through the ether. The results of the experiments were embarrassments which had to be dealt with.

THE MICHELSON-MORLEY EXPERIMENT

The Michelson-Morley experiment depends on observing the behavior of fringes when light beams are combined and allowed to interfere. While all such instruments are known as "interferometers," that term more and more is reserved for the particular interferometer of the Michelson-Morley experiment.

The apparatus is depicted, schematically, in Figure 11. Light is incident from the direction q onto the half-silvered mirror at 0. Part of the light is reflected from 0 to the fully-silvered mirror at A and is reflected back to 0, while part of the light incident on 0 is transmitted through 0 to another fully-silvered mirror at C whence it, too, is reflected back to 0. Part of the light arriving back from A is transmitted through 0 to the observer while part of the light arriving from C is reflected by 0 to the observer. It is thus possible for the observer (represented by the eye in Figure 11) to observe the fringe pattern resulting from the interference of the two beams of light.

Figure 11 depicts the interferometer in three different positions as it moves through space. The dotted lines represent the paths of the two beams of light as seen by an observer at rest with respect to absolute space. We have arranged the apparatus so that the arm OC of the instrument is parallel to the x axis, the presumed direction of the earth's motion through space.

* The most detailed account of the conditions under which the experiment took place are in L. S. Swenson, Jr., *The Ethereal Aether: A History of the Michelson-Morley-Miller Aether-Drift Experiments, 1880–1930* (Austin, 1972).

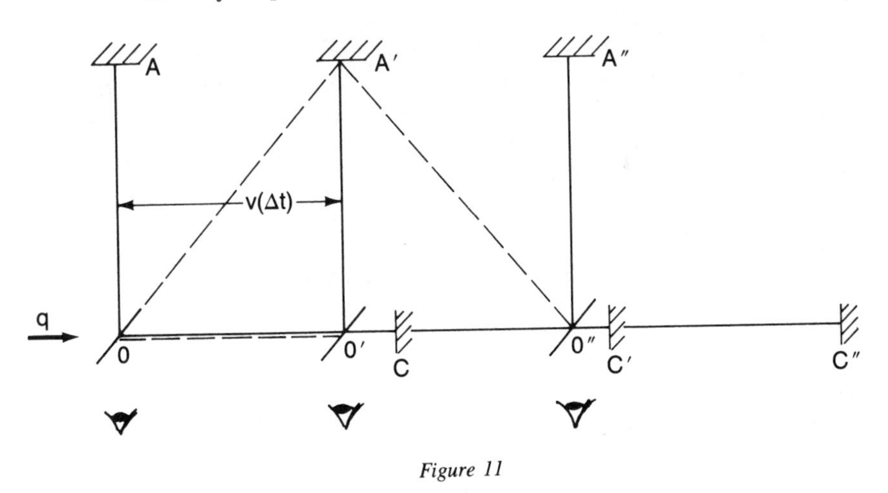

Figure 11

To detect some effect of the motion of the earth, Michelson and Morley set up the interferometer as shown and slowly rotated it until the arm 0A was parallel to the x axis, observing meanwhile, to see if a time difference in the paths of the two beams could be detected. This time difference would manifest itself as a shift in the fringe pattern. It would result from the fact that the roles of the arms 0A and 0C would replace one another. In other words, as we will show, the time required for the light to move back and forth in the x direction is different from the time required for the light to move back and forth in a perpendicular direction. Rotating the interferometer interchanges the role of the two arms of the instruments as 0A, originally in the perpendicular direction, rotates to the direction parallel to the x axis, whereas the arm 0C, originally parallel to the x axis, later becomes perpendicular to that direction.

As with the Fizeau experiment, there are three possible hypotheses: total drag, no drag and partial drag. We will analyze each in turn.

TOTAL DRAG HYPOTHESIS

If the earth drags the ether with it so that there is no relative velocity between them, the apparatus in Figure 11 is at rest with respect to the ether. The light path in the two arms of the interferometer is identical and there is no shift in the pattern of fringes as the instrument rotates.

NO DRAG HYPOTHESIS

In this hypothesis the ether is fixed and the relative speed between the interferometer and the ether is v, the speed of the earth through the ether. It is unlikely that the speed of the earth relative to the fixed ether is the same as the orbital speed of the earth around the sun since the sun is part of a huge galaxy of stars in motion. In a time Δt, the interferometer moves from A to A′ (Figure 11) in the ether. The distance AA′ equals $v\Delta t$. During that same time the light reflected from 0 proceeds to A′ with a velocity c. The distance 0A′ is given by $c\Delta t$. If we designate the length of the arm of the interferometer, 0A as L, it follows from the Pythagorean theorem that:

$$(0A')^2 = (A0)^2 + (AA')^2$$

or

$$c^2(\Delta t)^2 = L^2 + v^2(\Delta t)^2$$

collecting terms with $(\Delta t)^2$ in them,

$$(c^2 - v^2)(\Delta t)^2 = L^2$$

$$(\Delta t)^2 = \frac{L^2}{c^2 - v^2} \tag{5a}$$

$$\Delta t = \frac{L}{\sqrt{c^2 - v^2}}$$

Equation (5a) is the time required for the trip from 0 to A.′ Examining Figure 11, we see that, by the time the light returns to the half-silvered mirror, that mirror will be at 0.″ As the light path A′0″ is identical to the light path 0A,′ it follows that the total time for the trip is given by:

$$\Delta t = \frac{2L}{\sqrt{c^2 - v^2}}$$

It is convenient to divide the numerator and the denominator of this result by c. (In the denominator this is the same as dividing inside the square root by c^2). This gives:

$$\Delta t = \frac{2L/c}{\sqrt{1 - v^2/c^2}} \tag{5b}$$

At the instant that the light ray leaves 0 in the direction of the mirror at A, its partner (the remaining part of the beam split up by the half-silvered mirror) is transmitted toward the mirror at C. Let us calculate the time, ΔT required for the light to be reflected from the mirror originally at C, back to the half-silvered mirror originally at 0. In the time the light travels out, C moves to C.′ On reflection at C′ the light returns to the half silvered mirror which has advanced to 0.″ The light path is 0C′0.″ Let ΔT_1 represent the time required for light to travel from 0 to C.′

In that time, C must move to C.′ If v is the earth's velocity relative to the ether, $CC' = v\Delta T_1$ and $OC_1' = C\Delta T_1$. Therefore

$$0C' = L + CC'$$

or

$$c(\Delta T)_1 = L + v(\Delta T)_1 \tag{5c}$$

$$\Delta T_1 = \frac{L}{c - v}$$

Upon being reflected at C′ the light proceeds back toward 0,′ but in the time ΔT_2 the half-silvered mirror moves to 0″ where it encounters the beam reflected at C.′ Therefore, the light path for the return trip will be given by:

$$C'0'' = C(\Delta T)_2$$

All of the parts of the apparatus will have moved a distance $v(\Delta T)_2$, the distance C′C″ is given by $v(\Delta T)_2$. Examining the figure we see that:

$$0''C' = L - C'C''$$
$$C(\Delta T)_2 = L - v(\Delta T)_2$$

or

$$(\Delta T)_2 = \frac{L}{c + v} \tag{5d}$$

The total time for the light to travel from the half-silvered mirror at 0 to C′ and back to the half-silvered mirror at 0″ is given by the sum of (5c) and (5d):

$$\Delta T = \Delta T_1 + \Delta T_2 = \frac{L}{c + v} + \frac{L}{(c - v)} = \frac{2cL}{c^2 - v^2}$$

Dividing both the numerator and the denominator by c^2 yields:

$$\Delta T = \frac{2L/c}{1 - v^2/c^2} \tag{5e}$$

When the experiment is actually performed, any fringe shift that is observed depends on the time difference for the light to traverse the distance 0A0, compared to 0C0. This difference, designated by $\Delta T'$ is given by $\Delta T - \Delta t$. In other words, it is the difference between (5e) and (5b).

$$\Delta T' = \frac{\dfrac{2L}{c}}{1 - \dfrac{v^2}{c^2}} - \frac{\dfrac{2L}{c}}{\sqrt{1 - \dfrac{v^2}{c^2}}}$$

$$= \frac{2L}{c}\left[\frac{1}{1 - v^2/c^2} - \frac{1}{\sqrt{1 - v^2/c^2}}\right] \tag{5f}$$

At this point two mathematical approximations can be introduced to simplify the evaluation of (5f). To begin, let us refer to the ratio of the speed of an object to the speed of light, v/c as β. It is usually a very, very, small number. An expression like $1/(1 - \beta^2)$ can be approximated by a power series in β given by $(1 + \beta^2 + \ldots)$. Similarly, the quantity $\dfrac{1}{\sqrt{1 - \beta^2}}$ can be approximated by the quantity $(1 + \dfrac{\beta^2}{2} + \ldots)$.* Equation (5f) then becomes:

$$\Delta T' = \frac{2L}{c}\left(1 + \frac{v^2}{c^2} - 1^{-} \frac{v^2}{2c^2}\right)$$

$$= \frac{2L}{c}\left(\frac{v^2}{2c^2}\right) = \frac{L}{c}\left(\frac{v^2}{c^2}\right) \tag{5g}$$

In other words, there is a difference in the time required for light to move back and forth in the direction of motion when compared to the time required for light to move back and forth in a direction perpendicular to the direction of motion. As the arms of the interferometer are rotated ninety degrees and the positions of each arm replace each other, one will observe a shift in the fringe pattern in the instrument. The shift is proportional to the square of the ratio of v/c and thus, in contrast to previously discussed, first order, experiments, the Michelson-Morley experiment is a second order experiment.

THE PARTIAL DRAG HYPOTHESIS

In this case, as in the Fizeau experiment, equation (5g) becomes:

$$\Delta T' = \frac{L}{c}\left(\frac{f^2 v^2}{c^2}\right) \tag{5h}$$

---

* See footnote, p. 136.

f, the dragging coefficient, is less than one. Again the effect is a second order effect.

When Michelson first did the experiment and when it was repeated by Michelson and Morley, they and almost everyone else who knew about it expected to observe a fringe shift. Given the parameters of the apparatus (The instrument built by Michelson and Morley in 1887 had a light path of approximately eleven meters.), they expected a shift of up to a half fringe and were confident that with the sensitivity they could detect a shift of as little as 0.02 fringes. The entire apparatus was mounted on a large granite slab floating in mercury. Thus, after it was given a rotation, it continued to rotate at essentially the same rate for a long, long time. Although both men observed for hours at a time over a period of months, they never detected a significant shift in the fringe pattern. The result was null.

Since that time, the experiment has been repeated many, many times. The nature of the material from which the apparatus is constructed and the light source have been modified. The apparatus has been located on the tops of mountains, in cellars; the walls surrounding the apparatus have been changed from brick to glass. The results are always the same. There is no significant shift in the fringe pattern.

In terms of the traditional concept of the ether, there is only one possible interpretation: the apparatus is dragging the ether with it. There is no relative speed between the earth and the ether. However, such an interpretation cannot be made and, at the same time, allow for the partial-drag conclusion from the first order experiments which have been previously discussed. It cannot be both ways. The ether is either dragged by matter, partially dragged by matter, or not dragged at all. The impasse is not because the Michelson-Morley experiment showed that there was no ether. Rather, the impasse is because, if there is an ether, the properties possessed by that ether as suggested by the Michelson-Morley experiment are in sharp contradiction to the properties suggested by the Bradley, Fizeau, and Airy experiments.

Lorentz's solution was to suggest that the ether modified the properties of matter as it moved through it, thereby making it impossible to detect the absolute motion of the earth. As suggested in Chapter 3, Einstein solved the problem by reexamining the theory of measurement that had led to the impasse.

Appendix ·6·

Some Relativistic Derivations

THIS APPENDIX IS DEVOTED TO deriving some of the relationships between spatial and temporal coordinates in different inertial frames of reference by using the assumptions and postulates of special relativity. These relationships were only stated in the text.

THE LORENTZ TRANSFORMATIONS

In Chapter 3, it was pointed out that Lorentz had suggested a new set of transformation equations for relating spatial and temporal variables in different inertial frames of reference to replace the Galilean transformation equations. With the new transformation equations one was able to account satisfactorily for all of the results of first and second order ether drift experiments. Lorentz did not derive those equations from first principles. They *were* first principles. He postulated their use.

In contrast, Einstein *derived* the same equations from a different set of first principles that neither Lorentz, his close colleague Poincaré nor almost anyone else in the communities of world physics found acceptable or compatible with each other. Those principles were, of course, the principle of relativity and the principle of the invariance of the speed of light. The derivation provided here follows closely Einstein's derivation in his original paper of 1905. That proof was later reproduced by Einstein, in somewhat simplified form for a non-technical audience.*

* A. Einstein, *Relativity: The Special and the General Theory* (New York, 1921) Appendix A.

Appendix 6

THE PRINCIPLE OF RELATIVITY

There are many ways of stating this postulate. Here we give three equivalent statements:

a. The laws of physics have the same form in all inertial frames of reference.
b. The laws of physics are covariant in all inertial frames of reference.
c. No experiment can be performed to determine which of two different inertial frames of reference is moving.

THE SECOND POSTULATE

Again, there are many ways to state this postulate but recall that supporting it is Einstein's analysis of the problem of distant simultaneity and the conclusion that in any inertial frame of reference, it is necessary to stipulate the isoptropy of space for one signal speed. Einstein chose the speed of light. The second postulate represents one way of satisfying the need for such a stipulation. The following three statements of the second postulate are all equivalent:

a. The speed of light in empty space is an invariant.
b. The speed of light in empty space is the same in all directions and is independent of the motion of the source or the observer.
c. The speed of light in empty space has the same value in all inertial frames of reference.

COORDINATE POINTS ON AXES PERPENDICULAR TO THE DIRECTION OF MOTION

Let us begin by examining how coordinates transform between frames of reference in directions perpendicular to the direction of motion. Figure 1 depicts two three-dimensional frames of reference moving inertially with a speed v with respect to each other parallel to the x-x' axis. Earlier observers in the two frames of reference had communicated with each other and agreed to place marker pens at a certain point along the y (or y') axis so that when the origins of the two frames passed each other, the pen in one frame of reference would place a mark on the y-y' axis of the other frame of reference.

The situation has been redrawn in Figure 2 in a somewhat simplified form by removing the third dimension. Suppose that after the experiment it was found by observers in one or the other frame of reference that the marks were not at the same height. For example, suppose that it had been agreed before the experiment that the pens be placed so that they drew marks six units of length from the origin along the y or y' axis, and suppose further, that after the experiment, one of the observers found that the mark made by the pen from the other frame of reference

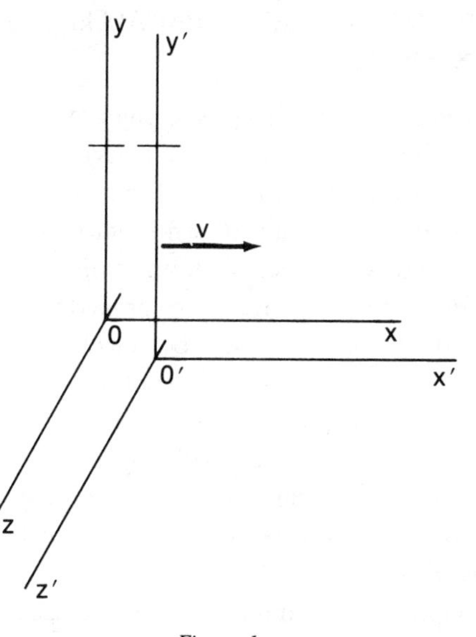

Figure 1

was not six units of length from the origin. That discrepancy could be used as a criterion for specifying which of the frames of reference was actually moving: The frame in which the mark was farther from or closer to the origin could be designated as "the frame in absolute motion."

Such a result would contradict the first postulate. It must therefore be the case that coordinate points in directions perpendicular to the direction of motion must be the same distance from the origin of the respective frames of reference. That is,

$$\begin{aligned} y &= y' \\ z &= z' \end{aligned} \tag{1a}$$

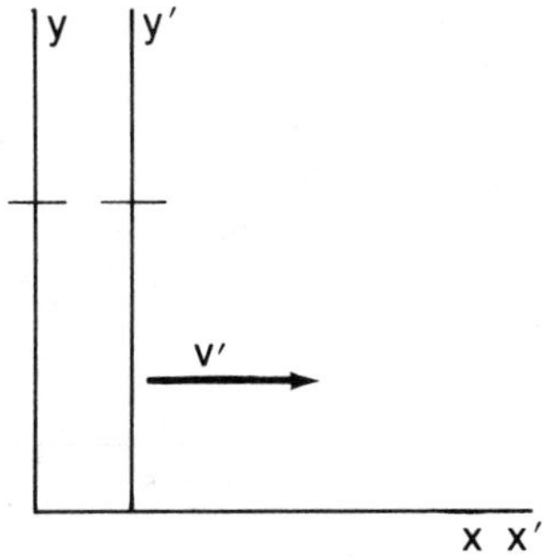

Figure 2

Appendix 6

COORDINATE POINTS ON AXES PARALLEL TO THE DIRECTION OF MOTION

According to the second postulate, the speed of light must have the same value for all inertial observers. Suppose that when the origins of the two inertial systems in Figure 3a coincide, clocks at the origins of both systems are set to zero and a light pulse is sent out, in the usual way, to synchronize clocks in both frames of reference. Observers in the two frames of reference will, of course, disagree that the clocks in the other frame of reference have been properly synchronized, but in both frames of reference, the pulse of light will be described as spreading out on the surface of a sphere whose radius is the product of the speed of light and the time elapsed from the moment when the origins of the two inertial frames coincided. Since the frames of reference are moving with respect to each other and they both describe the surface of the growing pulse of light as spherical, the graphical depiction for both frames cannot be made in one drawing. The situations are depicted in Figures 3b and 3c.

The radius of the sphere can be most easily characterized as a function of one or the other of the coordinate axes. Because we are interested in relating coordinate points between two frames of reference in the direction parallel to the direction in which the frames of reference are moving inertially, relative to each other, we chose

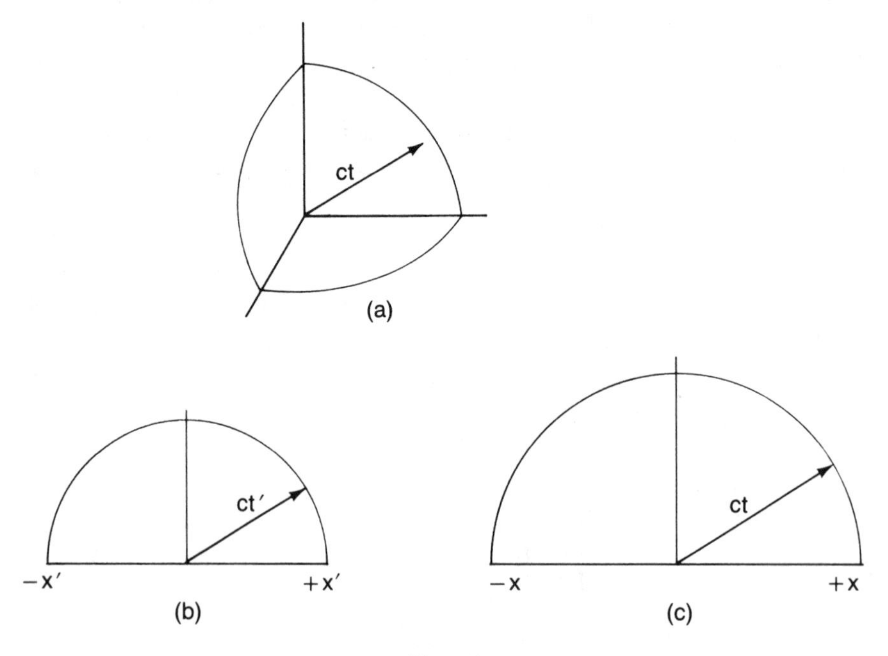

Figure 3

the x-x' axes. In the primed frame of reference, the point along the positive x' axis reached by the light pulse at any time t' after the origins of the two axes coincide is given by:

$$x' = ct'$$

or

$$x' - ct' = 0 \tag{1b}$$

Similarly, in the unprimed frame of reference the corresponding description is given by

$$x = ct$$

or

$$x - ct = 0 \tag{1c}$$

Equations (1b) and (1c) are describing the same events, and the right sides are equal. Therefore the terms on the left side of these equations can be related to each other by a constant. The constant must be linear. That expresses the fact that for each value of x and t there corresponds one, and only one, value of x' and t'.

$$(x' - ct') = m(x - ct) \tag{1}$$

A similar set of equations can be developed for the radius of the sphere of light in the negative x direction. This time, however:

$$-x = ct$$

and

$$x + ct = 0 \tag{2a}$$

$$-x' = ct'$$

and

$$x' + ct' = 0 \tag{2b}$$

$$(x' + ct') = n(x + ct) \tag{2}$$

Equations (1) and (2) may now be added or subtracted and rewritten so that values of x' and t' can be expressed as functions of x and t. We add the equations:

$$2x' = mx - mct + nx + nct$$

Collecting terms and dividing both sides of the equation by 2 gives:

$$x' = \frac{(m + n)x}{2} - \frac{(m - n)ct'}{2} \tag{3a}$$

If equations (1) and (2) are subtracted from each other we obtain:

$$ct' = \frac{(m+n)ct}{2} - \frac{(m-n)x}{2} \tag{3b}$$

In order to simplify further manipulation of equations (3a) and (3b) we introduce the following substitutions:

$$(m+n)/2 = a$$

and

$$(m-n)/2 = b \tag{3c}$$

Substituting the appropriate equation from (3c) into (3a) and (3b) gives the following:

$$x' = ax - bct \tag{4}$$

and

$$ct' = act - bx \tag{5}$$

It is at this point required to evaluate a and b. Consider the origin of the primed coordinate system. For that particular point, x' always equals zero and equation (5) becomes:

$$ax = bct$$

or

$$x = \frac{(bc)t}{a} \tag{6a}$$

This equation requires interpretation. It was arrived at by choosing a particular coordinate point, the origin of the primed system. It describes the position (x) of that origin in the unprimed system at any time (t). But in fact, another way of describing the position of the origin of the primed system in the unprimed frame of reference at any time after the origins coincide and clocks in each frame of reference have been synchronized is:

$$x = vt \tag{6b}$$

where v is the speed of the two frames of reference to each other. Comparing (6a) and (6b) we can now evaluate the constant bc/a:

$$v = bc/a. \tag{6}$$

Whereas we have been able to make a physical argument for evaluating the combination bc/a, our *task* is to evaluate b *and* a.

Note that the result of the measurement of a proper length in the unprimed system by observers in the primed system is the same as measurement of a proper length in the primed system by observers in the unprimed system. This is another way of stating the principle of relativity: The effects of the measuring process on systems in inertial motion relative to each other are symmetrical. We have seen (Chapter 3) that one way to measure the length of an object in uniform motion relative to another frame of reference is to simultaneously (that is, at the same time) note the coordinates in your frame of reference corresponding to the ends of the object.

Suppose the unprimed frame of reference were making such a measurement on a rod in the primed frame of reference at time $t = 0$. If one end of the rod were located at the origin of the primed system, from equation (4) the length of the rod would be given by

$$x' = ax$$

And if it were a unit length,

$$\begin{aligned} 1 &= ax \text{ or} \\ a &= 1/x \end{aligned} \tag{7a}$$

Now let us concentrate on measurement of a rod of unit length at rest in the unprimed frame of reference, at time $t' = 0$. We begin by eliminating the variable t from equations (4) and (5) by solving them simultaneously:

$$x' = ax - bct \tag{7b}$$

$$0 = act - bx \tag{7c}$$

From (7c) we get a value for t,

$$t = bx/ac$$

which we can substitute in (7b)

$$x' = ax - \frac{b^2x}{a} \tag{7d}$$

We can eliminate b from this equation by making use of equation (6). As

$$\begin{aligned} v &= bc/a \\ b &= va/c \end{aligned}$$

and

$$b^2 = v^2a^2/c^2 \tag{7e}$$

we can substitute the right side of (7e) in (7d), obtaining

$$x' = ax - axv^2/c^2.$$

The quantity ax can be factored from the right side of this last equation, yielding:

$$x' = ax(1 - v^2/c^2) \tag{7f}$$

Recall that in the unprimed system, we chose a unit length with one end located at the origin. We make the same choice of proper length in the unprimed system: $x = 1$. Equation (7e) then becomes

$$x' = a(1 - v^2/c^2) \tag{7g}$$

Because the principle of relativity requires that the measuring process between the two frames of reference is symmetrical it must be the case that x' of equation (7e) will have the same value as x of equation (7a). They are both the nonproper measurement of a proper unit length when the relative motion between the frames of reference is v. Combining those two equations gives:

$$1/a = a(1 - v^2/c^2).$$

Each side may be inverted and the result solved for a:

$$a^2 = 1/(1 - v^2/c^2) \tag{7h}$$

or

$$a = \sqrt{\frac{1}{1 - v^2/c^2}} \tag{7}$$

This gives us a value of a independent of b.

Our route to this end has been long and tortuous. We would not have taken it had we not had some inkling of how it would come out. Such inklings do not pop out of nowhere. They result from experience with the relationship between mathematical manipulation and physical meaning. They require the infusion of physical intuition. Consider for example how we identified in equation (7a) the quantity bc/a as being the inertial speed of relative motion between the primed and unprimed frames of reference.

We can now substitute this value for a into equation (6) to obtain a value for b. Since $v = bc/a$,

$$b = av/c \tag{8a}$$

Substituting for a from (7) yields:

$$b = \frac{v/c}{\sqrt{1 - v^2/c^2}} \tag{8}$$

We can now substitute for a and b in equations (4) and (5) from equations (7) and (8). Writing those equations with the equations with (1a) gives a complete set of transformation equations for spatial and temporal coordinates:

$$
\begin{aligned}
x' &= \frac{x - vt}{\sqrt{1 - v^2/c^2}} \\
y' &= y \\
z' &= z \\
t' &= \frac{t - vx/c^2}{\sqrt{1 - v^2/c^2}}
\end{aligned}
\tag{9}
$$

These will be recognized as the Lorentz transformations. Notice that we derived them by postulating the principle of relativity and the invariance of the speed of light and then following the physical and mathematical consequence of those postulates. We made no assumptions about the nature of matter or radiation or how they interact.

TIME DILATION

The measurement of spatial and temporal *coordinates* is not the same as determining a length or a time *interval.* A length is represented by the *difference* between two spatial coordinate points; a time interval by the *difference* between two temporal coordinate points. Since there is a unique set of spatial and temporal coordinates associated with each event, a given length is the distance between two particular events. Similarly, the difference between the time coordinate of two events represents what we commonly associate with the time interval. (In the Minkowskian, four dimensional formulation of special relativity, a time interval is proportional to the distance between two coordinates on the time axis.)

There are two ways to compare lengths and times as measured in different inertial frames of reference. We can utilize the postulates of our measurement theory, the special theory of relativity, directly, or we can employ the equations relating how coordinate points for each event compare in two inertial frames of reference. Those are the Lorentz transformation equations (9) which themselves were derived from the postulates of the theory. It is instructive to examine and compare both techniques.

MEASURING A TIME INTERVAL

Suppose we were to carry out the following thought experiment. You and I are in two different frames of reference moving inertially relative to each other with a speed v. Working at the origin of your frame of reference, you send a pulse of light to a mirror located in your frame of reference in a direction *perpendicular* to the direction of relative motion between us. It is important for this experiment that the line between your origin and the mirror be perpendicular to the direction of motion, because the principle of relativity insures that even though we are in

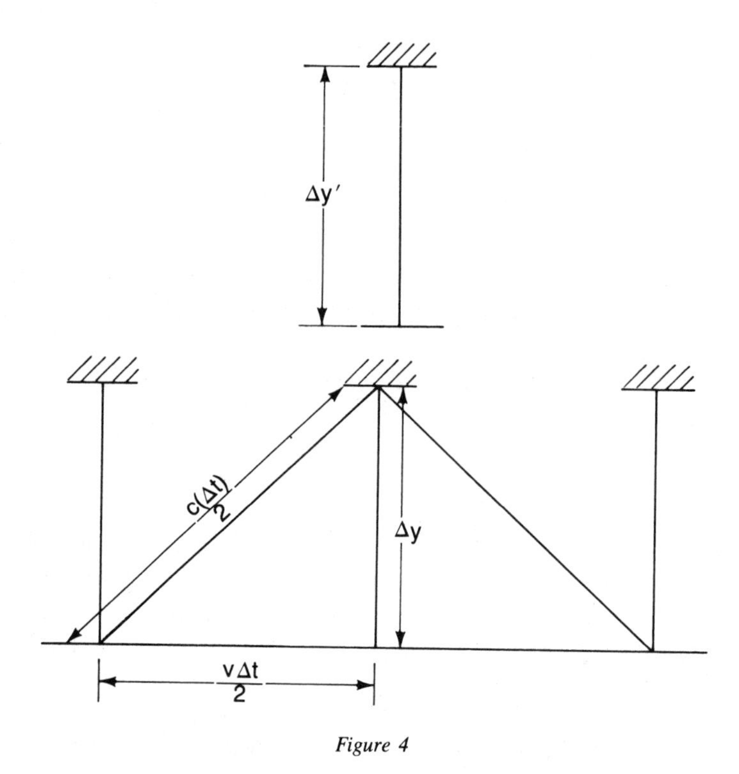

Figure 4

different inertial frames, we will agree on the magnitude of lengths in those directions. The situation is depicted in Figure 4. At the moment the pulse leaves, you activate a stopclock and measure the length of time for the pulse to travel to the mirror and return. The two events in question are:

Event {1} light pulse leaves your origin
Event {2} light pulse returns to your origin.

I also measure the time required for the light to travel to the mirror and return but, as can be seen from Figure 4, since events {1} and {2} occur at different points in my frame of reference, I will require more than one clock to measure the time interval between the two events. Because you use only one clock to make the measurement, your measurement is said to be *proper.* I require two clocks to measure the time interval between the same events and my measurement is *nonproper.* For this experiment, the frame of reference in which the proper time measurement is made has been designated the primed frame.

As you located the mirror at the coordinate point y', the distance traveled by the light is $\Delta y'$. Since the speed of light is c, your measurement of the time interval, $\Delta t'$, for the round trip journey of the light pulse is given by

$$\Delta t' = 2(\Delta y')/c \tag{10a}$$

Which value will *I* find that time interval to be? Note that there are a number of things you and I agree about concerning the parameters of this experiment. We agree that only you make the proper time measurement. We also agree that the speed of light in any direction is c (Second postulate) and we also agree that the relative speed between the frames of reference is v (First postulate) and that the distance from the x-x' axis to the mirror is $y = y'$. From my perspective, the round trip journey of the light pulse requires a time Δt but can be divided into two symmetrical halves: Going to the mirror and returning to the x-x' axis. Each half requires a time interval $(\Delta t)/2$. Referring to Figure 4, we see that in going to the mirror the light pulse follows the path $c(\Delta t)/2$. In the time required for the light pulse to reach the mirror, the primed frame of reference has moved a distance $v(\Delta t)/2$ along the x axis. Referring to Figure 4, the lengths $c(\Delta t)/2$, $v(\Delta t)/2$ and Δy form a right triangle which can be described by the Pythagorean theorem:

$$[c(\Delta t)/2]^2 = [\Delta y]^2 + [v(\Delta t)/2]^2$$

which can be solved for $(\Delta t)/2$:

$$[c^2 - v^2][(\Delta t)/2]^2 = (\Delta y)^2$$

$$(\Delta t)/2 = \frac{(\Delta y)}{\sqrt{(c^2 - v^2)}} \tag{10b}$$

As the trip from the source of the light pulse to the mirror is symmetrical with the trip from the mirror back to the source, the time for the total trip is obtained by multiplying (10b) by 2. For convenience we divide both the numerator and denominator by c:

$$(\Delta t) = \frac{2(\Delta y)/c}{\sqrt{1 - v^2/c^2}} \tag{10c}$$

The reason for dividing the numerator and the denominator by c now becomes apparent for if we compare equation (10c) with (10a), it is seen that since $(\Delta y) = (\Delta y)'$ the numerator of equation (10c) is the time interval between events $\{1\}$ and $\{2\}$ as measured in the primed frame of reference. In other words:

$$(\Delta t) = \frac{(\Delta t)'}{\sqrt{1 - v^2/c^2}} \tag{10}$$

The physical interpretation of these mathematical manipulations are as follows: The proper time interval between two events is less than any nonproper measurement of the same time interval, the discrepancy between the two measurements depending on the relative speed of the frames of reference between which the measurements are being compared.

Appendix 6

CALCULATING THE TIME INTERVAL FROM THE LORENTZ TRANSFORMATION EQUATIONS

The Lorentz transformation equations (9) express how coordinate points in the unprimed frame of reference would be measured in the primed frame of reference. In the unprimed frame of reference event {1} occurred at t_1 and event {2} occurred at t_2. The time interval is given by

$$\Delta t = t_2 - t_1 \tag{11a}$$

Let us now see how $\Delta t'$ relates to t by direct substitution from equation (9) into (11a):

$$\Delta t = \frac{(t_2' + vx_2'/c^2) - (t_1' + vx_1'/c^2)}{\sqrt{1 - v^2/c^2}} \tag{11b}$$

At this point it is important to remember that it is in the primed frame of reference that the proper measurement of the time interval is made. Since making a proper time measurement means using a single clock at the same place to time the interval between events at that place, it follows that in equation (11b) $x_2' = x_1'$. With that fact in mind, equation (11b) immediately reduces to

$$\begin{aligned} \Delta t &= \frac{t_2' - t_1'}{\sqrt{1 - v^2/c^2}} \\ &= \frac{\Delta t'}{\sqrt{1 - v^2/c^2}} \end{aligned} \tag{11}$$

which is identical to equation (10). In this last derivation, we did not make explicit reference to the postulates of relativity. Those postulates are implied by the transformation equations (9) themselves. Of course, there *are* other ways of getting the transformation equations. The result expressed by (10) or (11) does not, therefore, entail a unique theory about the nature of the measurement process, in fact, one can make the derivation without a thought to such matters. This accounts for the possibility of the variety of interpretations of the formal aspects of the theory of relativity detailed in Chapters 6–10.

MEASUREMENT OF LENGTH CONTRACTION

Observers in one and only one inertial frame of reference make a proper measurement of the time interval between two events. Similarly, observers in one and only one inertial frame of reference make a proper measurement of the length of an object. That inertial frame of reference is the inertial frame of reference in which the rod is at rest relative to measuring instruments. I again emphasize that in speaking about proper and nonproper measurements, one must pay attention to the events which specify the operations of measurement. A moment's thought will convince you that a proper length measurement will not be made in the frame of reference which makes a proper time measurement of the events associated with the proper length measurement.

In the thought experiment we performed earlier, the light pulse was sent from the origin of the primed frame of reference to a mirror at rest in that frame of reference. The line between the mirror and the origin was perpendicular to the direction of relative motion between the frames of reference. The result of that analysis was that proper and nonproper measurements of time intervals were related by

$$\text{(nonproper time interval)} = \frac{\text{(proper time interval)}}{\sqrt{1 - v^2/c^2}}$$

Note that since the denominator on the right side of this equation is less than one, a nonproper time interval measurement will always yield a value larger than that produced by the proper time interval measurement for the same two events.

We will now perform that thought experiment again. This time, however, the rod is at rest in the unprimed frame of reference and the mirror is located along a line parallel to the direction of relative motion between the two inertial frames of reference.

Figure 5 represents the new experimental arrangement. In the unprimed frame of reference, a proper measurement of the distance between the origin and the mirror has already been made (perhaps by using a ruler), and that length has been designated by L. An observer at the origin of the unprimed frame of reference now undertakes a proper time interval measurement for the events:

{1} light pulse leaves origin
{2} light pulse returns to origin.

Since the length of the resting rod is known to be L, this proper time interval is given by:

$$\Delta t = 2L/c \tag{12a}$$

In the primed frame of reference, in which observers do not make a proper time interval measurement of events {1} and {2}, the following relationships must apply. During the outward journey of the light pulse, $\Delta t_1'$ the light pulse travels a distance given by $c(\Delta t_1')$. In that time the mirror has been approaching the pulse with a speed v and has therefore traveled a distance $v(\Delta t_1')$. These quantities are related by:

$$c(\Delta t_1') = L' - v(\Delta t_1') \tag{12b}$$

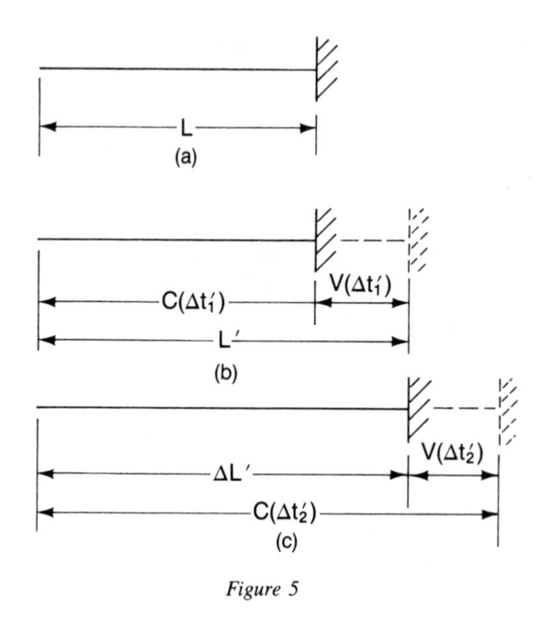

Figure 5

where L′ is the length of the rod in the primed frame of reference. This is shown in Figure 5b.

In the time required for the light to return from the mirror to the origin, $\Delta t_2'$, the origin has moved away from the approaching light pulse a distance $v\Delta t_2'$. Therefore,

$$c(\Delta t_2') = L' + v(\Delta t_2') \tag{12c}$$

The total time interval for events {1} and {2} as measured in the primed frame of reference is $\Delta t' = \Delta t_1' + \Delta t_2'$. From equations (12b) and (12c) we obtain the following values for $\Delta t_1'$ and $\Delta t_2'$:

$$\Delta t_1' = L'/(c+v)$$
$$\Delta t_2' = L'/(c-v)$$

When these are added together we obtain the following:

$$\Delta t' = \Delta t_1' + \Delta t_2' = \frac{L'}{(c+v)} + \frac{L'}{(c-v)}$$

When the terms on the right side of this last equation are rationalized and added together we obtain:

$$\Delta t' = \frac{2L'c}{(c^2 - v^2)}$$

The numerator and denominator of the right side of this equation can now be divided by c^2:

$$\Delta t' = \frac{2L'/c}{1 - v^2/c^2} \tag{12d}$$

Compare equations (12a) and (12d). $\Delta t'$ of (12d) is the nonproper measurement of the time interval between two events for which Δt of (12a) is the proper measurement. Since proper and nonproper time interval measurements are always related by

$$\text{(nonproper time interval)} = \frac{\text{(proper time interval)}}{\sqrt{1 - v^2/c^2}}$$

We can substitute directly from (12d) and (12a), values for $\Delta t'$ and Δt:

$$\frac{L'/c}{1 - v^2/c^2} = \frac{\mathrm{L/c}}{\sqrt{1 - v^2/c^2}}$$

Which is quickly simplified to:

$$L' = L\sqrt{1 - v^2/c^2}. \tag{12}$$

By now we recognize this as nothing more than the Lorentz contraction equation. But in terms of the analysis which has been done using the concepts of Einstein's special theory of relativity, the Lorentz contraction can be stated as a relationship between proper and nonproper length measurements:

$$\text{(nonproper length)} = \text{(proper length)}\ \sqrt{1 - v^2/c^2}.$$

In the Lorentz theory, the contraction was assumed to have been caused by an actual shrinking of the rod. Note that in the analysis given within the Einstein theory, nothing has been assumed about the nature of the material out of which the rod is made nor how such matter is affected by motion. And note also the direct manner in which the discrepancies between inertial frames of reference concerning the measurement of length intervals is affected by the disagreement of observers in the two frames of reference about the time interval between two specified events.

LENGTH CONTRACTION COMPUTED FROM THE LORENTZ TRANSFORMATIONS

Suppose that a proper measurement of the length of a rod is made in the primed frame of reference. The ends of the rod are at coordinates x_2' and x_1'. The length L' is given by

$$L' = x_2' - x_1' \tag{13a}$$

The proper length measurement was made when the rod was at rest relative to observers in that frame of reference. To determine the length of the rod in the unprimed frame of reference we use this information to transform the coordinates on the right side of (13a) with the Lorentz transformation equations (9)

$$x_1' = \frac{x_1 - vt_1}{\sqrt{1 - v^2/c^2}}$$

$$x_2' = \frac{x_2 - vt_2}{\sqrt{1 - v^2/c^2}} \qquad (13b)$$

The length of the rod in the unprimed frame of reference is determined by noting the position of the ends of the moving rod at the same time, $t_1 = t_2$

$$x_2' - x_1' = \frac{x_2 - x_1}{\sqrt{1 - v^2/c^2}}$$

Or, making use of equation (13a):

$$L' = \frac{L}{\sqrt{1 - v^2/c^2}}$$

$$L = L'\sqrt{1 - v^2/c^2} \qquad (13)$$

Once again the nonproper length is less than the proper length by a factor $\sqrt{1 - v^2/c^2}$.

In making use of the transformation equations to derive the length contraction, it was not necessary to pay much attention to the physics of the problem. The premises of the theory, the concepts of proper and nonproper measurements were imbedded implicitly within the equations themselves. But as we have seen so many times before in this book, there are any number of ways to arrive at the Lorentz transformation equations.

SYNCHRONIZING CLOCKS

As we have seen in Chapter 3, observers in different frames of reference will not agree on the simultaneity of distant events. It is that disagreement that undermined the classical theory of measurement and led Einstein to the replacement of the Galilean transformation equations with the Lorentz transformations. We now quantify the disagreement in simultaneity.

In Figure 6 we have illustrated the thought experiment which will be used to discuss the physical problem of synchronization. We suppose that two clocks located a distance L$'$ apart in the primed frame of reference are synchronized in a

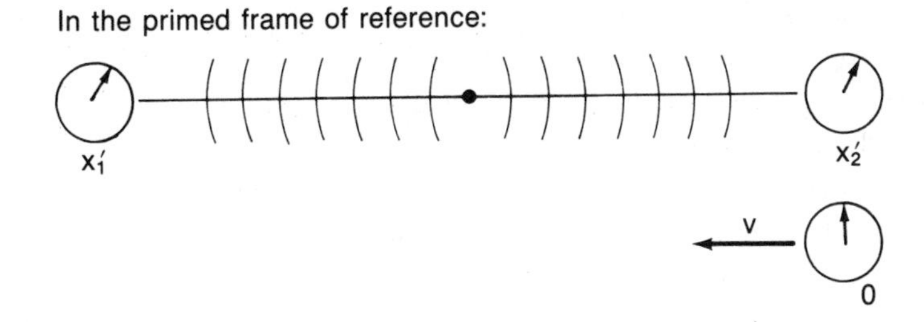

In the unprimed frame of reference:

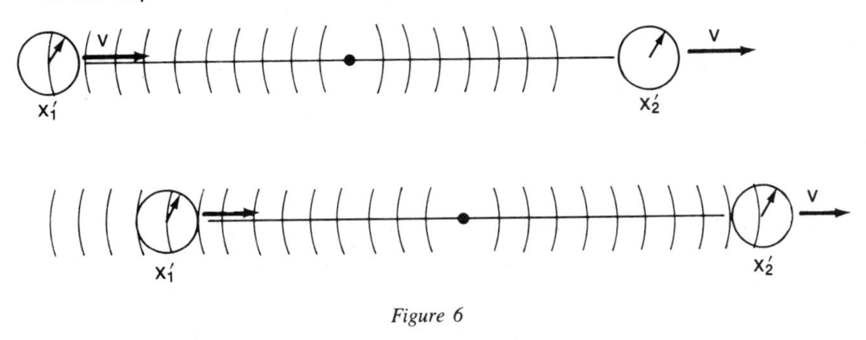

Figure 6

standard manner. For example, at the midpoint between the clocks, a flash bulb was exploded. As light moves uniformly in all directions, both clocks will start when a clock located at the midpoint and that was started when the light signals began propagating shows a time

$$t' = L'/2c$$

Not only will observers in the primed frame of reference not be able to make a proper time interval measurement of the interval between the events "light leaves flashbulb in primed frame of reference," and "light arrives at one of the clocks in the primed frame of reference," it is not possible to make a proper time interval measurement of those two events in any other frame of reference because nothing can keep up with a beam of light. That is why Einstein made the stipulation of isotropy of the speed of light in different directions.

Furthermore, it is not possible to make a proper time interval measurement of the events:

{1} light arrives at the clock located at x_1'

{2} light arrives at the clock located at x_2'.

In the primed frame of reference, that time interval is zero (that is what is meant by the fact that the clocks have been synchronized) and the clocks are located at two different places in the frame of reference. In other frames of reference, events {1} and {2} happen in a time interval less than the time required for light to go from one clock to the other. That is another way of saying that events {1} and {2} are not causally related.

Suppose that the observer in the unprimed frame of reference is located at 0 and starts his or her clock when he or she sees the flash from the flash bulb in the primed frame of reference. For the unprimed observer the distance the light travels to the clock located at x_1' is $(L'/2)\sqrt{1 - v^2/c^2}$. Since, according to the unprimed observer, the clock at x_1' is moving up to meet the light propagating outward from the flash bulb the total distance traveled by the light is given by

$$ct_1 = (L'/2)\sqrt{1 - v^2/c^2} - vt_1$$

which can be solved for t:

$$t_1 = \frac{(L'/2)\sqrt{1 - v^2/c^2}}{(c + v)} \tag{14a}$$

The clock at x_2 has been moving *away* from the on-rushing light pulse. From the point of view of the observer at 0 the total distance traveled by the light in this case is given by

$$ct_2 = (L'/2)\sqrt{1 - v^2/c^2} + vt_2$$

which can be solved for t:

$$t_2 = \frac{(L'/2)\sqrt{1 - v^2/c^2}}{c - v} \tag{14b}$$

The time difference between the arrival of the light signals at the clocks at x_1' and x_2' in the primed frame of reference is zero. In the unprimed frame of reference, the difference can be found by subtracting (14a) from (14b):

$$t_2 - t_1 = \frac{[L'(c + v)\sqrt{1 - v^2/c^2}]/2 - [L'(c - v)\sqrt{1 - v^2/c^2}]/2}{c^2 - v^2} \tag{14c}$$

Equation (14c) is simplified in the next three steps:

$$\begin{aligned} t_2 - t_1 &= \frac{L'v\sqrt{1 - v^2/c^2}}{(c^2 - v^2)} \\ &= \frac{(L'v/c^2)\sqrt{1 - v^2/c^2}}{1 - v^2/c^2} \\ &= \frac{L'v/c^2}{\sqrt{1 - v^2/c^2}} \end{aligned} \tag{14d}$$

Equation (14d) represents the time difference in the arrival of the synchronizing signals to the clocks at x_1' and x_2' as measured by observer in the unprimed frame of reference. At the instant the light arrives at the clock at x_2', that clock starts. Meanwhile, according to the unprimed observer, the clock at x_1' which had already been running, is "running slow." (The unprimed observer measurements of the rate at which the clock at x_1' runs are nonproper time interval measurements.) Therefore in the unprimed frame of reference the total time difference between the clocks at rest and synchronized in the primed frame of reference will be:

$$t_2 - t_1 = \frac{L'v/c^2}{\sqrt{1 - v^2/c^2}} \cdot \sqrt{(1 - v^2/c^2)}$$

$$t_2 - t_1 = L'v/c^2 \tag{14}$$

Note that not only will the unprimed observer *not* agree that the clocks in the primed frame of reference were properly synchronized, the farther apart the clocks are, the more serious the discrepancy. And the greater the relative speed between the frames of reference, the more seriously out of phase will be the clocks in the primed frame of reference.

DETERMINING THE SYNCHRONIZATION ERROR FROM THE LORENTZ TRANSFORMATIONS

From the equations in (9) the transformation for temporal coordinates is given by

$$t = \frac{t' + vx'/c^2}{\sqrt{1 - v^2/c^2}} \tag{15a}$$

From (15a)

$$t_2 - t_1 = \frac{t_2' + vx_2'/c^2 - t_1' - vx_1'/c^2}{\sqrt{1 - v^2/c^2}} \tag{15b}$$

But because we are speaking of clocks that are synchronized in the primed frame of reference,

$$t_2' = t_1'$$

Equation (15b) then becomes:

$$t_2 - t_1 = \frac{(v/c^2)(x_2' - x_1')}{\sqrt{1 - v^2/c^2}} \tag{15c}$$

But $x_2' - x_1'$ is nothing more than the distance between the two clocks (Figure 6) in the primed frame of reference, L.′ In the unprimed frame of reference, that distance is $L'\sqrt{1 - v^2/c^2}$. Equation (15c) then becomes:

$$t_2 - t_1 = L'v/c^2 \tag{15}$$

in agreement with (14).

THE VELOCITY ADDITION LAW

In the classical theory of measurement, the Galilean transformation equations were used to derive the classical velocity addition law. We will now apply the theory of measurement represented by the special theory of relativity to derive a new velocity addition law.

Two familiar frames of reference are depicted in Figure 7. One, the unprimed frame of reference, represents the frame of reference of the railway embankment. The primed frame of reference represents the frame of reference of a train moving along the x axis with a speed v. In the primed frame of reference a person is walking from the back to the front of the train. The person's speed is u_x when measured in the primed frame of reference.

$$u_x = (\Delta x')/(\Delta t') = (x_2' - x_1')/(t_2' - t_1')$$

In order to simplify matters we rewrite the last equation as

$$u_x(t_2' - t_1') = x_2' - x_1'. \tag{16a}$$

And to simplify still further, we assume that $x_1 = x_1' = t_1 = t_1' = 0$. Equation (16a) then becomes:

$$u_x t' = x' \tag{16b}$$

Now we apply the Lorentz transformations (9) to equation (16b):

$$\frac{u_x(t - vx/c^2)}{\sqrt{1 - v^2/c^2}} = \frac{x - vt}{\sqrt{1 - v^2/c^2}} \tag{16c}$$

This can be immediately simplified to:

$$x(1 + u_x v/c^2) = (u_x + v)t$$

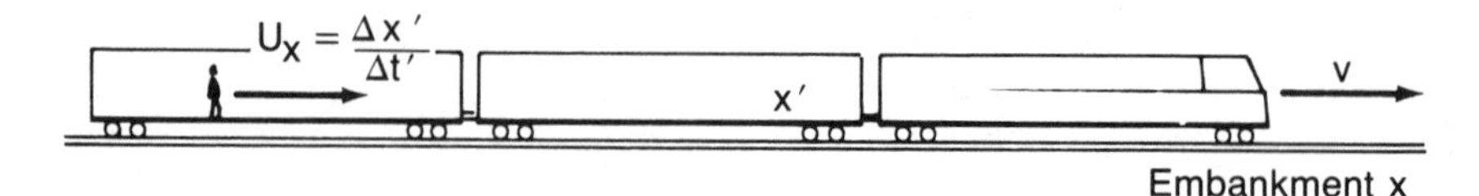

Figure 7

which can be rewritten as

$$x/t = (u_x + v)/(1 + u_x v/c^2) \tag{16d}$$

The left side of (16d) should be recognized as the speed of the walking passenger as determined by observers in the unprimed frame of reference. If we call this speed w_x then 16d becomes:

$$w_x = (u_x + v)/(1 + u_x vc^2) \tag{16}$$

which is the relativistic velocity addition law. Notice that it reduces to the classical velocity addition law of Chapter 2 for the cases where the product of $u_x v$ is a small fraction of the square of the speed of light. Notice also that we did not use the contraction or time dilation formulas in this case. Those relationships relate proper and nonproper measurements of length and time. Think through the measurement process in this thought experiment again, making explicit who makes proper time and distance measurements.

Let us consider one more situation. This time the passenger is moving in a direction perpendicular to the direction of relative speed between the frame and the embankment (The train is very wide). The speed of the passenger as measured on the train is u_y where

$$u_y = (y_2' - y_1')/(t_2' - t_1') \tag{17a}$$

As before, we rewrite this equation, set $y_1' = y_1 = t_1' = t_1 = 0$, and apply the Lorentz transformation equations:

$$y' = u_y t'$$
$$y = \frac{u_y(t - vx/c^2)}{\sqrt{1 - v^2/c^2}} \tag{17b}$$

Notice that we have used the transformation $y = y'$. Equation (17b) can be rewritten as

$$y\sqrt{1 - v^2/c^2} = u_y(t - vx/c^2) \tag{17c}$$

Let us now divide both sides of this equation by t:

$$(y/t)\sqrt{1 - v^2/c^2} = u_y(1 - v\{x/t\}/c^2)$$

If we call the speed of the object as measured on the embankment w_y, this last equation can be rewritten as

$$w_y = \frac{u_y(1 - y^2/c^2)}{1 - u_x v/c^2} \tag{17}$$

Where we have used the term u_x to represent x/t in the denominator on the right side of the equation.

It is a small matter to show that the equation for velocities in the z direction has exactly the same form as equation (17). And if we make the speed in the primed system the dependent variable, equation (17) becomes

$$u_y = \frac{w_y(1 - v^2/c^2)}{1 + u_x v/c^2} \qquad (18)$$

The physical meaning of these velocity addition equations is more complicated than many of the relationships we have examined in this book. Note that the observed speed of the passenger walking on the train in the y direction depends on the speed of the passenger relative to the train in the x direction.

SUMMARY

In this appendix, we have derived the Lorentz transformation equations and explored the relationship between proper and nonproper measurements of distance and time intervals. We applied those concepts to the addition of velocities. It is now possible to make use of the relationships derived here to calculate how masses relate to each other in different inertial frames of reference, to calculate the relationship between energy and mass, and to calculate the time relationships in the clock paradox problems. The physics of those situations is discussed in Chapters 3 and 4. While the intellectual puzzle of working out the details of the analysis should not be made light of, those details should present little difficulty if you have followed the argument to this point. Keep before you the following questions:

What are the events for which temporal and spatial coordinates must be compared?

Who (if anyone) makes a proper temporal or spatial measurement?

Bibliographic Essay: On Understanding Relativity

THIS BIBLIOGRAPHIC ESSAY IS PROVIDED as a guide to those who are interested in pursuing further work in the history of the reception of relativity, or in further study of science as a social institution. It is not a catalogue of the research behind this book nor a compendium of primary sources. It is not written for professional historians of science nor is there any pretense that it is exhaustive.

PART I: THE THEORY OF RELATIVITY

The best popular account of the theory of relativity is still Einstein's, *Relativity: The Special and the General Theory* (New York: Crown rpr., 17th ed., 1961). The reader should not be misled by Einstein's disarmingly simple and spare prose. Each sentence must be studied with great care. As I have noted in the text, almost all American treatments of the theory of relativity assume that the postulates have been tested, directly, by experiment. Bearing that in mind, the reader will be interested in the treatment of the theory by Lillian Lieber and Hugh Lieber, *The Einstein Theory of Relativity* (New York, 1st ed., 1936; 2nd ed., 1945). The second edition is recommended because it includes a treatment of general relativity. The Lieber and Lieber book is one of a series in which the authors undertake to explain painstakingly and to explicate physics and mathematics by unpacking the intricacies of the language of mathematics itself. These books are also noteworthy for their blank verse format. As far as I know all the books by

Lieber and Lieber, including *Non Euclidean Geometry, Galois and the Theory of Groups, The Education of T. C. Mits,* and *Infinity*, are out of print. (T. C. Mits is an acronym for "The Common Man In The Streets".) They are unique and deserve to be made available once again.

Another excellent book is Max Talmay's *The Relativity Theory Simplified and the Formative Period of Its Inventor* (New York, 1932). Talmay was a boarder in Einstein's house during Einstein's youth and they formed a close and lasting friendship. At a somewhat more advanced level, the most outstanding treatment of the theory is A. P. French's *Special Relativity* (London/Camden: Thomas Nelson Press, 1968). The book is written for students who have completed an introductory course in physics, and French assumes an understanding of the notation of elementary calculus. French's exploration of the relationship between the experimental evidence, the postulates and conclusions of the theory is clear and insightful. Dennis Sciama's *The Physical Foundations of General Relativity* (Garden City, 1969) is recommended as an elementary introduction to the physics which Einstein's general theory of relativity addresses.

DRAMATIS PERSONAE

Accounts of the lives and work of almost all of the individual scientists mentioned in this book can be found in *The Dictionary of Scientific Biography* (16 Vols., New York: Scribners, 1970–1980). The essays in that biographical dictionary contain bibliographies of relevant primary and secondary literature. There have been many biographies of Einstein. Einstein's own biographical sketch can be found in P. A. Schillp (ed.), *Albert Einstein: Philosopher-Scientist* (2 Vols., New York, 1949), Vol. 1, pp. 1–95. Schillp provided the reader with Einstein's original German text and an English translation. The sketch gives wonderful insights into Einstein's self-image and a sense of his assessment of the importance of individual contributions to the stream of human understanding. The reader who wants to know more about Einstein's views on these subjects is encouraged to read Einstein's other writings, which have been collected in *The World as I See It* (London, 1935) and *Out of My Later Years* (London, 1950). Many of Einstein's occasional writings have been excerpted and collected in *Ideas and Opinions* (New York, 1954).

The most trusted biographies of Einstein are Philipp Frank, *Einstein: His Life and His Times* (London, 1948), Banesh Hoffman and Helen Dukas, *Albert Einstein: Creator and Rebel* (New York, 1972), and Jeremy Bernstein's *Einstein* (New York, 1973). The Frank biography is somewhat limited by not maintaining distance from its subject. On the other hand it contains some wonderful anecdotal material. The most sensitive portrayal of Einstein is the 1982 account by Abraham Pais, *Subtle is the Lord: The Science and Life of Albert Einstein* (New York/Oxford: Oxford University Press, 1982). In dealing with Einstein's work, Pais makes no compromise for the readers who are not familiar with mathematics. On the other hand, Pais' interpretations of physical theory are infused with enthusiasm. The book lacks an appreciation of historical nuance.

THE ORIGINS OF RELATIVITY THEORY

For an account of the seventeenth century background to current physical thought, Richard S. Westfall's *The Construction of Modern Science* (Cambridge: Cambridge University Press, 1977) is highly recommended. For the eighteenth century, the account in John Randall's *The Making of the Modern Mind* (Cambridge, Mass., 1940) has never been eclipsed. A. R. Hall's *The Scientific Revolution, 1500–1800* (Boston, 1954) not only deals with physical thought of the seventeenth and eighteenth century, it contains a summary account of the emergence of the social institutions of science. An overview of the major themes in nineteenth century physics can be obtained from P. M. Harman's *Energy, Force, and Matter: The Conceptual Development of Nineteenth Century Physics* (Cambridge: Cambridge University Press, 1982). Those trained in physics might turn to Edmund Whittaker's *History of the Theories of Aether and Electricity* (2 Vols., New York, 1951–1952). However, the reader is warned that Whittaker's account lacks a sense of historical nuance and gives a badly distorted picture of how the works of one individual relate to those of another. Whittaker not only modernized the notation of works on which he reports, in doing so he often obscured the motivations and intent of his subjects. A better source would be J. T. Merz's *A History of European Thought in the Nineteenth Century* (4 Vols., Chicago, 1902–1912, repr. New York: Dover Publications, 1965), especially volumes 1 and 2. The difficulty with Merz's account is that, having been written at the end of the nineteenth century, it has no distance from its subject.

An excellent account of the spirit of nineteenth century physical thought can be obtained from the 1981 novel by Russell McCormmach, *Night Thoughts of a Classical Physicist* (Cambridge: Harvard University Press, 1981). Another excellent source is McCormmach's prefaces to *Historical Studies in Physical Sciences* Vols. 1–7, 1969–1976. The reader may also want to explore many of the individual articles contained in that journal, now under the general editorship of John Heilbron. Heilbron's own synoptic history of twentieth century physics is soon to be published by Cambridge University Press.

Einstein himself gave several different, though compatible, accounts of the influences that led him to formulate the theory of relativity. The most accessible are those in *Relativity: The Special and the General Theories,* and the informal account delivered on December 14, 1922 at Kyoto University. The speech was delivered in German without notes and simultaneously translated into Japanese by the Japanese physicist J. Ishiwara. Ishiwara then wrote out his recollection of the speech which he published in Japanese. Those notes have subsequently been translated and published in English. See Albert Einstein, "How I Created The Theory of Relativity," tr. Y. A. Ono, *Physics Today,* 1982, *35*:45–48. Despite the tortuous path by which this account found its way into the English language, not only is it compatible with other accounts, it makes sense in the light of Einstein's professed epistemological credo and is in close agreement with the account found in Chapter 3 of this book. This book was written before I knew of the existence of the Ishiwara-transcribed account.

In the absence of definitive statements by Einstein and with no primary source material to document the process, there have been *many* conflicting secondary accounts of how Einstein created the special theory of relativity. These accounts (including mine) reflect more their authors' views of the nature of science and the relationship of evidence to theory than they reflect how Einstein "really did it." Those who are interested can consult the following: Jacques Hadamard, *The Psychology of Invention in the Mathematical Field* (New York, 1954); Max Wertheimer, *Productive Thinking* (New York, 1964); G. Holton, "What Precisely is Thinking? . . . Einstein's Answer," *The Physics Teacher,* 1979, *17*:157–164; G. Holton, "Einstein's Model for Constructing a Scientific Theory," Bernulf Kanitscheider, "Einstein's Treatment of Theoretical Concepts," and A. I. Miller's "On the History of the Special Theory of Relativity" all to be found in P. C. Aichelburg and R. U. Sexl (eds.) *Albert Einstein: His Influence on Physics, Philosophy and Politics* (Braunschweig, 1978); G. Holton, "Einstein's Scientific Program: The Formative Years" in Harry Woolf (ed.); *Some Strangeness in the Proportion: A Centennial Symposium to Celebrate the Achievements of Albert Einstein* (Reading: Addison-Wesley, 1980); the essays on the history of relativity theory by Holton in *Thematic Origins of Scientific Thought* (Cambridge: Harvard University Press, 1973); A. I. Miller, "The Special Theory of Relativity: Einstein's Response to the Physics of 1905," in Gerald Holton and Yehuda Elkana (eds.) *Albert Einstein: Historical and Cultural Perspectives* (Princeton: Princeton University Press, 1982); and Stanley Goldberg, "Albert Einstein and the Creative Act" in R. Aris et al, *The Springs of Scientific Creativity* (Minneapolis: The University of Minnesota Press, 1983).

The path breaking work on the history of special relativity is by Gerald Holton, "On the Origins of the Special Theory of Relativity," *American Journal of Physics,* 1960, *28*:627–636. That essay and others by Holton related to the early history of the theory of relativity may be found in *Thematic Origins of Scientific Thought.*

The account by Pais (see above) of Einstein's formulation of relativity theory is excellent. Another excellent source of the relationship of the ideas in relativity theory to the ambiance of ideas in physics at the turn of the century can be found in a work coauthored by Einstein and Leopold Infeld, *Evolution of Physics* (New York, 1936). The first part of A. I. Miller's *Albert Einstein's Special Theory of Relativity: Emergence and Early Interpretation (1905–1911)* (Reading: Addison-Wesley, 1981) is also useful although it should be noted that Miller relies too heavily on uninterpreted formalisms in the place of historical analysis. For readers who want to compare the approaches of the works thus far mentioned to the approach of those who are committed to rational reconstruction, E. Zahar's "Why did Einstein's Research Programme Supercede Lorentz'?" *The British Journal for the Philosophy of Science*, 1973, *24*:95–123, 223–262 or Adolph Gruenbaum's *Philosophical Problems of Space and Time* (Dordrect, 2nd ed., 1973) should prove interesting.

PART II AND PART III

Part II of this book was based in large part on my doctoral dissertation *Early Response to Einstein's Theory of Relativity, 1905–1911: A Case Study in National Differences* (Harvard University, Unpublished PhD. Dissertation, 1969). That work, largely the result of research in the primary literature, is available in the Harvard University Archives. A. I. Miller's *Albert Einstein's Special Theory of Relativity: Emergence (1905) and Early Interpretation (1905–1911)* examines the same material from a different point of view. As mentioned earlier, Miller dwells on recapitulation of formal, mathematical arguments.

The most useful journal literature in journals about questions of the reception and assimilation of scientific ideas, and of science as a social organization, is found in *Historical Studies in Physical Science, Minerva, History of Science, Social Studies of Science* and recent volumes of *Isis*.

The most useful source for beginning the examination of differences in the science of European cultures is J. T. Merz' *A History of European Thought* . . . The first volume of that work appeared more than eighty years ago, and yet there has been nothing like it since. Russell McCormmach's *Night Thoughts of a Classical Physicist* presents a supposition in fictionalized form of how the established physics community in Germany received new ideas like those of Einstein. Fritz Ringer's *The Decline of the Mandarins: The German Academic Community. 1890–1933* (Cambridge: Harvard University Press, 1969) is the best available source for understanding the intellectual and social ferment of the German academic community in this century.

Published collections of letters and reminiscences by scientists are other rich sources of information for the ambiance of various physics communities. Of those now readily available the most useful for our purposes are *The Born-Einstein*

Letters (London, 1971). For physics in Germany, Max Born's *Physics in My Generation* (New York, 1969) as well as his *My Life and Views* (New York, 1968) are useful. The recollections of the German biochemist Richard Willstaeter, *From My Life* (New York, 1958), are good sources for details on the German scientific community during the first part of this century. Chapter 4 of Pierre Duhem's *Aim and Structure of Physical Theory* (Princeton, 1954) gives an interesting insight into the contrast between the character of French physics and that of English physics near the turn of the century. J. J. Thomson's *Recollections and Reflections* (London, 1936) is the most useful British account of the character of the early twentieth century British physics community and the training required for admission to that community. One of the most remarkable documents in the history of science is *The Autobiography of Robert A. Millikan* (New York, 1950), for the insights it provides not only about Millikan but also about understanding the dynamics of the organization of the American scientific establishment in the first few decades of this century. Millikan's account should be read in conjunction with Robert Kargon's recent *The Rise of Robert Millikan: Portrait of a Life in American Science* (Ithaca: Cornell University Press, 1982). But the most useful material of this kind for acquiring insight into the changing character of American science in this century is *Robert Oppenheimer: Letters and Recollections* edited by Alice Kimball Smith and Charles Weiner (Cambridge: Harvard University Press, 1980). The editors have very artfully used the primary material to document the shift in the American scholarly enterprise in this century from a profession of the social and economic elite to a profession for the upwardly mobile intellectual.

There is an ever growing literature on science as a social institution. The essays collected and reprinted by Bernard Barber and Walter Hirsch in *The Sociology of Science* (Glencoe, 1962) provide insight into the different orientations that have been used in addressing questions about the social institutions of science.

Essays by Charles E. Rosenberg, "Science and American Social Thought"; Charles Weiner, "Science and Higher Education"; Howard Miller, "Science and Private Agencies"; and Carrol W. Pursell Jr., "Science and Government Agencies"; can be found in the collection edited by David D. Van Tassell and Michael G. Hall, *Science and Society in the United States* (Homewood: The Dorsey Press, 1966). In addition to an excellent set of articles, the editors have provided very valuable bibliographic information. Nathan Reingold's *Science in Nineteenth Century America: A Documentary History* (New York, 1964) and Nathan and Ida Reingold's *Science in America: A Documentary History, 1900–1939* (Chicago, University of Chicago Press, 1981) contain excellent, insightful introductory essays by the editors.

Daniel J. Kevles's *The Physicists: The History of a Scientific Community in Modern America* (New York: Alfred A. Knopf, 1978) is an important source of information. Kevles has included a bibliographic essay that is required reading for anyone who is undertaking serious research on science in American culture.

It goes without saying that the interpretations in this book are often at odds with the interpretations to be found in the literature I have cited. The following articles and books are cited, not so much because they are directly relevant to the subject matter of this book, but because the approach and point of view of the authors have either directly influenced my work or are compatible with views I have developed independently.

The notion of case study that this book represents has its roots in James Conant's *On Understanding Science* (New Haven, 1947). Although the details of that work are dated, Conant's idea of using the history of science, both as a heuristic introduction to substantive scientific questions for individuals untrained in science, and as an organizing tool for introducing the study of science as a social institution, continues to motivate me.

Burton Bledstein's *The Culture of Professionalism: The Middle Class and the Development of Higher Education in America* (New York: W. W. Norton, 1976) and Clifford Geertz's *The Interpretation of Cultures* (New York: Basic Books, 1973) have also influenced my thinking.

The writings of Jacob Bronowski are a constant echo to my own thoughts. A summary of his views can be found in the recently reprinted *Science and Human Values* (New York: Harper and Row, 1975) and *Science and Common Sense* (Cambridge: Harvard University Press, 1975). I share a great deal in point of view with David Noble as represented in his wonderful *America by Design* (New York: Knopf, 1977). Steven Shapin's work in the sociology of knowledge only recently came to my attention. Had I known of it earlier, his "Social Uses of Science," in G. S. Rousseau and Roy Porter (eds.), *The Ferment of Knowledge* (Cambridge: Cambridge University Press, 1980) or his "History of Science and its Sociological Reconstruction," *History of Science,* 1982, *15*:157–211, would have played a more prominent role in the analyses presented in this book. The recent controversies between American social and intellectual historians as represented by the articles in the volume edited by John Higham and Paul Conkin, *New Directions in American Intellectual History* (Baltimore: Johns Hopkins University Press, 1979) has been a wonderful stimulant to my own thinking about the nature of historical investigation.

Index